Stochastic Analysis, Stochastic Systems, and Applications to Finance

Stochastic Analysis, Stochastic Systems, and Applications to Finance

Edited by

Allanus Tsoi
University of Missouri, Columbia, USA

David Nualart
University of Kansas, USA

George Yin
Wayne State University, Michigan, USA

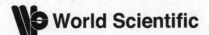
World Scientific

NEW JERSEY · LONDON · SINGAPORE · BEIJING · SHANGHAI · HONG KONG · TAIPEI · CHENNAI

Published by

World Scientific Publishing Co. Pte. Ltd.

5 Toh Tuck Link, Singapore 596224

USA office: 27 Warren Street, Suite 401-402, Hackensack, NJ 07601

UK office: 57 Shelton Street, Covent Garden, London WC2H 9HE

British Library Cataloguing-in-Publication Data
A catalogue record for this book is available from the British Library.

ISBN-13 978-981-4355-70-4
ISBN-10 981-4355-70-4

Printed in Singapore.

Contents

Preface

This volume contains 11 chapters. It is an expanded version of the papers presented at the first Kansas–Missouri Winter School of Applied Probability, which was organized by Allanus Tsoi and was held at the University of Missouri, February 14 and 15, 2008. It brought together researchers from different parts of the country to review and to update the recent advances, and to identify future directions in the areas of applied probability, stochastic processes, and their applications.

After the successful conference was over, there was a strong support of publishing the papers delivered in the conference as an archival volume. Based on the support, we began the preparation on this project. In addition to papers reported at the conference, we have invited a number of colleagues to contribute additional papers.

As an archive, this volume presents some of the highlights of the conference, as well as some of most recent developments in stochastic systems and applications. This book is naturally divided into two parts. The first part contains some recent results in stochastic analysis, stochastic processes and related fields. It explores the Itô formula for multidimensional Gaussian processes using the Wick integral, introduces the notion of fractional white noise multiplication, and discusses the LaSalle type of invariance principles for hybrid switching diffusions. The second part of the book is devoted to financial mathematics, insurance models, and applications. Included here are optimal investment policies for irreversible capital investment projects under uncertainty in monopoly and Stackelberg leader-follower environments, finding expectations of monotone functions of binary random variables by simulation, with applications to reliability, finance, and round robin tournaments, jump bond markets with general models in applications to hedging and utility problems, algorithm and weak convergence for recombining tree in a regime-switching model, applications of counting processes and martingales in survival analysis, extended filtering micro-movement model with counting process observations and applications to bond price tick data, optimal reinsurance for a jump diffusion model, recursive algorithms and numerical studies for mean-reverting asset trading.

Without the encouragement and assistance of many colleagues, this volume would have never come into being. We thank all the authors of this volume, and all of the speakers of the conference for their contributions. The financial support provided by the University of Missouri for this conference is also greatly acknowledged.

Allanus Tsoi
Columbia, Missouri
David Nualart
Lawrence, Kansas
George Yin
Detroit, Michigan

Contributors and Addresses

- Alain Bensoussan, School of Management, University of Texas at Dallas, Richardson, TX 75083-0688, USA. & The Hong Kong Polytechnic University, Hong Kong. Email: alain.bensoussan@utdallas.edu
- Mark Brown, Department of Mathematics, City College, CUNY, New York, NY, USA. Email: cybergarf@aol.com
- J. David Diltz, Department of Finance and Real Estate, The University of Texas at Arlington, Arlington, TX 76019, USA. Email: diltz@uta.edu
- SingRu Hoe, School of Management, University of Texas at Dallas, Richardson, TX 75083-0688, USA. Email: celinehoe@utdallas.edu
- Xing Hu, Department of Economics, Princeton University, Princeton, 08544, USA. Email: xinghu@princeton.edu
- Michael Kohlmann, Department of Mathematics and Statistics, University of Konstanz, D-78457, Konstanz, Germany. Email: michael.kohlmann@uni-konstanz.de
- David R. Kuipers, Department of Finance, Henry W. Bloch School of Business and Public Administration, University of Missouri at Kansas City, Kansas City, MO 64110, USA. Email: kuipersd@umkc.edu
- Ruihua Liu, Department of Mathematics, University of Dayton, 300 College Park, Dayton, OH 45469-2316, USA. Email: ruihua.liu@notes.udayton.edu

- Shangzhen Luo, Department of Mathematics, University of Northern Iowa, Cedar Falls, Iowa, 50614-0506, USA. Email: luos@uni.edu
- David Nualart, Department of Mathematics, University of Kansas, Lawrence, KS 66045, USA. Email: nualart@math.ku.edu
- Salvador Ortiz-Latorre, Departament de Probabilitat, Lògica i Estadística, Universitat de Barcelona, Gran Via 585, 08007 Barcelona, Spain. Email: sortiz@ub.edu
- Erol A. Pekoz, School of Management, Boston University, 595 Commonwealth Avenue, Boston, MA 02215, USA. Email: pekoz@bu.edu
- Sheldon M. Ross, Department of Industrial and Systems Engineering, University of Southern California, Los Angeles, CA 90089, USA. Email: smross@usc.edu
- Jianguo Sun, Department of Statistics, University of Missouri, USA. Email: sunj@missouri.edu
- Allanus Hak-Man Tsoi, Department of Mathematics, University of Missouri, Columbia, MO 65211, USA. Email: tsoia@missouri.edu
- Dewen Xiong, Department of Mathematics, Shanghai Jiaotong University, Shanghai 200240, People's Republic of China. Email: xiongdewen@sjtu.edu.cn
- George Yin, Department of Mathematics, Wayne State University, Detroit, MI 48202, USA. Email: gyin@math.wayne.edu
- Yong Zeng, Department of Mathematics and Statistics, University of Missouri at Kansas City, Kansas City, MO 64110, USA. Email: zengy@umkc.edu
- Qing Zhang, Department of Mathematics, University of Georgia, Athens, GA 30602, USA. Email: qingz@math.uga.edu
- Chao Zhu, Department of Mathematical Sciences, University of Wisconsin-Milwaukee, Milwaukee, WI 53201, USA. Email: zhu@uwm.edu
- Chao Zhuang, Marshall School of Business, University of Southern California, Los Angeles, CA 90089, USA. Email: czhuang@usc.edu

PART I

STOCHASTIC ANALYSIS AND SYSTEMS

Multidimensional Wick-Itô Formula for Gaussian Processes

D. Nualart*

Department of Mathematics, University of Kansas
Lawrence, KS 66045, USA
E-mail: nualart@math.ku.edu

S. Ortiz-Latorre

Departament de Probabilitat, Lògica i Estadística, Universitat de Barcelona
Gran Via 585, 08007 Barcelona, Spain
Email: sortiz@ub.edu

An Itô formula for multidimensional Gaussian processes using the Wick integral is obtained. The conditions allow us to consider processes with infinite quadratic variation. As an example we consider a correlated heterogenous fractional Brownian motion. We also use this Itô formula to compute the price of an exchange option in a Wick-fractional Black-Scholes model.

Keywords: Wick-Itô formula; Gaussian processes; Malliavin calculus.

1. Introduction

The classical stochastic calculus and Itô's formula can be extended to semi-martingales. There has been a recent interest in developing a stochastic calculus for Gaussian processes which are not semimartingales such as the fractional Brownian motion (fBm for short). These developments are motivated by the fact that fBm and other related processes are suitable input noises in practical problems arising in a variety of fields including finance, telecommunications and hydrology (see, for instance, Mandelbrot and Van Ness[7] and Sottinen[13]).

A possible definition of the stochastic integral with respect to the fBm is based on the divergence operator appearing in the stochastic calculus of variations. This approach to define stochastic integrals started from the work by Decreusefond and Üstünel[3] and was further developed by Carmona

---

*Supported by the NSF Grant DMS-0604207

and Coutin[2] and Duncan, Hu and Pasik-Duncan[4] (see also Hu[5] and Nualart[9] for a general survey papers on the stochastic calculus for the fBm). The divergence integral can be approximated by Riemman sums defined using the Wick product, and it has the important property of having zero expectation.

Nualart and Taqqu[11,12] have proved a Wick-Itô formula for general Gaussian processes. In 11 they have considered Gaussian processes with finite quadratic variation, which includes the fBm with Hurst parameter $H > 1/2$. The paper 12 deals with the change-of-variable formula for Gaussian processes with infinite quadratic variation, in particular the fBm with Hurst parameter $H \in (1/4, 1/2)$. The lower bound for H is a natural one, see Alòs, Mazet and Nualart.[1]

The aim of this paper is to generalize the results of Nualart and Taqqu[12] to the multidimensional case. We introduce the multidimensional Wick-Itô integral as a limit of forward Riemann sums and prove a Wick-Itô formula under conditions similar to those in Nualart and Taqqu,[12] allowing infinite quadratic variation processes.

The paper is organized as follows. In Section 2, we introduce the conditions that the multidimensional Gaussian process must satify and state the Itô formula. Section 3 contains some definitions in order to introduce the Wick integral. In Section 4 we prove some technical lemmas using extensively the integration by parts formula for the derivative operator. The convergence results used in the proof of the main theorem are proved in Section 5. Section 6 is devoted to study two examples related to the multidimensional fBm with parameter $H > 1/4$. Finally, in section 7 we use our Itô formula to compute the price of an exchange option on a Wick-fractional Black-Scholes market.

2. Preliminaries

Let $X = \{X_t, t \in [0, T]\}$ be a d-dimensional centered Gaussian process with continuous covariance function matrix $R(s, t)$, that is,

$$R^{i,j}(s, t) = \mathbb{E}[X_s^i X_t^j],$$

for $i, j = 1, \ldots, d$. For $s = t$, we have the covariance matrix $V_t = R(t, t)$.

We denote by $\mathcal{H}$ be the space obtained as the completion of the set of step functions in $A = [0, T] \times \{1, \ldots, d\}$ with respect the scalar product

$$\langle \mathbf{1}_{[0,s]}^i, \mathbf{1}_{[0,t]}^j \rangle_{\mathcal{H}} = R^{i,j}(s, t), \quad 0 \leq s, t \leq T, \quad 1 \leq i, j \leq d,$$

where

$$1^i_{[0,s]} = 1_{[0,s] \times \{i\}} (x, k), \quad (x, k) \in A.$$

The mapping $1^i_{[0,t]} \mapsto X^i_t$ can be extended to a linear isometry between the space $\mathcal{H}$ and the Gaussian Hilbert space generated by the process X. We denote by let $I_1 : h \mapsto X(h), h \in \mathcal{H}$ this isometry.

Let $\mathcal{H}^{\otimes m}$ denote the mth tensor power of $\mathcal{H}$, equipped with the following scalar product

$$\langle h_1 \otimes \cdots \otimes h_m, g_1 \otimes \cdots \otimes g_m \rangle_{\mathcal{H}^{\otimes m}} = \prod_{i=1}^{m} \langle h_i, g_i \rangle_{\mathcal{H}},$$

where $h_1, \ldots, h_m, g_1, \ldots, g_m \in \mathcal{H}$. The subspace of mth symmetric tensors will be denoted by $\mathcal{H}^{\odot m}$. In $\mathcal{H}^{\odot m}$ we introduce the modified scalar product given by $\langle \cdot, \cdot \rangle_{\mathcal{H}^{\odot m}} = m! \langle \cdot, \cdot \rangle_{\mathcal{H}^{\otimes m}}$. In this way, the multiple stochastic integral I_m is an isometry between $\mathcal{H}^{\odot m}$ and the mth Wiener chaos (see Nualart[10] and also Janson[6] for a more detailed discussion of tensor products of Hilbert spaces). We denote by $h_1 \odot \cdots \odot h_m$ the symmetrization of the tensor product $h_1 \otimes \cdots \otimes h_m$.

Now consider the set of smooth random variables $\mathcal{S}$. A random variable $F \in \mathcal{S}$ has the form

$$F = f(X(h_1), \ldots, X(h_n)), \tag{1}$$

with $h_1, \ldots, h_n \in \mathcal{H}$, $n \geq 1$, and $f \in C_b^\infty(\mathbb{R}^n)$ (f and all its partial derivatives are bounded). In $\mathcal{S}$ one can define the derivative operator D as

$$DF = \sum_{i=1}^{n} \partial_i f(X(h_1), \ldots, X(h_n)) h_i,$$

which is an element of $L^2(\Omega; \mathcal{H})$. By iteration one obtains

$$D^m F = \sum_{i_1, \ldots, i_m = 1}^{n} \frac{\partial^m f}{\partial x_{i_1} \cdots \partial x_{i_m}} (X(h_1), \ldots, X(h_n)) h_{i_1} \otimes \cdots \otimes h_{i_m},$$

which is an element of $L^2(\Omega; \mathcal{H}^{\odot m})$.

Definition 2.1. For $m \geq 1$, the space $\mathbb{D}^{m,2}$ is the completion of $\mathcal{S}$ with respect to the norm $\|F\|_{m,2}$ defined by

$$\|F\|^2_{m,2} = \mathbb{E}[F^2] + \sum_{i=1}^{m} \mathbb{E}[\|D^i F\|^2_{\mathcal{H}^{\otimes i}}].$$

The Wick product $F \diamond X\left(h\right)$ between a random variable $F \in \mathbb{D}^{1,2}$ and the Gaussian random variable $X\left(h\right)$ is defined as follows.

Definition 2.2. Let $F \in \mathbb{D}^{1,2}$ and $h \in \mathcal{H}$. Then the Wick product $F \diamond X\left(h\right)$ is defined by

$$F \diamond X\left(h\right) = FX\left(h\right) - \langle DF, h\rangle_{\mathcal{H}}.$$

Actually, the Wick product coincides with the divergence (or the Skorohod integral) of Fh, and by the properties of the divergence operator we can write

$$\mathbb{E}\left[FX\left(h\right)\right] = E\left[\langle DF, h\rangle_{\mathcal{H}}\right]. \tag{2}$$

The Wick integral of a stochastic process u with respect to X is defined as the limit of Riemann sums constructed using the Wick product. For this we need some notation. Denote by $\mathcal{D}$ the set of all partitions of $[0, T]$

$$\pi = \{0 = t_0 < t_1 < \cdots < t_n = T\}$$

such that

$$\frac{|\pi|}{|\pi|_{\inf}} \leq D,$$

where $|\pi| = \max_{0 \leq i \leq n-1}\left(t_{i+1} - t_i\right)$, $|\pi|_{\inf} = \min_{0 \leq i \leq n-1}\left(t_{i+1} - t_i\right)$, and D is a positive constant.

Definition 2.3. Let $u = \{u_t, t \in [0, T]\}$ be a d-dimensional stochastic process such that $u_t^i \in \mathbb{D}^{1,2}$ for all $t \in [0, T]$ and $i = 1, \ldots, d$. The Wick integral

$$\int_0^T u_t \diamond dX_t = \sum_{j=1}^{d} \int_0^T u_t^j \diamond dX_t^j$$

is defined as the limit in probability, if it exists, of the forward Riemann sums

$$\sum_{j=1}^{d} \sum_{i=0}^{n-1} u_{t_i}^j \diamond \left(X_{t_{i+1}}^j - X_{t_i}^j\right)$$

as $|\pi|$ tends to zero, where π runs over all the partitions of the interval $[0, T]$ in the class $\mathcal{D}$.

3. Main Result

We will make use of the following assumptions.

Assumptions.

(A1) For all $j, k \in \{1, \ldots, d\}$ the function $t \mapsto V_t^{j,k}$ has bounded variation on $[0, T]$.

(A2) For all $k, l \in \{1, \ldots, d\}$

$$\sum_{i,j=0}^{n-1} \left| \mathbb{E}[\Delta_i X^k \Delta_j X^l] \right|^2 \to 0, \quad \text{as} \quad |\pi| \to 0.$$

(A3) For all $j, k \in \{1, \ldots, d\}$

$$\sum_{i=0}^{n-1} \sup_{0 \le s \le t} \left| \mathbb{E}[X_s^j \Delta_i X^k] \right|^2 \to 0, \quad \text{as} \quad |\pi| \to 0,$$

where $\Delta_i X^j = X_{t_{i+1}}^j - X_{t_i}^j$, and π runs over all partitions of $[0, T]$ in the class $\mathcal{D}$.

Our purpose is to derive a change-of-variable formula for the process $f(X_t)$, where $f : \mathbb{R}^d \to \mathbb{R}$ if a function satisfying the following condition.

(A4) For every multi-index $\alpha = (\alpha_1, \ldots, \alpha_d) \in \mathbb{N}^d$ with $|\alpha| := \alpha_1 + \cdots + \alpha_d \le 7$, the iterated derivatives

$$\partial^\alpha f(x) = \frac{\partial^{|\alpha|} f}{\partial x_1^{\alpha_1} \cdots \partial x_d^{\alpha_d}}(x)$$

exist, are continuous, and satisfy

$$\sup_{t \in [0,T]} \mathbb{E}\left[|\partial^\alpha f(X_t)|^2 \right] < \infty. \tag{3}$$

Condition (3) holds if $\det V_t > 0$ for all $t \in (0, T]$, and the partial derivatives $\partial^\alpha f$ satisfy the exponential growth condition

$$|\partial^\alpha f(x)| \le C_T e^{c_T |x|^2}, \tag{4}$$

for all $t \in [0, T], x \in \mathbb{R}^d$, where $C_T > 0$ and c_T are such that

$$0 < c_T < \frac{1}{4} \inf_{\substack{0 < t \le T \\ x \in \mathbb{R}^d, |x| > 0}} \frac{x^T V_t^{-1} x}{|x|^2} < \infty \tag{5}$$

(see Lemma 4.5).

Besides the multi-index notation for the derivatives, we will also use the following notation for iterated derivatives. Let $f(x_1, \ldots, x_d)$ be a sufficiently smooth function, then

$$\partial_i f = \frac{\partial f}{\partial x_i}, \quad i \in \{1, \ldots, d\}$$

$$\partial_{i_1,\ldots,i_m}^m f = \partial_{i_m} \left(\partial_{i_{m-1}} \left(\cdots \partial_{i_2} (\partial_{i_1} f) \right) \right), \quad i_k \in \{1, \ldots, d\}, k = 1, \ldots, m.$$

The next theorem is the main result of the paper.

Theorem 3.1. *Suppose that the Gaussian process X and the function f satisfy the preceding assumptions (A1) to (A4). Then the forward integrals (see Definition 2.3)*

$$\int_0^t \partial_j f(X_s) \diamond dX_s^j, \quad 0 \le t \le T, \quad j = 1, \ldots, d$$

exist and the following Wick-Itô formula holds:

$$f(X_t) = f(X_0) + \sum_{j=1}^d \int_0^t \partial_j f(X_s) \diamond dX_s^j + \frac{1}{2} \sum_{j,k=1}^d \int_0^t \partial_{j,k}^2 f(X_s) dV_s^{j,k}.$$

Proof. Using the Taylor expansion of f up to fourth order in two consecutive points of a partition $\pi = \{0 = t_0 < t_1 < \cdots < t_n = t\}$ in the class $\mathcal{D}$ we obtain

$$f\left(X_{t_{i+1}}\right) = f\left(X_{t_i}\right) + \sum_{j=1}^d \partial_j f\left(X_{t_i}\right) \Delta_i X^j + \frac{1}{2} \sum_{j,k=1}^d \partial_{j,k}^2 f\left(X_{t_i}\right) \Delta_i X^j \Delta_i X^k$$
$$+ \frac{1}{3!} T_3^\pi(i) + \frac{1}{4!} T_4^\pi(i),$$

where

$$T_3^\pi(i) = \sum_{j,k,l=1}^d \partial_{j,k,l}^3 f\left(X_{t_i}\right) \Delta_i X^j \Delta_i X^k \Delta_i X^l,$$

$$T_4^\pi(i) = \sum_{j,k,l,m=1}^d \partial_{j,k,l,m}^4 f(\overline{X}_i) \Delta_i X^j \Delta_i X^k \Delta_i X^l \Delta_i X^m,$$

and

$$\overline{X}_i = \lambda X_{t_i} + (1 - \lambda) X_{t_{i+1}}, \quad 0 \le \lambda \le 1.$$

By the definition of the Wick product, see Definition 2.2, one has

$$\partial_j f\left(X_{t_i}\right) \Delta_i X^j = \partial_j f\left(X_{t_i}\right) \diamond \Delta_i X^j + \langle D\left(\partial_j f\left(X_{t_i}\right)\right), \mathbf{1}_{\delta_i}^j \rangle_{\mathcal{H}},$$

where $\delta_i = (t_i, t_{i+1}]$. Taking into account that

$$D\left(\partial_j f\left(X_{t_i}\right)\right) = \sum_{k=1}^{d} \partial_{j,k}^2 f\left(X_{t_i}\right) \mathbf{1}_{[0,t_i]}^{k},$$

one gets

$$\sum_{j=1}^{d} \partial_j f\left(X_{t_i}\right) \Delta_i X^j = \sum_{j=1}^{d} \partial_j f\left(X_{t_i}\right) \diamond \Delta_i X^j + \sum_{j,k=1}^{d} \partial_{j,k}^2 f\left(X_{t_i}\right) \langle \mathbf{1}_{[0,t_i]}^{k}, \mathbf{1}_{\delta_i}^{j} \rangle_{\mathcal{H}}.$$

Using the definition of $\langle \cdot, \cdot \rangle_{\mathcal{H}}$ and adding and subtracting $\frac{1}{2}\mathbb{E}\left[\Delta_i X^j \Delta_i X^k\right]$ we have

$$\langle \mathbf{1}_{[0,t_i]}^{k}, \mathbf{1}_{\delta_i}^{j} \rangle_{\mathcal{H}} = \mathbb{E}[X_{t_i}^k (X_{t_{i+1}}^j - X_{t_i}^j)] = \frac{1}{2}\varphi_i^{j,k} - \frac{1}{2}\mathbb{E}\left[\Delta_i X^j \Delta_i X^k\right],$$

where

$$\varphi_i^{j,k} = \mathbb{E}\left[\left(X_{t_{i+1}}^j - X_{t_i}^j\right)\left(X_{t_{i+1}}^k + X_{t_i}^k\right)\right].$$

This gives

$$f\left(X_{t_{i+1}}\right) = f\left(X_{t_i}\right) + \sum_{j=1}^{d} \partial_j f\left(X_{t_i}\right) \diamond \Delta_i X^j$$

$$+ \frac{1}{2} \sum_{j,k=1}^{d} \partial_{j,k}^2 f\left(X_{t_i}\right) \left\{ \Delta_i X^j \Delta_i X^k - \mathbb{E}\left[\Delta_i X^j \Delta_i X^k\right] \right\}$$

$$+ \frac{1}{2} \sum_{j,k=1}^{d} \partial_{j,k}^2 f\left(X_{t_i}\right) \varphi_i^{j,k} + T_3^\pi(i) + T_4^\pi(i).$$

Hence,

$$f\left(X_t\right) = f\left(X_0\right) + \sum_{i=0}^{n-1} \left[f\left(X_{t_{i+1}}\right) - f\left(X_{t_i}\right)\right]$$

$$= f\left(X_0\right) + \sum_{i=0}^{n-1} \sum_{j=1}^{d} \partial_j f\left(X_{t_i}\right) \diamond \Delta_i X^j$$

$$+ \frac{1}{2} \sum_{i=0}^{n-1} \sum_{j,k=1}^{d} \partial_{j,k}^2 f\left(X_{t_i}\right) \varphi_i^{j,k} + \frac{1}{2} R_2^\pi + \frac{1}{3!} R_3^\pi + \frac{1}{4!} R_4^\pi,$$

where

$$R_2^\pi = \sum_{i=0}^{n-1} \sum_{j,k=1}^{d} \partial_{j,k}^2 f\left(X_{t_i}\right) \left\{ \Delta_i X^j \Delta_i X^k - \mathbb{E}\left[\Delta_i X^j \Delta_i X^k\right]\right\}$$

$$R_3^\pi = \sum_{i=0}^{n-1} T_3^\pi(i) = \sum_{i=0}^{n-1} \sum_{j,k,l=1}^{d} \partial_{j,k,l}^3 f\left(X_{t_i}\right) \Delta_i X^j \Delta_i X^k \Delta_i X^l,$$

$$R_4^\pi = \sum_{i=0}^{n-1} T_4^\pi(i) = \sum_{i=0}^{n-1} \sum_{j,k,l,m=1}^{d} \partial_{j,k,l,m}^4 f(\overline{X}_i) \Delta_i X^j \Delta_i X^k \Delta_i X^l \Delta_i X^m.$$

Note that

$$\frac{1}{2} \sum_{j,k=1}^{d} \partial_{j,k}^2 f\left(X_{t_i}\right) \varphi_i^{j,k} = \frac{1}{2} \sum_{j=1}^{d} \partial_{j,j}^2 f\left(X_{t_i}\right) \varphi_i^{j,j}$$

$$+ \frac{1}{2} \sum_{k>j=1}^{d} \partial_{j,k}^2 f\left(X_{t_i}\right) \left(\varphi_i^{j,k} + \varphi_i^{k,j}\right)$$

$$= \frac{1}{2} \sum_{j=1}^{d} \partial_{j,j}^2 f\left(X_{t_i}\right) \left(V_{t_{i+1}}^{j,j} - V_{t_i}^{j,j}\right)$$

$$+ \sum_{k>j=1}^{d} \partial_{j,k}^2 f\left(X_{t_i}\right) \left(V_{t_{i+1}}^{j,k} - V_{t_i}^{j,k}\right)$$

$$= \frac{1}{2} \sum_{j,k=1}^{d} \partial_{j,k}^2 f\left(X_{t_i}\right) \left(V_{t_{i+1}}^{j,k} - V_{t_i}^{j,k}\right).$$

Using Assumption (A1) it is easy to show the almost sure convergence

$$\lim_{|\pi|\to 0} \frac{1}{2} \sum_{i=0}^{n-1} \sum_{j,k=1}^{d} \partial_{j,k}^2 f\left(X_{t_i}\right) \left(V_{t_{i+1}}^{j,k} - V_{t_i}^{j,k}\right) = \frac{1}{2} \sum_{j,k=1}^{d} \partial_{j,k}^2 f\left(X_s\right) dV_s^{j,k}$$

as $|\pi| \to 0$. The convergences of R_2^π and R_3^π to zero in $L^2(\Omega)$ as $|\pi| \to 0$ are proved in Propositions 5.1 and 5.2. The convergence of R_4^π to zero in $L^1(\Omega)$ as $|\pi| \to 0$ is proved in Proposition 5.3. This clearly implies the convergence in probability

$$\lim_{|\pi|\to 0} \sum_{i=0}^{n-1} \sum_{j=1}^{d} \partial_j f\left(X_{t_i}\right) \diamond \Delta_i X^j = \sum_{j=1}^{d} \int_0^t \partial_j f\left(X_s\right) \diamond dX_s^j,$$

and the result follows. $\square$

Remark 3.1. We can also consider a function $f(t, x)$ depending on time such that the partial derivative $\frac{\partial f}{\partial t}(t, x)$ exists and is continuous. In this case we obtain the additional term $\int_0^t \frac{\partial f}{\partial t}(s, X_s)ds$.

In order to prove Propositions 5.1, 5.2 and 5.3 we need to introduce some technical concepts and prove a number of lemmas.

4. Technical Lemmas

In this section we establish some preliminary lemmas. The first one is trivial.

Lemma 4.1. *Let* $F \in \mathbb{D}^{m+n,2}$ *and* $h_1, \ldots, h_m, g_1, \ldots, g_n \in \mathcal{H}$. *Then*

$$\langle D^n \langle D^m F, h_1 \odot \cdots \odot h_m \rangle_{\mathcal{H}^{\otimes m}}, g_1 \odot \cdots \odot g_n \rangle_{\mathcal{H}^{\otimes n}}$$
$$= \langle D^{m+n} F, h_1 \odot \cdots \odot h_m \odot g_1 \odot \cdots \odot g_n \rangle_{\mathcal{H}^{\otimes m+n}}.$$

The next lemmas are based on the integration by parts formula.

Lemma 4.2. *Let* $F \in \mathbb{D}^{2,2}$ *and* $h, g \in \mathcal{H}$. *Then*

$$\mathbb{E}\left[FX(h)X(g)\right] = \mathbb{E}[\langle D^2 F, h \odot g \rangle_{\mathcal{H}^{\otimes 2}}] + \mathbb{E}\left[F\right]\langle h, g \rangle_{\mathcal{H}}.$$

Proof. See Nualart and Taqqu[12], Lemma 6. □

Lemma 4.3. *Let* $F \in \mathbb{D}^{2,2}$, $h, g \in \mathcal{H}, \xi = X(h)X(g) - \langle h, g \rangle_{\mathcal{H}}$. *Then*

$$\mathbb{E}\left[F\xi\right] = \mathbb{E}[\langle D^2 F, h \odot g \rangle_{\mathcal{H}^{\otimes 2}}].$$

Proof. It is an immediate consequence of the preceding lemma. □

Lemma 4.4. *Let* $F \in \mathbb{D}^{4,2}$, $h_1, h_2, g_1, g_2 \in \mathcal{H}, \xi_1 = X(h_1)X(g_1) - \langle h_1, g_1 \rangle_{\mathcal{H}}$ *and* $\xi_2 = X(h_2)X(g_2) - \langle h_2, g_2 \rangle_{\mathcal{H}}$. *Then*

$$\mathbb{E}\left[F\xi_1\xi_2\right] = \mathbb{E}[\langle D^4 F, h_2 \odot g_2 \odot h_1 \odot g_1 \rangle_{\mathcal{H}^{\otimes 4}}] + \mathbb{E}[\langle D^2 F, h_2 \odot g_1 \rangle_{\mathcal{H}^{\otimes 2}}]\langle h_1, g_2 \rangle_{\mathcal{H}}$$
$$+ \mathbb{E}[\langle D^2 F, g_1 \odot g_2 \rangle_{\mathcal{H}^{\otimes 2}}]\langle h_1, h_2 \rangle_{\mathcal{H}} + \mathbb{E}[\langle D^2 F, h_1 \odot h_2 \rangle_{\mathcal{H}^{\otimes 2}}]\langle g_1, g_2 \rangle_{\mathcal{H}}$$
$$+ \mathbb{E}[\langle D^2 F, h_1 \odot g_2 \rangle_{\mathcal{H}^{\otimes 2}}]\langle h_2, g_1 \rangle_{\mathcal{H}} + 2\mathbb{E}\left[F\right]\langle h_1 \odot g_1, h_2 \odot g_2 \rangle_{\mathcal{H}^{\otimes 2}}.$$

Proof. Applying the last lemma with F replaced by $F\xi_1$ and ξ by ξ_2, we get

$$\mathbb{E}\left[F\xi_1\xi_2\right] = \mathbb{E}[\langle D^2 (F\xi_1), h_2 \odot g_2 \rangle_{\mathcal{H}^{\otimes 2}}].$$

Now, by the Leibniz rule for the derivative operator,

$$D^2 (F\xi_1) = \left(D^2 F\right)\xi_1 + 2DF \odot D\xi_1 + FD^2\xi_1,$$

where

$$D\xi_1 = h_1 X(g_1) + X(h_1) g_1,$$
$$D^2\xi_1 = 2(h_1 \odot g_1),$$

and thus

$$D^2(F\xi_1) = (D^2 F)\xi_1 + 2X(g_1)(DF \odot h_1) + 2X(h_1)(DF \odot g_1)$$
$$+ 2F(h_1 \odot g_1) = A_1 + 2A_2 + 2A_3 + 2A_4.$$

Then,

$$\mathbb{E}\left[\langle A_1, h_2 \odot g_2\rangle_{\mathcal{H}^{\otimes 2}}\right] = \mathbb{E}[\xi_1 \langle D^2 F, h_2 \odot g_2\rangle_{\mathcal{H}^{\otimes 2}}]$$
$$= \mathbb{E}[\langle D^2 \langle D^2 F, h_2 \odot g_2\rangle_{\mathcal{H}^{\otimes 2}}, h_1 \odot g_1\rangle_{\mathcal{H}^{\otimes 2}}]$$
$$= \mathbb{E}[\langle D^4 F, h_2 \odot g_2 \odot h_1 \odot g_1\rangle_{\mathcal{H}^{\otimes 4}}],$$

where we have applied Lemmas 4.3 and 4.1 in the second and third equalities respectively. For the term B, we have

$$\mathbb{E}\left[\langle A_2, h_2 \odot g_2\rangle_{\mathcal{H}^{\otimes 2}}\right]$$
$$= \mathbb{E}\left[X(g_1)\langle DF \odot h_1, h_2 \odot g_2\rangle_{\mathcal{H}^{\otimes 2}}\right]$$
$$= \frac{1}{2}\left[X(g_1)\langle DF, h_2\rangle_{\mathcal{H}}\right]\langle h_1, g_2\rangle_{\mathcal{H}} + \frac{1}{2}\left[X(g_1)\langle DF, g_2\rangle_{\mathcal{H}}\right]\langle h_1, h_2\rangle_{\mathcal{H}}$$
$$= \frac{1}{2}\mathbb{E}[\langle D^2 F, h_2 \odot g_1\rangle_{\mathcal{H}^{\otimes 2}}]\langle h_1, g_2\rangle_{\mathcal{H}} + \frac{1}{2}\mathbb{E}[\langle D^2 F, g_1 \odot g_2\rangle_{\mathcal{H}^{\otimes 2}}]\langle h_1, h_2\rangle_{\mathcal{H}}.$$

Where we have used the integration by parts formula (2) and Lemma 4.1. Analogously, for A_3 we obtain

$$\mathbb{E}\left[\langle A_3, h_2 \odot g_2\rangle_{\mathcal{H}^{\otimes 2}}\right]$$
$$= \frac{1}{2}\mathbb{E}[\langle D^2 F, h_1 \odot h_2\rangle_{\mathcal{H}^{\otimes 2}}]\langle g_1, g_2\rangle_{\mathcal{H}} + \frac{1}{2}\mathbb{E}[\langle D^2 F, h_1 \odot g_2\rangle_{\mathcal{H}^{\otimes 2}}]\langle h_2, g_1\rangle_{\mathcal{H}}.$$

Finally,

$$\mathbb{E}\left[\langle A_4, h_2 \odot g_2\rangle_{\mathcal{H}^{\otimes 2}}\right] = \mathbb{E}[F]\langle h_1 \odot g_1, h_2 \odot g_2\rangle_{\mathcal{H}^{\otimes 2}}.$$

Adding up all the terms the result follows.   □

Lemma 4.5. *The exponential growth condition (4) implies (3).*

Proof. The exponential growth assumption (4) implies

$$\mathbb{E}[|\partial^\alpha f(X_t)|^2] \leq C_T^2 \sup_{0 \leq t \leq T} \mathbb{E}[e^{2c_T |X_t|^2}]. \tag{6}$$

For any symmetric and positive definite matrix A we have

$$\int_{\mathbb{R}^d} e^{-\langle x, Ax \rangle} dx = \left(\frac{\pi^d}{|A|} \right)^{1/2},$$

where $|A| = \det(A)$. As a consequence,

$$\mathbb{E}[e^{2c_T |X_t|^2}] = \frac{1}{(2\pi)^{d/2} |V_t|^{1/2}} \int_{\mathbb{R}^d} e^{-\langle x, Ax \rangle} dx = \frac{1}{2^{d/2} |V_t|^{1/2} |A|^{1/2}},$$

with

$$A = \frac{1}{2} V_t^{-1} - 2c_T I_d = 2c_T \left(\frac{1}{4c_T} V_t^{-1} - I_d \right),$$

and this gives

$$\mathbb{E}[e^{2c_T |X_t|^2}] = |I_d - 4c_T V_t|^{-1/2},$$

provided A is symmetric and positive definite. This matrix is positive definite if and only if for all $x \in \mathbb{R}^d$ with $|x| > 0$

$$x^T \left(\frac{1}{4c_T} V_t^{-1} - I_d \right) x = \frac{1}{4c_T} x^T V_t^{-1} x - |x|^2 > 0,$$

which is implied by (5). Therefore,

$$\mathbb{E}\left[|\partial^\alpha f(X_t)|^2 \right] \leq C_T^2 \sup_{0 \leq t \leq T} |I_d - 4c_T V_t|^{-1/2} =: a_T,$$

which is finite by condition (5). $\qquad \square$

5. Convergence Results

From now on, C will denote a finite positive constant that may change from line to line.

Proposition 5.1. *Let*

$$R_2^\pi = \sum_{i=0}^{n-1} \sum_{j,k=1}^{d} \partial_{j,k}^2 f(X_{t_i}) \left\{ \Delta_i X^j \Delta_i X^k - \mathbb{E}\left[\Delta_i X^j \Delta_i X^k \right] \right\}.$$

Then

$$\lim_{|\pi| \to 0} \mathbb{E}[(R_2^\pi)^2] = 0.$$

Proof. Set $F_i^{j,k} = \partial_{j,k}^2 f(X_{t_i})$ and

$$\varphi_i^{j,k} = \Delta_i X^j \Delta_i X^k - \mathbb{E}\left[\Delta_i X^j \Delta_i X^k\right] = X(1_{\delta_i}^j) X(1_{\delta_i}^k) - \langle 1_{\delta_i}^j, 1_{\delta_i}^k \rangle_{\mathcal{H}}.$$

Then

$$\mathbb{E}[(R_2^\pi)^2] = \sum_{i_1,i_2=0}^{n-1} \sum_{j_1,j_2,k_1,k_2=1}^{d} \mathbb{E}[F_{i_1}^{j_1,k_1} F_{i_2}^{j_2,k_2} \varphi_{i_1}^{j_1,k_1} \varphi_{i_2}^{j_2,k_2}],$$

and by Lemma 4.4 we get the decomposition

$$E[F_{i_1}^{j_1,k_1} F_{i_2}^{j_2,k_2} \varphi_{i_1}^{j_1,k_1} \varphi_{i_2}^{j_2,k_2}] = B_1 + B_2 + B_3 + B_4 + B_5 + B_6,$$

where

$$B_1 = \langle \mathbb{E}[D^4(F_{i_1}^{j_1,k_1} F_{i_2}^{j_2,k_2})], 1_{\delta_{i_1}}^{j_1} \odot 1_{\delta_{i_2}}^{j_2} \odot 1_{\delta_{i_1}}^{k_1} \odot 1_{\delta_{i_2}}^{k_2} \rangle_{\mathcal{H}^{\otimes 4}},$$

$$B_2 = \langle \mathbb{E}[D^2(F_{i_1}^{j_1,k_1} F_{i_2}^{j_2,k_2})], 1_{\delta_{i_2}}^{j_2} \odot 1_{\delta_{i_1}}^{k_1} \rangle_{\mathcal{H}^{\otimes 2}} \langle 1_{\delta_{i_1}}^{j_1}, 1_{\delta_{i_2}}^{k_2} \rangle_{\mathcal{H}},$$

$$B_3 = \langle \mathbb{E}[D^2(F_{i_1}^{j_1,k_1} F_{i_2}^{j_2,k_2})], 1_{\delta_{i_1}}^{k_1} \odot 1_{\delta_{i_2}}^{k_2} \rangle_{\mathcal{H}^{\otimes 2}} \langle 1_{\delta_{i_1}}^{j_1}, 1_{\delta_{i_2}}^{j_2} \rangle_{\mathcal{H}},$$

$$B_4 = \langle \mathbb{E}[D^2(F_{i_1}^{j_1,k_1} F_{i_2}^{j_2,k_2})], 1_{\delta_{i_1}}^{j_1} \odot 1_{\delta_{i_2}}^{j_2} \rangle_{\mathcal{H}^{\otimes 2}} \langle 1_{\delta_{i_1}}^{k_1}, 1_{\delta_{i_2}}^{k_2} \rangle_{\mathcal{H}},$$

$$B_5 = \langle \mathbb{E}[D^2(F_{i_1}^{j_1,k_1} F_{i_2}^{j_2,k_2})], 1_{\delta_{i_1}}^{j_1} \odot 1_{\delta_{i_2}}^{k_2} \rangle_{\mathcal{H}^{\otimes 2}} \langle 1_{\delta_{i_2}}^{j_2}, 1_{\delta_{i_1}}^{k_1} \rangle_{\mathcal{H}},$$

$$B_6 = 2\mathbb{E}[F_{i_1}^{j_1,k_1} F_{i_2}^{j_2,k_2}] \langle 1_{\delta_{i_1}}^{j_1} \odot 1_{\delta_{i_1}}^{k_1}, 1_{\delta_{i_2}}^{j_2} \odot 1_{\delta_{i_2}}^{k_2} \rangle_{\mathcal{H}^{\otimes 2}}.$$

Notice that the terms B_h, $h = 1, \ldots, 6$, depend on the indices i_1, i_2, j_1, j_2, k_1, and k_2. We omit this dependence to simplify the notation and we set

$$\mathcal{B}_h = \sum_{i_1,i_2=0}^{n-1} \sum_{j_1,j_2,k_1,k_2=1}^{d} B_h,$$

so $\mathbb{E}[(R_2^\pi)^2] = \sum_{h=1}^{6} \mathcal{B}_h$. We have that

$$B_1 = \sum_{p=0}^{4} \binom{4}{p} \langle \mathbb{E}[D^p(F_{i_1}^{j_1,k_1}) \odot D^{4-p}(F_{i_2}^{j_2,k_2})], 1_{\delta_{i_1}}^{j_1} \odot 1_{\delta_{i_2}}^{j_2} \odot 1_{\delta_{i_1}}^{k_1} \odot 1_{\delta_{i_2}}^{k_2} \rangle_{\mathcal{H}^{\otimes 4}}.$$

On the other hand

$$D^p(F_i^{j,k}) = \sum_{\substack{u_1,\ldots,u_d=0 \\ u_1+\cdots+u_d=p}}^{p} \frac{p!}{u_1! \cdots u_d!} \partial^u(\partial_{j,k}^2 f(X_{t_i}))(1_{[0,t_i]}^1)^{\odot u_1} \odot \cdots \odot (1_{[0,t_i]}^d)^{\odot u_d}.$$

Hence,

$$
B_1 = \sum_{p=0}^{4} \binom{4}{p} \sum_{\substack{u_1,\ldots,u_d=0 \\ u_1+\cdots+u_d=p}}^{p} \sum_{\substack{v_1,\ldots,v_d=0 \\ v_1+\cdots+v_d=4-p}}^{4-p} \cdot \frac{p!}{u_1!\cdots u_d!} \frac{(4-p)!}{v_1!\cdots v_d!}
$$

$$
\mathbb{E}\left[\partial^{\boldsymbol{u}}(\partial^2_{j_1,k_1}f(X_{t_{i_1}}))\partial^{\boldsymbol{v}}(\partial^2_{j_2,k_2}f(X_{t_{i_2}}))\right]
$$

$$
\times \langle ((1^1_{[0,t_{i_1}]})^{\odot u_1} \odot \cdots \odot (1^d_{[0,t_{i_1}]})^{\odot u_d} \odot (1^1_{[0,t_{i_2}]})^{\odot v_1} \odot \cdots \odot (1^d_{[0,t_{i_2}]})^{\odot v_d},
$$

$$
\mathbf{1}^{j_1}_{\delta_{i_1}} \odot \mathbf{1}^{j_2}_{\delta_{i_2}} \odot \mathbf{1}^{k_1}_{\delta_{i_1}} \odot \mathbf{1}^{k_2}_{\delta_{i_2}} \rangle_{\mathcal{H}^{\otimes 4}}.
$$

Notice that

$$
(1^1_{[0,t_{i_1}]})^{\odot u_1} \odot \cdots \odot (1^d_{[0,t_{i_1}]})^{\odot u_d} \odot (1^1_{[0,t_{i_2}]})^{\odot v_1} \odot \cdots \odot (1^d_{[0,t_{i_2}]})^{\odot v_d}
$$

$$
= \mathbf{1}^{w_1}_{[0,s_1]} \odot \mathbf{1}^{w_2}_{[0,s_2]} \odot \mathbf{1}^{w_3}_{[0,s_3]} \odot \mathbf{1}^{w_4}_{[0,s_4]},
$$

where $w_k \in \{1,\ldots,d\}$, $s_k \in \{t_{i_1}, t_{i_2}\}$, $k = 1,\ldots,4$. But for any $0 \le s_k \le t$, $w_k \in \{1,\ldots,d\}$, $k = 1,\ldots,4$,

$$
|\langle \mathbf{1}^{w_1}_{[0,s_1]} \odot \mathbf{1}^{w_2}_{[0,s_2]} \odot \mathbf{1}^{w_3}_{[0,s_3]} \odot \mathbf{1}^{w_4}_{[0,s_4]}, \mathbf{1}^{j_1}_{\delta_{i_1}} \odot \mathbf{1}^{j_2}_{\delta_{i_2}} \odot \mathbf{1}^{k_1}_{\delta_{i_1}} \odot \mathbf{1}^{k_2}_{\delta_{i_2}} \rangle_{\mathcal{H}^{\otimes 4}}|
$$

$$
\le \frac{1}{4!} \sum_{\sigma \in \Sigma_4} |\langle \mathbf{1}^{w_{\sigma(1)}}_{[0,s_{\sigma(1)}]}, \mathbf{1}^{j_1}_{\delta_{i_1}} \rangle_{\mathcal{H}} \langle \mathbf{1}^{w_{\sigma(2)}}_{[0,s_{\sigma(2)}]}, \mathbf{1}^{j_2}_{\delta_{i_2}} \rangle_{\mathcal{H}} \langle \mathbf{1}^{w_{\sigma(3)}}_{[0,s_{\sigma(3)}]}, \mathbf{1}^{k_1}_{\delta_{i_1}} \rangle_{\mathcal{H}} \langle \mathbf{1}^{w_{\sigma(4)}}_{[0,s_{\sigma(4)}]}, \mathbf{1}^{k_2}_{\delta_{i_2}} \rangle_{\mathcal{H}}|
$$

$$
\le \sup_{\substack{0 \le s \le t \\ 1 \le w \le d}} |\langle \mathbf{1}^{w}_{[0,s]}, \mathbf{1}^{j_1}_{\delta_{i_1}} \rangle_{\mathcal{H}}| |\langle \mathbf{1}^{w}_{[0,s]}, \mathbf{1}^{j_2}_{\delta_{i_2}} \rangle_{\mathcal{H}}| |\langle \mathbf{1}^{w}_{[0,s]}, \mathbf{1}^{k_1}_{\delta_{i_1}} \rangle_{\mathcal{H}}| |\langle \mathbf{1}^{w}_{[0,s]}, \mathbf{1}^{k_2}_{\delta_{i_2}} \rangle_{\mathcal{H}}|
$$

$$
= \sup_{\substack{0 \le s \le t \\ 1 \le w \le d}} |\mathbb{E}[X^w_s \Delta_{i_1} X^{j_1}]| |\mathbb{E}[X^w_s \Delta_{i_2} X^{j_2}]| |\mathbb{E}[X^w_s \Delta_{i_1} X^{k_1}]| |\mathbb{E}[X^w_s \Delta_{i_2} X^{k_2}]|.
$$

Furthermore, by Assumption (A4), we have

$$
\mathbb{E}[|\partial^{\boldsymbol{u}}(\partial^2_{j_1,k_1}f(X_{t_i}))\partial^{\boldsymbol{v}}(\partial^2_{j_2,k_2}f(X_{t_{i_2}}))|] \le a_T < \infty.
$$

Hence, using Cauchy-Schwartz inequality,

$$
\mathcal{B}_1 \le C a_T \left(\sum_{i=0}^{n-1} \sum_{j,k=1}^{d} \sup_{\substack{0 \le s \le t \\ 1 \le w \le d}} |\mathbb{E}[X^w_s \Delta_i X^j]| |\mathbb{E}[X^w_s \Delta_i X^k]| \right)^2
$$

$$
\le C a_T \left(\sum_{j,k=1}^{d} \sum_{i=0}^{n-1} \sup_{\substack{0 \le s \le t \\ 1 \le w \le d}} |\mathbb{E}[X^w_s \Delta_i X^j]|^2 \right)^2.
$$

The last expression tends to zero as $|\pi| \to 0$ by Assumption (A3). Analogously

$$
B_2 = \sum_{p=0}^{2} \binom{2}{p} \sum_{\substack{u_1,\ldots,u_d=0 \\ u_1+\cdots+u_d=p}}^{p} \sum_{\substack{v_1,\ldots,v_d=0 \\ v_1+\cdots+v_d=2-p}}^{2-p} \frac{p!}{u_1!\cdots u_d!} \frac{(2-p)!}{v_1!\cdots v_d!}
$$

$$
\mathbb{E}[\partial^{\boldsymbol{u}}(\partial^2_{j_1,k_1}f(X_{t_{i_1}}))\partial^{\boldsymbol{v}}(\partial^2_{j_2,k_2}f(X_{t_{i_2}}))]
$$

$$
\times \langle (\mathbf{1}^1_{[0,t_{i_1}]})^{\odot u_1} \odot \cdots \odot (\mathbf{1}^d_{[0,t_{i_1}]})^{\odot u_d} \odot (\mathbf{1}^1_{[0,t_{i_2}]})^{\odot v_1} \odot \cdots \odot (\mathbf{1}^d_{[0,t_{i_2}]})^{\odot v_d},
$$

$$
\mathbf{1}^{j_2}_{\delta_{i_2}} \odot \mathbf{1}^{k_1}_{\delta_{i_1}} \rangle_{\mathcal{H}^{\otimes 2}} \langle \mathbf{1}^{j_1}_{\delta_{i_1}}, \mathbf{1}^{k_2}_{\delta_{i_2}} \rangle_{\mathcal{H}}
$$

$$
\leq Ca_T \sup_{\substack{0 \leq s \leq t \\ 1 \leq w \leq d}} \left| \mathbb{E}[X^w_s \Delta_{i_2} X^{j_2}] \right| \left| \mathbb{E}[X^w_s \Delta_{i_1} X^{k_1}] \right| \left| \mathbb{E}[\Delta_{i_1} X^{j_1} \Delta_{i_2} X^{k_2}] \right|.
$$

Therefore, by Cauchy-Schwartz inequality

$$
\mathcal{B}_2 \leq Ca_T \left(\sum_{i=0}^{n-1} \sum_{j=1}^{d} \sup_{\substack{0 \leq s \leq t \\ 1 \leq w \leq d}} \left| \mathbb{E}[X^w_s \Delta_i X^j] \right|^2 \right)
$$

$$
\times \left(\sum_{i_1,i_2=0}^{n-1} \sum_{j,k=1}^{d} \left| \mathbb{E}[\Delta_{i_1} X^j \Delta_{i_2} X^k] \right|^2 \right)^{1/2}
$$

which tends to zero as $|\pi| \to 0$ by Assumptions (A2) and (A3). The proof for the terms $\mathcal{B}_3, \mathcal{B}_4$ and $\mathcal{B}_5$ is almost the same as for the term $\mathcal{B}_2$. Finally,

$$
B_6 = \mathbb{E}[F^{j_1,k_1}_{i_1} F^{j_2,k_2}_{i_2}] \langle \mathbf{1}^{j_1}_{\delta_{i_1}}, \mathbf{1}^{j_2}_{\delta_{i_2}} \rangle_{\mathcal{H}} \langle \mathbf{1}^{k_1}_{\delta_{i_1}}, \mathbf{1}^{k_2}_{\delta_{i_2}} \rangle_{\mathcal{H}}
$$

$$
+ \mathbb{E}[F^{j_1,k_1}_{i_1} F^{j_2,k_2}_{i_2}] \langle \mathbf{1}^{j_1}_{\delta_{i_1}}, \mathbf{1}^{k_2}_{\delta_{i_2}} \rangle_{\mathcal{H}} \langle \mathbf{1}^{k_1}_{\delta_{i_1}}, \mathbf{1}^{j_2}_{\delta_{i_2}} \rangle_{\mathcal{H}}
$$

$$
\leq a_T \left| \mathbb{E}[\Delta_{i_1} X^{j_1} \Delta_{i_2} X^{j_2}] \right| \left| \mathbb{E}[\Delta_{i_1} X^{k_1} \Delta_{i_2} X^{k_2}] \right|
$$

$$
+ a_T \left| \mathbb{E}[\Delta_{i_1} X^{j_1} \Delta_{i_2} X^{k_2}] \right| \left| \mathbb{E}[\Delta_{i_1} X^{k_1} \Delta_{i_2} X^{j_2}] \right|.
$$

Hence,

$$
\mathcal{B}_6 \leq Ca_T \sum_{i_1,i_2=0}^{n-1} \left(\sum_{j,k=1}^{d} \mathbb{E}[\Delta_{i_1} X^j \Delta_{i_2} X^k] \right)^2
$$

$$
\leq Ca_T \sum_{j,k=1}^{d} \sum_{i_1,i_2=0}^{n-1} \left| \mathbb{E}[\Delta_{i_1} X^j \Delta_{i_2} X^k] \right|^2,
$$

which tends to zero as $|\pi| \to 0$ by Assumption (A2). $\qquad \square$

Proposition 5.2. *If*

$$
R^\pi_3 = \sum_{i=0}^{n-1} \sum_{j,k,l=1}^{d} \partial^3_{j,k,l}f(X_{t_i}) \Delta_i X^j \Delta_i X^k \Delta_i X^l,
$$

then

$$\lim_{|\pi| \to 0} \mathbb{E}[(R_3^\pi)^2] = 0.$$

Proof. Setting

$$\Delta_i X^j \Delta_i X^k \Delta_i X^l = \left\{ \Delta_i X^j \Delta_i X^k - \mathbb{E}\left[\Delta_i X^j \Delta_i X^k\right] \right\} \Delta_i X^l$$
$$+ \mathbb{E}\left[\Delta_i X^j \Delta_i X^k\right] \Delta_i X^l,$$

one gets

$$\mathbb{E}[(R_3^\pi)^2]$$

$$\leq 2\mathbb{E}\left[\left(\sum_{i=0}^{n-1} \sum_{j,k,l=1}^{d} \partial_{j,k,l}^3 f\left(X_{t_i}\right) \Delta_i X^l \left\{\Delta_i X^j \Delta_i X^k - \mathbb{E}\left[\Delta_i X^j \Delta_i X^k\right]\right\}\right)^2\right]$$

$$+ 2\mathbb{E}\left[\left(\sum_{i=0}^{n-1} \sum_{j,k,l=1}^{d} \partial_{j,k,l}^3 f\left(X_{t_i}\right) \Delta_i X^l \mathbb{E}\left[\Delta_i X^j \Delta_i X^k\right]\right)^2\right]$$

$$= 2C_1 + 2C_2.$$

To prove the convergence of C_1 to zero, observe that

$$C_1 \leq C \sum_{l=1}^{d} \mathbb{E}\left[\left(\sum_{i=0}^{n-1} \sum_{j,k=1}^{d} \partial_{j,k,l}^3 f\left(X_{t_i}\right) \Delta_i X^l \left\{\Delta_i X^j \Delta_i X^k - \mathbb{E}\left[\Delta_i X^j \Delta_i X^k\right]\right\}\right)^2\right].$$

So it suffices to fix l and apply Proposition 5.1 with the term $\partial_{j,k}^2 f\left(X_{t_i}\right)$ replaced by $\partial_{j,k,l}^3 f\left(X_{t_i}\right) \Delta_i X^l =: g\left(X_{t_i}, X_{t_{i+1}}\right)$ whose exact form does not matter because it satisfies the exponential condition (4). Using Lemma 4.2, we obtain that

$$C_2 = \sum_{i_1,i_2=0}^{n-1} \sum_{j_1,k_1,l_1,j_2,k_2,l_2=1}^{d} \mathbb{E}[\partial_{j_1,k_1,l_1}^3 f(X_{t_{i_1}}) \partial_{j_2,k_2,l_2}^3 f(X_{t_{i_2}}) \Delta_{i_1} X^{l_1} \Delta_{i_2} X^{l_2}]$$

$$\times \mathbb{E}\left[\Delta_{i_1} X^{j_1} \Delta_{i_1} X^{k_1}\right] \mathbb{E}\left[\Delta_{i_2} X^{j_2} \Delta_{i_2} X^{k_2}\right]$$

$$= \mathcal{E}_1 + \mathcal{E}_2,$$

where $\mathcal{E}_h = \sum_{i_1,i_2=0}^{n-1} \sum_{j_1,k_1,l_1,j_2,k_2,l_2=1}^{d} E_h$, for $h = 1, 2$, and

$$E_1 = \mathbb{E}[\langle D^2(\partial_{j_1,k_1,l_1}^3 f(X_{t_{i_1}}) \partial_{j_2,k_2,l_2}^3 f(X_{t_{i_2}})), 1_{\delta_{i_1}}^{l_1} \odot 1_{\delta_{i_2}}^{l_2} \rangle_{\mathcal{H}^{\odot 2}}]$$

$$\times \mathbb{E}\left[\Delta_{i_1} X^{j_1} \Delta_{i_1} X^{k_1}\right] \mathbb{E}\left[\Delta_{i_2} X^{j_2} \Delta_{i_2} X^{k_2}\right],$$

$$E_2 = \mathbb{E}[\partial_{j_1,k_1,l_1}^3 f(X_{t_{i_1}}) \partial_{j_2,k_2,l_2}^3 f(X_{t_{i_2}})] \langle 1_{\delta_{i_1}}^{l_1}, 1_{\delta_{i_2}}^{l_2} \rangle_{\mathcal{H}}$$

$$\times \mathbb{E}\left[\Delta_{i_1} X^{j_1} \Delta_{i_1} X^{k_1}\right] \mathbb{E}\left[\Delta_{i_2} X^{j_2} \Delta_{i_2} X^{k_2}\right].$$

Similarly to the preceding proposition, the term E_1 can be bounded by

$$E_1 \leq Ca_T \sup_{\substack{0 \leq s \leq t \\ 1 \leq w \leq d}} \left| \mathbb{E}[X_s^w \Delta_{i_1} X^{l_1}] \right| \left| \mathbb{E}[X_s^w \Delta_{i_2} X^{l_2}] \right|$$
$$\times \mathbb{E} \left[\Delta_{i_1} X^{j_1} \Delta_{i_1} X^{k_1} \right] \mathbb{E} \left[\Delta_{i_2} X^{j_2} \Delta_{i_2} X^{k_2} \right].$$

As a consequence, we obtain

$$\mathcal{E}_1 \leq Ca_T \left(\sum_{i=0}^{n-1} \sum_{j=1}^{d} \sup_{\substack{0 \leq s \leq t \\ 1 \leq w \leq d}} \left| \mathbb{E}[X_s^w \Delta_i X^j] \right|^2 \right) \left(\sum_{i=0}^{n-1} \sum_{j,k=1}^{d} \left| \mathbb{E}\left[\Delta_i X^j \Delta_i X^k \right] \right|^2 \right),$$

where we have used the Cauchy-Schwartz inequality. This term tends to zero as $|\pi| \to 0$ by Assumptions (A2) and (A3). For the therm E_2, we have

$$E_2 \leq a_T \left| \mathbb{E}\left[\Delta_{i_1} X^{l_1} \Delta_{i_2} X^{l_2} \right] \right| \left| \mathbb{E}\left[\Delta_{i_1} X^{j_1} \Delta_{i_1} X^{k_1} \right] \right| \left| \mathbb{E}\left[\Delta_{i_2} X^{j_2} \Delta_{i_2} X^{k_2} \right] \right|,$$

and by Cauchy-Schwartz inequality,

$$\mathcal{E}_2 \leq Ca_T \left(\sum_{i,j=0}^{n-1} \sum_{k,l=1}^{d} \left| \mathbb{E}\left[\Delta_i X^k \Delta_j X^l \right] \right|^2 \right) \left(\sum_{i=0}^{n-1} \sum_{j,k=1}^{d} \left| \mathbb{E}\left[\Delta_i X^j \Delta_i X^k \right] \right|^2 \right)$$

which converges to 0 as $|\pi| \to 0$ by Assumption (A2). $\qquad \square$

Proposition 5.3. *Let $\overline{X}_i$ be a point in the straight line that joins X_{t_i} and $X_{t_{i+1}}$ and*

$$R_4^\pi = \sum_{i=0}^{n-1} \sum_{j,k,l,m=1}^{d} \partial_{j,k,l,m}^4 f(\overline{X}_i) \Delta_i X^j \Delta_i X^k \Delta_i X^l \Delta_i X^m,$$

then

$$\lim_{|\pi| \to 0} \mathbb{E}[|R_4^\pi|] = 0.$$

Proof. We have

$$\|R_4^\pi\|_{L^1(\Omega)}$$
$$\leq \sum_{i=0}^{n-1} \sum_{j,k,l,m=1}^{d} \left\| \partial_{j,k,l,m}^4 f(\overline{X}_i) \Delta_i X^j \Delta_i X^k \Delta_i X^l \Delta_i X^m \right\|_{L^1(\Omega)}$$
$$\leq \sum_{i=0}^{n-1} \sum_{j,k,l,m=1}^{d} \left(\mathbb{E}[(\partial_{j,k,l,m}^4 f(\overline{X}_i))^2] \right)^{1/2}$$
$$\times \left(\mathbb{E}[\left(\Delta_i X^j \right)^2 \left(\Delta_i X^k \right)^2 \left(\Delta_i X^l \right)^2 \left(\Delta_i X^m \right)^2] \right)^{1/2}.$$

Appliying iteratively the Cauchy-Schwartz inequality one obtains

$$\left(\mathbb{E}[\left(\Delta_i X^j\right)^2 \left(\Delta_i X^k\right)^2 \left(\Delta_i X^l\right)^2 \left(\Delta_i X^m\right)^2] \right)^{1/2}$$

$$\leq (\mathbb{E}[\left(\Delta_i X^j\right)^8])^{1/8}(\mathbb{E}[\left(\Delta_i X^k\right)^8])^{1/8}(\mathbb{E}[\left(\Delta_i X^l\right)^8])^{1/8}(\mathbb{E}[\left(\Delta_i X^m\right)^8])^{1/8}.$$

Hence, by Assumption (A4)

$$\|R_4^\pi\|_{L^1(\Omega)} \leq (a_T)^{1/2} \sum_{i=0}^{n-1} \left(\sum_{j=1}^d (\mathbb{E}[\left(\Delta_i X^j\right)^8])^{1/8} \right)^4$$

$$= (a_T)^{1/2} \sum_{i=0}^{n-1} \left(k\left(8\right) \sum_{j=1}^d (\mathbb{E}[\left(\Delta_i X^j\right)^2])^{1/2} \right)^4$$

$$\leq C\left(a_T\right)^{1/2} k\left(8\right)^4 \sum_{i=0}^{n-1} \sum_{j=1}^d \left| \mathbb{E}[\left(\Delta_i X^j\right)^2] \right|^2 \tag{7}$$

where we have used that for all $p > 0$ if ξ is a centered Gaussian variable one has

$$\|\xi\|_{L^p(\Omega)} = \kappa\left(p\right) \|\xi\|_{L^2(\Omega)},$$

where

$$\kappa\left(p\right) = \sqrt{2} \left(\frac{\Gamma\left(\frac{p+1}{2}\right)}{\sqrt{\pi}} \right)^{1/p}, \quad p > 0.$$

And the last term in equation (7) converges to zero as $|\pi| \to 0$ by Assumption (A2). $\qquad\square$

6. Examples

6.1. *Correlated Heterogeneous Fractional Brownian Motion*

In this section we give an example where the theory previously developed applies. Let $B^H = \{(B_t^{1,H_1}, \ldots, B_t^{d,H_d}), t \in [0,T]\}$ be a d-dimensional heterogeneous fractional Brownian motion with Hurst parameter $H = (H_1, \ldots, H_d) \in (0,1)^d$ and $H_1 \leq \cdots \leq H_d$. That is, B^H is a d-dimensional centered Gaussian process with covariance function matrix $R_H(s,t)$ given by

$$R_H^{i,j}(s,t) = \delta_{ij} R_{H_i}(s,t) = \frac{\delta_{ij}}{2} \{s^{2H_i} + t^{2H_i} - |s-t|^{2H_i}\},$$

for $i,j = 1, \ldots, d$. Set $X_t = AB_t^H$, where $A = (a_{i,j})_{i,j=1,\ldots,d}$ is a $d \times d$ matrix. We call X a correlated heterogeneous fractional Brownian motion. X is a d-dimensional Gaussian process, with the following correlation function matrix

$$R^{i,j}(s,t) = \mathbb{E}[X_s^i X_t^j] = \mathbb{E}\left[\left(\sum_{k=1}^d a_{i,k} B_s^{k,H_k}\right)\left(\sum_{l=1}^d a_{j,l} B_t^{l,H_l}\right)\right]$$

$$= \sum_{k=1}^d a_{i,k} a_{j,k} R_{H_k}(s,t).$$

Proposition 6.1. *The process $X = AB_t^H$ with $1/4 < \min_i H_i < 1$ satisfies Assumptions (A1) to (A3). Therefore, the Wick-Itô formula applies to X.*

Proof. We have that

$$V_t^{j,k} = R^{j,k}(t,t) = \sum_{m=1}^d a_{j,m} a_{k,m} t^{2H_m},$$

so Assumption (A1) is fulfilled. Let's check Assumption (A2). For any k, l we have

$$\mathbb{E}\left[\Delta_i X^k \Delta_j X^l\right]$$
$$= R^{k,l}(t_{i+1}, t_{j+1}) - R^{k,l}(t_{i+1}, t_j) - R^{k,l}(t_i, t_{j+1}) + R^{k,l}(t_i, t_j)$$
$$= \sum_{m=1}^d a_{k,m} a_{l,m}$$
$$\times \left(R_{H_m}^{k,l}(t_{i+1}, t_{j+1}) - R_{H_m}^{k,l}(t_{i+1}, t_j) - R_{H_m}^{k,l}(t_i, t_{j+1}) + R_{H_m}^{k,l}(t_i, t_j)\right)$$
$$= \frac{1}{2} \sum_{m=1}^d a_{k,m} a_{l,m}$$
$$\times \left(|t_{j+1} - t_i|^{2H_m} + |t_j - t_{i+1}|^{2H_m} - |t_j - t_i|^{2H_m} - |t_{j+1} - t_{i+1}|^{2H_m}\right).$$

Therefore

$$A_n = \sum_{i,j=0}^{n-1} \left|\mathbb{E}\left[\Delta_i X^k \Delta_j X^l\right]\right|^2 = B_n + C_n + D_n,$$

where

$$B_n = \sum_{i=0}^{n-1} \left(\sum_{m=1}^{d} a_{k,m} a_{l,m} \, |t_{i+1} - t_i|^{2H_m} \right)^2,$$

$$C_n = \frac{1}{2} \sum_{i=0}^{n-2} \left(\sum_{m=1}^{d} a_{k,m} a_{l,m} \left(|t_{i+2} - t_i|^{2H_m} - |t_{i+1} - t_i|^{2H_m} - |t_{i+2} - t_{i+1}|^{2H_m} \right) \right)^2,$$

$$D_n = \frac{1}{2} \sum_{i=0}^{n-3} \sum_{j=i+2}^{n-1} \left(\sum_{m=1}^{d} a_{k,m} a_{l,m} \right.$$

$$\left. \times \left(|t_{j+1} - t_i|^{2H_m} + |t_j - t_{i+1}|^{2H_m} - |t_j - t_i|^{2H_m} - |t_{j+1} - t_{i+1}|^{2H_m} \right) \right)^2.$$

We have

$$B_n \leq C \sum_{i=0}^{n-1} \sum_{m=1}^{d} |t_{i+1} - t_i|^{4H_m} \leq CT \sum_{m=1}^{d} |\pi|^{4H_m-1},$$

which converges to zero as $|\pi| \to 0$ if $H_1 > 1/4$. By a similar argument we obtain the same result for C_n. The term D_n is more complicated. In Nualart and Taqqu[11] it is proved that, when $j \neq i$ and $j \neq i + 1$,

$$|t_{j+1} - t_i|^{2H_m} + |t_j - t_{i+1}|^{2H_m} - |t_j - t_i|^{2H_m} - |t_{j+1} - t_{i+1}|^{2H_m}$$
$$\leq C \, (j - i - 1)^{2H_m-2} \, |\pi|^{2H_m}.$$

Then,

$$D_n \leq C \sum_{m=1}^{d} |\pi|^{4H_m} \sum_{h=1}^{n} \sum_{k=1}^{h} k^{4H_m-4}.$$

As $n \to +\infty$, one has the following asymptotics

$$\sum_{h=1}^{n} \sum_{k=1}^{h} k^{4H_m-4} \sim \begin{cases} Cn^{4H_m-2} & \text{if} \quad H_m > 3/4 \\ Cn \ln n & \text{if} \quad H_m = 3/4 \\ n & \text{if} \quad H_m < 3/4 \end{cases}.$$

Since our partitions are in the class $\mathcal{D}$ we have $n \leq C \, |\pi|^{-1}$. Therefore,

$$D_n \leq \begin{cases} C \, |\pi|^2 & \text{if} \quad H_1 > 3/4 \\ C \, |\pi|^2 \ln(|\pi|^{-1}) & \text{if} \quad H_1 = 3/4 \\ C \, |\pi|^{4H_1-1} & \text{if} \quad H_1 < 3/4 \end{cases},$$

and Assumption (A2) is fulfilled. Finally, let us check Assumption (A3). One has

$$
\left(\mathbb{E}\left[X_t^k \Delta_j X^l\right]\right)^2 = \left(R^{k,l}\left(t, t_{j+1}\right) - R^{k,l}\left(t, t_j\right)\right)^2
$$

$$
= \frac{1}{4}\left(\sum_{m=1}^{d} a_{k,m} a_{l,m}(t_{i+1}^{2H_m} - t_i^{2H_m} + |t - t_i|^{2H_m} - |t - t_{i+1}|^{2H_m})\right)^2
$$

$$
\leq C \sum_{m=1}^{d} \left(t_{i+1}^{2H_m} - t_i^{2H_m}\right)^2 + \left(|t - t_i|^{2H_m} - |t - t_{i+1}|^{2H_m}\right)^2.
$$

As before, the convergence to zero of $\left(\mathbb{E}\left[X_t^k \Delta_j X^l\right]\right)^2$ as $|\pi| \to 0$ is controlled by the term with H_1. If $1/4 < H_1 \leq 1/2$, one has that $t_{i+1}^{2H_1} - t_i^{2H_1}$ and $|t - t_i|^{2H_1} - |t - t_{i+1}|^{2H_1}$ are both bounded by $(t_{i+1} - t_i)^{2H_1}$, and the sum of their squares is bounded by $C|\pi|^{4H_1-1}$, which converges to zero as $|\pi| \to 0$. If $H_1 > 1/2$, either term is bounded by $C(t_{i+1} - t_i)$. Hence, the sum of their squares is bounded by $C|\pi|$, which converges to zero as $|\pi| \to 0$. So the proof is concluded. $\qquad \square$

6.2. Multidimensional Fractional Brownian Motion

The d-dimensional fractional brownian motion with Hurst parameter $H \in (0, 1)$ is the centered d-dimensional Gaussian process B^H with the following covariance function matrix $R_H(s, t)$

$$
R_H^{i,j}\left(s, t\right) = \delta_{ij} R_H\left(s, t\right) = \frac{\delta_{ij}}{2}\{s^{2H} + t^{2H} - |s - t|^{2H}\}.
$$

Obviously, B^H is the process X considered in the previous section with parameter $(H, \ldots, H)$ and $A = I_d$, therefore we have the following result.

Proposition 6.2. *The process B^H with $1/4 < H < 1$ satisfies Assumptions (A1) to (A3). Therefore, the Wick-Itô formula applies to B^H.*

7. Application to the Pricing of an Exchange Option

The market consists in two risky assets S^1, S^2 and a risk free asset B. Assume the following form for their dynamics

$$
S_t^1 = S_0^1 \exp\left\{\mu_1 t + \sigma_1 X_t^1 - \frac{\sigma_1^2}{2} V_t^{1,1}\right\}, \quad S_0^1 > 0,
$$

$$
S_t^2 = S_0^2 \exp\left\{\mu_2 t + \sigma_2 X_t^2 - \frac{\sigma_2^2}{2} V_t^{2,2}\right\}, \quad S_0^1 > 0,
$$

$$
B_t = B_0 \exp\left\{rt\right\}, \quad B_0 > 0,
$$

where $X = (X^1, X^2)$ is the following correlated heterogeneous fractional Brownian

$$X_t^1 = B_t^{1,H_1},$$
$$X_t^2 = \rho B_t^{1,H_1} + \sqrt{1 - \rho^2} B_t^{2,H_2},$$

where $\rho \in (0, 1)$ and $H_1 \leq H_2$. Note that

$$V_t^{1,1} = \mathbb{E}[(X_t^1)^2] = t^{2H_1},$$
$$V_t^{2,2} = \mathbb{E}[(X_t^2)^2] = \rho^2 t^{2H_1} + (1 - \rho^2) t^{2H_2},$$
$$V_t^{1,2} = \mathbb{E}[X_t^1 X_t^2] = \rho t^{2H_1}.$$

Suppose that $H_1 > 1/4$, hence the Wick-Itô formula applies to X and we obtain that

$$dS_t^1 = \mu_1 S_t^1 dt + \sigma_1 S_t^1 \diamond dX_t^1,$$
$$dS_t^2 = \mu_2 S_t^2 dt + \sigma_2 S_t^2 \diamond dX_t^2.$$

Our aim is to price at time $t \in [0, T]$ the contingent claim $(S_T^1 - S_T^2)^+$, which is known as an exchange option. Assume that the price process for this option has the form $C(t, S_t^1, S_t^2)$, where $C(t, x, y)$ is a function of class $C^{1,2,2}$ and satisfies the exponential growth condition (4). Then the Wick-Itô formula yields

$$C(t, S_t^1, S_t^2) = C(0, S_0^1, S_0^2) + \int_0^t \frac{\partial C}{\partial u}(u, S_u^1, S_u^2) du$$

$$+ \mu_1 \int_0^t \frac{\partial C}{\partial x}(u, S_u^1, S_u^2) S_u^1 du + \sigma_1 \int_0^t \frac{\partial C}{\partial x}(u, S_u^1, S_u^2) S_u^1 \diamond dX_u^1$$

$$+ \mu_2 \int_0^t \frac{\partial C}{\partial y}(u, S_u^1, S_u^2) S_u^2 du + \sigma_2 \int_0^t \frac{\partial C}{\partial y}(u, S_u^1, S_u^2) S_u^2 \diamond dX_u^2$$

$$+ \frac{1}{2}\sigma_1^2 \int_0^t \frac{\partial^2 C}{\partial x^2}(u, S_u^1, S_u^2) (S_u^1)^2 (V_u^{1,1})' du$$

$$+ \frac{1}{2}\sigma_2^2 \int_0^t \frac{\partial^2 C}{\partial y^2}(u, S_u^1, S_u^2) (S_u^2)^2 (V_u^{2,2})' du$$

$$+ \sigma_1 \sigma_2 \int_0^t \frac{\partial^2 C}{\partial x \partial y}(u, S_u^1, S_u^2) S_u^1 S_u^2 (V_u^{1,2})' du. \tag{8}$$

The price $C(t, S_t^1, S_t^2)$ should coincide with the value at time t of a portfolio which replicates the contingent claim $(S_T^1 - S_T^2)^+$. Let Π_t denote the amuount of this portfolio invested in the risk free asset B_t an h_t^1, h_t^2 the amount of stocks S^1 and S^2, respectively. Then,

$$C(t, S_t^1, S_t^2) = \Pi_t + h_t^1 S_t^1 + h_t^2 S_t^2.$$

We will consider portfolios satisfying the following Wick self-financing type condition

$$C\left(t, S_t^1, S_t^2\right) = C\left(0, S_0^1, S_0^2\right) + r \int_0^t \Pi_u du$$

$$+ \mu_1 \int_0^t h_u^1 S_u^1 du + \sigma_1 \int_0^t (h_u^1 S_u^1) \diamond dX_u^1 \qquad (9)$$

$$+ \mu_2 \int_0^t h_u^2 S_u^2 du + \sigma_2 \int_0^t (h_u^2 S_u^2) \diamond dX_u^2.$$

We also suppose that the portfolio is admissible, that is, $\int_0^T |\Pi_t| \, dt < \infty$, $\int_0^T |h_t^i| \, dt < \infty$ and $\{h_u^i S_u^i\}$ is Wick forward integrable on any interval $[0, t]$, $i = 1, 2$. Choosing $h_t^1 = \frac{\partial C}{\partial x}\left(t, S_t^1, S_t^2\right)$ and $h_t^2 = \frac{\partial C}{\partial y}\left(t, S_t^1, S_t^2\right)$ and comparing equations (8) and (9) we get that $C\left(t, x, y\right)$ must satisfy the partial differential equation

$$rC = \frac{\partial C}{\partial t} + r\frac{\partial C}{\partial x}x + r\frac{\partial C}{\partial y}y$$

$$+ \frac{\sigma_1^2}{2}\frac{\partial^2 C}{\partial x^2}x^2(V_t^{1,1})' + \frac{\sigma_2^2}{2}\frac{\partial^2 C}{\partial y^2}y^2(V_t^{2,2})' + \sigma_1\sigma_2\frac{\partial^2 C}{\partial x \partial y}xy(V_t^{1,2})', \quad (10)$$

with terminal condition

$$C\left(T, x, y\right) = (x - y)^+$$

and boundary conditions

$$C\left(t, 0, y\right) = 0,$$
$$C\left(t, x, 0\right) = x.$$

Reasoning as Margrabe,[8] $C\left(t, x, y\right)$ is homogeneous of degree 1 in x and y. Therefore, thanks to Euler's theorem for homogeneous functions, we have that

$$C\left(t, x, y\right) = x\frac{\partial C\left(t, x, y\right)}{\partial x} + y\frac{\partial C\left(t, x, y\right)}{\partial y}$$

and equation (10) simplifies to

$$\frac{\partial C}{\partial t} + \frac{\sigma_1^2}{2}\frac{\partial^2 C}{\partial x^2}x^2(V_t^{1,1})' + \frac{\sigma_2^2}{2}\frac{\partial^2 C}{\partial y^2}y^2(V_t^{2,2})' + \sigma_1\sigma_2\frac{\partial^2 C}{\partial x \partial y}xy(V_t^{1,2})' = 0.$$

Using again the homogeneity of $C\left(t, x, y\right)$ we can define $\overline{C}(t, z) := C\left(t, x, y\right)/y$ where $z = x/y$ and find the following partial differential equation for $\overline{C}$

$$\frac{\partial \overline{C}}{\partial t} + \frac{z^2}{2}\{\sigma_1^2(V_t^{1,1})' + \sigma_2^2(V_t^{2,2})' - 2\sigma_1\sigma_2(V_t^{1,2})'\}\frac{\partial^2 \overline{C}}{\partial z^2} = 0, \qquad (11)$$

with terminal condition

$$\overline{C}(T,z) = (z-1)^{+}$$

and boundary condition

$$\overline{C}(t,0) = 0.$$

Define

$$\theta(t) := \sigma_1^2 (V_t^{1,1})' + \sigma_2^2 (V_t^{2,2})' - 2\sigma_1 \sigma_2 (V_t^{1,2})',$$

then the solution to equation (11) is

$$\overline{C}(t,z) = zN(\overline{d}_1) - N(\overline{d}_2),$$

where

$$\overline{d}_1 := \frac{\ln z + \frac{1}{2}\int_t^T \theta(s)\,ds}{\sqrt{\int_t^T \theta(s)\,ds}}, \quad \overline{d}_2 := \overline{d}_1 - \sqrt{\int_t^T \theta(s)\,ds},$$

and $N(x)$ is the $\mathcal{N}(0,1)$ cumulative distribution function.

Finally, taking into account the values of $V_t^{1,1}, V_t^{2,2}$ and $V_t^{1,2}$, we get

$$C\left(t, S_t^1, S_t^2\right) = S_t^1 N(d_1) - S_t^2 N(d_2),$$

where d_1 and d_2 are obtained from $\overline{d}_1$ and $\overline{d}_2$ making $z = S_t^1/S_t^2$ and

$$\int_t^T \theta(s)\,ds = \left(\sigma_1^2 + \sigma_2^2 - 2\rho\sigma_1\sigma_2\right)\left(T^{2H_1} - t^{2H_1}\right)$$
$$+ \sigma_2^2\left(1 - \rho^2\right)\left(T^{2H_2} - t^{2H_2}\right).$$

References

1. E. Alòs, O. Mazet and D. Nualart, *Stochastic calculus with respect to Gaussian processes*, Ann. Probab. **29**, 766–801 (2001).
2. P. Carmona and L. Coutin, *Stochastic integration with respect to fractional Brownian motion*, Ann. Inst. H. Poincaré **39**, 27–68 (2003).
3. L. Decreusefond and A. S. Üstünel, *Stochastic analysis of the fractional Brownian motion*, Potential Analysis **10**, 177–214 (1998).
4. T. E. Duncan, Y. Hu and B. Pasik-Duncan, *Stochastic calculus for fractional Brownian motion I. Theory*, SIAM J. Control Optim. **38**, 582–612 (2000).
5. Y. Hu, *Integral transformations and anticipative calculus for fractional Brownian motions*, Memoirs of the AMS **175** (2005).
6. S. Janson, *Gaussian Hilbert Spaces* (Cambridge University Press, Cambridge, 1997).
7. B. B. Mandelbrot and J. W. Van Ness, *Fractional Brownian motions, fractional noises and applications*, SIAM Review **10**, 422–437 (1968).

8. W. Margrabe, *The value of an option to exchange one asset for another*, The Journal of Finance **33**, 177-186 (1978).
9. D. Nualart, *Stochastic integration with respect to fractional Brownian motion and applications*, Contemporary Mathematics **336**, 3–39 (2003).
10. D. Nualart, *The Malliavin Calculus and Related Topics*, (Springer-Verlag, Berlin, 2006).
11. D. Nualart and M.S. Taqqu, *Wick-Itô formula for regular processes and applications to the Black and Scholes formula*, Stochastics and Stochastics Reports, to appear.
12. D. Nualart and M.S. Taqqu, *Wick-Itô formula for Gaussian processes*, J. Stoch. Anal. Appl. **24**, 599–614 (2006).
13. T, Sottinen, Fractional Brownian motion in finance and queueing, Ph.D. Thesis, University of Helsinki (2003).

Fractional White Noise Multiplication

Allanus H. Tsoi

Department of Mathematics, University of Missouri
Columbia, MO 65211, USA
Email: tsoia@missouri.edu

Fractional Brownian motion can be viewed as a generalization of the classical Brownian motion process, which can be described through the associated Hurst parameter. The Hurst parameter is a description of the memory of the fractional Brownian motion. The range of the Hurst parameter is between 0 and 1. The classical Brownian motion is a special version of the fractional Brownian motion when the Hurst parameter equal 1/2. In this paper we study fractional Brownian motion in terms of the standard white noise, with Hurst parameter $\frac{1}{2} < H < 1$. We consider fractional white noise as the fractional integral of the standard white noise. We construct a fractional differential operator and study some of its properties. Finally we formulate the notion of fractional white noise multiplication.

1. Introduction

White noise multiplication has been discussed in the book by H.-H. Kuo.[12] One of its advantage over the Ito differential is that it generalizes the diffusion coefficient term $b(t, X_t)$ in the classical Ito differential equation:

$$dX_t = a(t, X_t)dt + b(t, X_t)dB_t, \tag{1}$$

in the sense that the $b(t, X_t)dB_t$ is replaced by the white noise term $\dot{B}_t \cdot b(t, X_t)dt$ in such a way that the term b comes from a comparatively much wider class of functions.

The application of white noise calculus to solve a class of Cauchy problems were presented in the work by Chung, Ji and Saitô.[3] On the other hand, the classical investment and consumption problem can be formulated and solved in terms of solving a Cauchy problem, as can be seen in the book by I. Karatzas and S.E. Shreve.[11] As a consequence, our fractional white noise multiplication results can be used to: (1) generate and solve a wide class of fractional Cauchy problems, which, in turn, (2) can be applied to solve the fractional white noise counterpart of the classical investment and consumption problem.

A fractional Brownian motion is an extension of the notion of the classical Brownian motion, in the sense that a fractional Brownian motion is a Gaussian process which possesses "memory". The study of fractional Brownian motion started in the sixties, when Mandelbrot and Van Ness published their paper in SIAM.[14] Ever since then applied mathematicians and econometricians have been employing the concept of fractional Brownian motion to study and analyse financial and economic data in the case when their data exhibit autocorrelation phenomena.

In this paper we study fractional Brownian motion in terms of white noise analysis. Physically speaking, white noise is the time derivative of Brownian motion. Since the sample paths of a Brownian motion is of unbounded variation, we need to study the derivative of the sample paths of Brownian motion in terms of generalized function theory. The analysis of Brownian motion sample paths in terms of generalized function theory is called white noise analysis. The first study of white noise analysis appeared in the work of T. Hida.[6] Ever since then numerous applications of white noise analysis on quantum physics and mathematical finance emerged. See Refs. 2, 14, and 18. In this paper we first present some background of the standard Gaussian white noise calculus which can be found in Hida, Kuo, Potthoff and Streit,[8] Kuo,[12] or Saitô and Tsoi.[19]

Consider the Gel'fand triple:

$$\mathcal{S}(\mathbf{R}) \subset L^2(\mathbf{R}) \subset \mathcal{S}'(\mathbf{R}) \tag{2}$$

on the real line $\mathbf{R}$. We use $(L^2) = L^2(\mathcal{S}'(\mathbf{R}), \mu_B)$ to denote the standard white noise space of square integrable functionals. We recall the Wiener-Itô theorem which says that any function $\phi \in (L^2)$ can be decomposed uniquely as a sum of multiple Wiener integrals:

$$\phi = \sum_{n=0}^{\infty} \mathcal{I}_n(f^n), \quad f^n \in \hat{L}_c^2(\mathbf{R}^n), \tag{3}$$

where $\mathcal{I}_n$ is the multiple Wiener integral of order n with respect to the Brownian motion, and $\hat{L}_c^2(\mathbf{R}^n)$ denotes the space of symmetric complex-valued L^2-funtions on $\mathbf{R}^n$.

The (L^2)-norm $\|\phi\|_0$ of ϕ is given by:

$$\|\phi\|_0 = \left(\sum_{n=0}^{\infty} n! |f^n|_0^2 \right)^{1/2}, \tag{4}$$

where $|\cdot|_0$ denotes the $L_c^2(\mathbf{R}^n)$-norm for any n.

Let A denote the operator $A = -(d/du)^2 + u^2 + 1$. For ϕ given by the expression Eq. (3) above, and which satisfies the condition:

$$\sum_{n=0}^{\infty} n! |A^{\otimes n} f^n|_0^2 < \infty, \tag{5}$$

we define $\Gamma(A)\phi \in (L^2)$ by:

$$\Gamma(A)\phi = \sum_{n=0}^{\infty} \mathcal{I}_n(A^{\otimes n} f^n). \tag{6}$$

The operator $\Gamma(A)$ is called the second quantization operator of A.

For each $p \geq 0$, define

$$\|\phi\|_p = \|\Gamma(A)^p \phi\|_0, \tag{7}$$

where $\|\cdot\|_0$ is the (L^2)-norm. Let

$$(\mathcal{S}_p) = \{\phi \in (L^2) : \|\phi\|_p < \infty\}. \tag{8}$$

Then $(\mathcal{S}_p)$ is a Hilbert space with norm $\|\cdot\|_p$. Define

$$(\mathcal{S}) = \text{ projective limit of } \{(\mathcal{S}_p); \ p \geq 0\}. \tag{9}$$

Then $(\mathcal{S})$ is a nuclear space, which is called the space of test functions. The dual space $(\mathcal{S})'$ of $(\mathcal{S})$ is called a space of generalized functions (or Hida distributions). Thus we have the Gel'fand triple:

$$(\mathcal{S}) \subset (L^2) \subset (\mathcal{S})'. \tag{10}$$

The bilinear pairing of $(\mathcal{S})'$ and $(\mathcal{S})$ will be denoted by $\langle\langle \cdot, \cdot \rangle\rangle$.

The **S**-transform $\mathbf{S}\Phi$ of a generalized function $\Phi \in (\mathcal{S})'$ is defined to be the function:

$$\mathbf{S}\Phi(\xi) = \langle\langle \Phi, : e^{\langle \cdot, \xi \rangle} : \rangle\rangle, \ \xi \in \mathcal{S}_c(\mathbf{R}), \tag{11}$$

or equivalently,

$$\mathbf{S}\Phi(\xi) = e^{-\frac{1}{2}\langle \xi, \xi \rangle} \langle\langle \Phi, e^{\langle \cdot, \xi \rangle} \rangle\rangle. \ \xi \in \mathcal{S}_c(\mathbf{R}). \tag{12}$$

Here $\mathcal{S}_c(\mathbf{R})$ denotes the complexification of $\mathcal{S}(\mathbf{R})$.

If $\Phi(x) = \sum_{n=0}^{\infty} \langle : x^{\otimes n} :, F^n \rangle \in (\mathcal{S})'$, then its **S**-transform is given by:

$$\mathbf{S}\Phi(\xi) = \sum_{n=0}^{\infty} \langle F^n, \xi^{\otimes n} \rangle, \ \xi \in \mathcal{S}_c(\mathbf{R}). \tag{13}$$

Further properties of the **S**-transform can be found in Hida et.al.[8] or Kuo.[12]

We note that similar results hold if we replace $\mathbf{R}$ by the half-line $(-\infty, T]$, where $T > 0$ is a fixed positive constant throughout this paper.

The fractional Brownian motion (FBM) $B^H(t), t \in \mathbf{R}$, with Hurst parameter $H, 0 < H < 1$, and starting value $B^H(0) = 0$, is given by:

$$B^H(t) = \frac{1}{\Gamma(H + \frac{1}{2})} \{ \int_{-\infty}^{0} [(t - v)^{H-1/2} - (-v)^{H-1/2}] dB(v) \\ + \int_{0}^{t} (t - v)^{H-1/2} dB(v) \}, \tag{14}$$

for $t > 0$, and is defined similarly for $t < 0$.

See Ref. 14 for the above definitions of fractional Brownian motions.

FBM has received much attention in a variety of applications for several decades. See, for example, Refs. 1, 2, 4, 13, 14, 15, and 18. When $\frac{1}{2} < H < 1$, FBM has a long range dependence in the sense that if we let

$$r(n) = cov(B^H(1), (B^H(n + 1) - B^H(n))), \tag{15}$$

then

$$\sum_{n=1}^{\infty} r(n) = \infty. \tag{16}$$

In this paper we study the fractional Brownian motion with Hurst parameter $\frac{1}{2} < H < 1$ in terms of the standard Gaussian white noise. We express the FBM as a fractional integral of Gaussian white noise in the sense of generalized functionals. We construct a fractional differential operator in Sec. 3 and study some of its properties. In Sec. 4 we formulate the notion of fractional white noise multiplication, while in Sec. 5 we present an application of the above concept on a Binomial type finance model.

2. Fractional Integral Representation of Fractional Brownian Motion

We recall some results about generalized funtions according to Gel'fand and Shilov.[5]

First we discuss the direct product of two generalized functions $f(x)$ and $g(y)$ of one variable defined on the space of infinitely differentiable functions with bounded supports. Consider a test function $\phi(x, y), (x, y) \in \mathbf{R}^2$. The direct product $f(x) \times g(y)$ is the generalized function on the space of test functions $\phi(x, y)$ defined as

$$< f(x) \times g(y), \phi(x, y) > = < f(x), < g(y), \phi(x, y) >> . \tag{17}$$

Now suppose that both the generalized functions $f(x)$ and $g(y)$ have support on the same interval $(-\infty, T]$ for some constant $T > 0$. Consider $(f(x), \phi(x+y))$ where ϕ has bounded support, which is an infinitely differentiable function of y. For sufficiently large negative values of y, the support of $\phi(x+y)$ does not intersect with that of $f(x)$. Thus for such y the function $(f(x), \phi(x+y))$ vanishes, and its support is therefore bounded on the left. But the support of $g(y)$ is bounded on the right by assumption. Hence for large enough $a > 0$, the strip $|x+y| \le a$ that contains the support of $\phi(x+y)$ has a bounded intersection with the support of the product $f \times g$. Thus in $|x+y| \le a$ we can replace $\phi(x+y)$ by a function $\phi(x, y)$ with bounded support having the same values in the intersection of the strip and the support of $f \times g$. Now we can define the convolution $f * g$ of two such generalized functions f and g by:

$$< f*g, \phi >=< f(x) \times g(y), \phi(x+y) >=< g(y), < f(x), \phi(x+y) >> . \quad (18)$$

Similarly, if the supports of both f and g are bounded on the left side (i.e., $f = 0$ for $x < a$ and $g = 0$ for $y < b$), then the product $f \times g$ and the convolution $f * g$ can be defined. See page 103, Gel'fand and Shilov.[5]

Now we consider generalized functions g concentrated on the interval $(-\infty, T]$, for fixed constant $T > 0$. Define, for $\lambda > 0$,

$$\gamma_T^\lambda = x^\lambda$$

if $x \in (-\infty, T]$, and

$$\gamma_T^\lambda(x) = 0$$

otherwise.

Next we define the fractional integral of g of order $\lambda > 0$ as the convolution:

$$I_\lambda(g)(x) = g_\lambda(x) = (g * \frac{\gamma_T^{\lambda-1}}{\Gamma(\lambda)})(x). \quad (19)$$

Note that in the above definition, the function $\gamma_T^{\lambda-1}$ is locally summable. If $\lambda \le 0$, we have to view $\frac{\gamma_T^{\lambda-1}}{\Gamma(\lambda)}$ as a generalized function, and regularization and normalization are involved. See Chapters 3 and 5 in Gel'fand and Shilov.[5]

Since Eq. (19) can also be expressed as:

$$I_\lambda(g)(t) = \frac{1}{\Gamma(\lambda)} \int_{-\infty}^t (t-v)^{\lambda-1} g(v) dv, \quad (20)$$

we see that if g is an ordinary continuous function with bounded support, and if $\lambda = n$ is a positive integer, then Eq. (20) is just the usual n-folded integral of g given by the Cauchy formula (see Ref. 5, pg. 115) or through integration-by-parts.

Next, for $\lambda > 0$ the fractional derivative of order λ of g is defined as the convolution

$$D_\lambda(g)(x) = g * \frac{\gamma_T^{-\lambda-1}}{\Gamma(-\lambda)}. \tag{21}$$

We also use the convention that, if $\lambda > 0$, we write

$$I_{-\lambda}(g) = D_\lambda(g),$$

and

$$D_{-\lambda}(g) = I_\lambda(g).$$

Proposition 2.1. *For $\alpha > 0$ and $\lambda > 0$ and generalized function g with support $(-\infty, T]$, we have:*

$$I_\alpha I_\lambda(g)) = I_{\alpha+\lambda}(g). \tag{22}$$

Proof.

$$I_\alpha(I_\lambda(g))(t) = \frac{1}{\Gamma(\alpha)} \int_{-\infty}^t (t-z)^{\alpha-1} \frac{1}{\Gamma(\lambda)} \int_{-\infty}^z (z-v)^{\lambda-1} g(v) dv dz$$

$$= \frac{1}{\Gamma(\alpha)\Gamma(\lambda)} \int_{-\infty}^t g(v) \int_v^t (t-z)^{\alpha-1} (z-v)^{\lambda-1} dz dv.$$

Make a change of variable $w = z - v$ so that the above quantity equals

$$\frac{1}{\Gamma(\alpha)\Gamma(\lambda)} \int_{-\infty}^t g(v) \int_0^{t-v} (t-w-v)^{\alpha-1} w^{\lambda-1} dw dv.$$

Next let $u = \frac{w}{t-v}$ so that the above integral becomes

$$\frac{1}{\Gamma(\alpha)\Gamma(\lambda)} \int_{-\infty}^t (t-v)^{\alpha+\lambda-1} g(v) \int_0^1 (1-u)^{\alpha-1} u^{\lambda-1} du dv$$

which is the same as

$$\frac{1}{\Gamma(\alpha+\lambda)} \int_{-\infty}^t (t-v)^{\alpha+\lambda-1} g(v) dv = I_{\alpha+\lambda}(g)(t),$$

since

$$\int_0^1 (1-u)^{\alpha-1} u^{\lambda-1} du = \frac{\Gamma(\alpha)\Gamma(\lambda)}{\Gamma(\alpha+\lambda)}.$$

See also Ref. 16, page 72. □

For the rest of this paper we shall concentrate on the Hurst parameter H, with $\frac{1}{2} < H < 1$. Recall that the fractional Brownian motion(FBM) with Hurst parameter H, $\frac{1}{2} < H < 1$, can be expressed in the form given on the right hand side of Eq. (14).

Now we consider the white noise space $(\mathcal{S}'(\mathbf{R}), \mu_B)$, where μ_B is the standard white noise measure and B is the standard Brownian motion given in Sec. 1.

Denote $\dot{B}(t)(x) = \frac{dB(t)(x)}{dt} = x(t)$ as the sample path of the Brownian noise (time derivative of the Brownian motion) evaluated at the sample path $x \in \mathcal{S}'(\mathbf{R})$.

We would like to express B^H as

$$B^H(t) = \int_0^t \dot{B}^H(\tau)d\tau \tag{23}$$

for some generalized function $\dot{B}^H \in \mathcal{S}'(\mathbf{R})$.

According to R. Barton,[2] let

$$\dot{B}^H(t) = \frac{1}{\Gamma(H-\frac{1}{2})} \int_{-\infty}^t |t-\tau|^{H-\frac{3}{2}} \dot{B}(\tau)d\tau$$
$$= I_{H-\frac{1}{2}}(\dot{B})(t). \tag{24}$$

If we substitute this into Eq. (23), then, upon changing the order of integration, Eq. (23) becomes

$$B^H(t) = \frac{1}{\Gamma(H-\frac{1}{2})} \int_0^t \int_{-\infty}^\tau |\tau-s|^{H-\frac{3}{2}} \dot{B}(s)dsd\tau$$
$$= \frac{1}{\Gamma(H-\frac{1}{2})} \left[\int_{-\infty}^0 \left[\int_0^t |\tau-s|^{H-\frac{3}{2}} d\tau \right] \dot{B}(s)ds + \int_0^t \left[\int_s^t |\tau-s|^{H-\frac{3}{2}} d\tau \right] \dot{B}(s)ds \right]$$
$$= \frac{1}{\Gamma(H+\frac{1}{2})} \left[\int_{-\infty}^0 \left(|t-s|^{H-\frac{1}{2}} - |s|^{H-\frac{1}{2}} \right) \dot{B}(s)ds + \int_0^t |t-s|^{H-\frac{1}{2}} \dot{B}(s)ds \right]$$

which agrees with the definition of fractiona Brownian motion given in expression Eq. (14). We call $\dot{B}^H$ in Eq. (24) the fractional white noise of Hurst parameter H.

We also have

$$
\begin{aligned}
B^H(t) &= \frac{1}{\Gamma(H - \frac{1}{2})} \int_0^t \int_{-\infty}^{\tau} |\tau - s|^{H - \frac{3}{2}} \dot{B}(s) ds d\tau \\
&= \int_0^t \dot{B}^H(\tau) d\tau \\
&= \int_0^t I_{H - \frac{1}{2}}(\dot{B})(\tau) d\tau \\
&= \langle \chi_{[0,t]}, I_{H - \frac{1}{2}}(\dot{B}) \rangle \\
&= \int_{\mathbf{R}} \chi_{[0,t]}(\tau) I_{H - \frac{1}{2}}(\dot{B})(\tau) d\tau.
\end{aligned}
$$

Remark 2.1.

$$
\begin{aligned}
x_H(t) &= \langle x_H, \delta_t \rangle \\
&= \dot{B}^H(t)(x) \tag{25} \\
&= I_{H - \frac{1}{2}}(x)(t).
\end{aligned}
$$

Note that in the above Eq. (25) the integral $\int_0^t I_{H - \frac{1}{2}}(s)$ is understood to be $\langle I_{H - \frac{1}{2}}(\dot{B}), I_{[0,t]} \rangle$, which is the generalized function $I_{H - \frac{1}{2}}(\dot{B}) \in \mathcal{S}^*$ acting the $L^{(}\mathbf{R})$ function $I_{[0,t]}$, which is interpreted as the $L^2(\mathcal{S}^*, \mu)$ limit of a sequence $\langle I_{H - \frac{1}{2}}(\dot{B}), \xi_n \rangle$, where the sequence $\{\xi_n\} \subset \mathcal{S}(\mathbf{R})$ converging in $L^2(\mathbf{R})$ to $I_{[0,t]}$.

3. Fractional Differential Operator

Definition 3.1. For $y \in \mathcal{S}'((-\infty, T])$, $\Phi_n(x) = \langle : x^{\otimes n} :, f^n \rangle \in (\mathcal{S})$, with $f^n \in \hat{L}_{c,CS}^2(\mathbf{R}^n)$ denoting functions in $\hat{L}_c^2(\mathbf{R}^n)$ with compact supports, define the fractional differential operator as:

$$
\mathbf{D}_y^H \Phi_n(x) := n \langle : x^{\otimes (n-1)} :, \langle y_H, f^n \rangle \rangle, \tag{26}
$$

where

$$
\langle y_H, f^n \rangle(\cdot) = \int_{\mathbf{R}} f(\cdot, t) y_H(t) dt \in L_c^2(\mathbf{R}^{n-1}), \tag{27}
$$

and

$$
y_H = I_{H - \frac{1}{2}}(y) \in \mathcal{S}'(\mathbf{R}). \tag{28}
$$

Consequently,

$$\mathbf{D}_y^H(: e^{\langle \cdot, \xi \rangle} :) = \sum_{n=1}^{\infty} n \frac{1}{n!} \langle : x^{\otimes(n-1)} :, \langle y_H, \xi \rangle \xi^{\otimes(n-1)} \rangle \tag{29}$$
$$= \langle \xi, y_H \rangle : e^{\langle \cdot, \xi \rangle} :,$$

and

$$\partial_t^H(: e^{\langle \cdot, \xi \rangle} :) = D_{\delta_t}^H(: e^{\langle \cdot, \xi \rangle} :)$$
$$= \langle \xi, \delta_t^H \rangle : e^{\langle \cdot, \xi \rangle} :.$$

By definition,

$$\mathbf{S}(\mathbf{D}_y^{H^*}\Phi)(\xi) = \langle\langle \mathbf{D}_y^{H^*}\Phi, : e^{\langle \cdot, \xi \rangle} : \rangle\rangle$$
$$= \langle\langle \Phi, D_y^H(: e^{\langle \cdot, \xi \rangle} :) \rangle\rangle \tag{30}$$
$$= \langle \xi, y_H \rangle \mathbf{S}\Phi(\xi).$$

That is,

$$\mathbf{D}_y^{H^*}\Phi(x) = \langle x, y_H \rangle \Phi(x). \tag{31}$$

Thus

$$\mathbf{D}_y^{H^*}(\langle : x^{\otimes n} :, f^n \rangle) = \langle x, y_H \rangle \langle : x^{\otimes n} :, f^n \rangle \tag{32}$$
$$= \langle : x^{\otimes(n+1)} :, f^n \hat{\otimes} y_H \rangle.$$

Remark 3.1.

$$\partial_t^{H^*}\Phi(x) = \langle x, \delta_t^H \rangle \Phi(x). \tag{33}$$

In particular,

$$\partial_t^{H^*}(\langle : x^{\otimes n} :, f^n \rangle) = \langle x, \delta_t^H \rangle \langle : x^{\otimes n} :, f^n \rangle \tag{34}$$
$$= \langle : x^{\otimes(n+1)} :, f^n \hat{\otimes} \delta_t^H \rangle.$$

Proposition 3.1.

$$\langle x, y_H \rangle = \langle x_H, y \rangle. \tag{35}$$

In particular,

$$\langle x, \delta_{t_0}^H \rangle = \langle x_H, \delta_{t_0} \rangle.$$

Proof. Note that

$$\langle x, y_H \rangle = \int_{\mathbf{R}} x(t) I_{H-\frac{1}{2}} y(t) dt$$

$$= \int_{\mathbf{R}} x(t) \frac{1}{\Gamma(H-\frac{1}{2})} \int_{-\infty}^{t} (t-v)^{H-\frac{3}{2}} y(v) dv dt$$

$$= \int_{\mathbf{R}} y(v) \int_{v}^{\infty} (t-v)^{H-\frac{3}{2}} x(t) dt dv.$$

However, letting $w = -(t-v)$, we have

$$\int_{v}^{\infty} (t-v)^{H-\frac{3}{2}} x(t) dt = \int_{0}^{-\infty} (-w)^{H-\frac{3}{2}} x(v-w) d(-w)$$

$$= -\int_{\infty}^{0} w^{H-\frac{3}{2}} x(v-w) dw$$

$$= \int_{-\infty}^{v} (v-u)^{H-\frac{3}{2}} x(u) du,$$

where the last equality is obtained by letting $u = v - w$. Thus from the above, we have

$$\langle x, y_H \rangle = \int_{\mathbf{R}} y(v) \int_{-\infty}^{v} (v-u)^{H-\frac{3}{2}} x(u) du dv$$

$$= \langle I_{H-\frac{1}{2}}(x), y \rangle$$

$$= \langle x_H, y \rangle. \qquad \square$$

Remark 3.2. For $\phi =: e^{\langle \cdot, \xi \rangle} :,\ \psi =: e^{\langle \cdot, \eta \rangle} :$,

$$D_\nu^H \phi(x) = \langle \nu_H, \xi \rangle (: e^{\langle \cdot, \xi \rangle} :); \tag{36}$$

$$D_\nu^H \psi(x) = \langle \nu_H, \eta \rangle (: e^{\langle \cdot, \eta \rangle} :). \tag{37}$$

Thus

$$\langle\langle D_\nu^H \phi, \psi \rangle\rangle + \langle\langle \phi, D_\nu^H \psi \rangle\rangle = \langle\langle \langle \nu_H, \xi \rangle : e^{\langle \cdot, \xi \rangle} :, \psi \rangle\rangle + \langle\langle \phi, \langle \nu_H, \eta \rangle \psi \rangle\rangle$$

$$= \langle \nu_H, \xi + \eta \rangle \langle\langle : e^{\langle \cdot, \xi \rangle} :, : e^{\langle \cdot, \eta \rangle} : \rangle\rangle \tag{38}$$

$$= \langle \nu_H, \xi + \eta \rangle e^{\langle \eta, \xi \rangle}.$$

4. Fractional White Noise Multiplication

Definition 4.1. For $\Phi_n = \langle : x^{\otimes n} :, f^n \rangle$, $f^n \in \mathcal{S}(\mathbf{R}^n)$ with compact support, define

$$K_H \Phi_n := \langle : x^{\otimes n} :, f_H^n \rangle, \tag{39}$$

where

$$f_H^n(t_1, \ldots, t_n) = I_{H-\frac{1}{2}}^{(n)}(t_1, \ldots, t_n)$$

$$= \int_{-\infty}^{t_1} \cdots \int_{-\infty}^{t_n} (t_1 - v_1)^{H-\frac{3}{2}} \cdots (t_n - v_n)^{H-\frac{3}{2}} f^n(v_1, \ldots, v_n) dv_n \ldots dv_1.$$

(40)

Definition 4.2. For $\Phi \in (\mathcal{S})'$, $\phi, \psi \in (\mathcal{S})$, define the fractional multiplication $\phi \circ_H \Phi$ of order H by:

$$\langle\langle \phi \circ_H \Phi, \psi \rangle\rangle := \langle\langle \phi(K_H \Phi), \psi \rangle\rangle$$
$$= \langle\langle K_H \Phi, \phi \psi \rangle\rangle.$$

(41)

Remark 4.1. From Definition 4.1, for $\Phi = \langle \cdot, \nu \rangle$ and $\nu \in \mathcal{S}(\mathbf{R})$ with compact support, we obtain $K_H \Phi(x) = \langle \cdot, \nu_H \rangle$.

Next, for $\phi =: e^{\langle \cdot, \xi \rangle} :$, $\psi =: e^{\langle \cdot, \eta \rangle} :$, where $\xi, \eta \in \mathcal{S}(\mathbf{R})$, we have

$$\langle\langle \langle \cdot, \nu \rangle \circ_H \phi, \psi \rangle\rangle = \langle\langle \langle \cdot, \nu_H \rangle, \phi \psi \rangle\rangle$$

$$= e^{-\frac{1}{2}(|\xi|_0^2 + |\eta|_0^2)} \int \langle \cdot, \nu_H \rangle e^{\langle \cdot, \xi + \eta \rangle} d\mu(x)$$

(42)

$$= \langle \nu_H, \xi + \eta \rangle e^{\langle \eta, \xi \rangle},$$

where the last equality is a consequence of Lemma 9.16 on page 113 from Kuo.[12]

Theorem 4.1. *The fractional white noise multiplication is given by:*

$$\dot{B}^H(t) \circ_H \phi = (\partial_t^H + \partial_t^{H*})\phi, \ \phi \in (\mathcal{S}).$$

(43)

Proof. First by Eq. (28),

$$\langle\langle \langle \cdot, \nu \rangle \circ_H \phi, \psi \rangle\rangle = \langle\langle D_\nu^H \phi, \psi \rangle\rangle + \langle\langle \phi, D_\nu^H \psi \rangle\rangle.$$

(44)

Now,

$$\langle\langle \langle \cdot, \nu \rangle \circ_H \phi, \langle : \cdot^{\otimes n} :, f^n \rangle \rangle\rangle = \langle\langle \langle \cdot, \nu_H \rangle \langle : \cdot^{\otimes n} :, f^n \rangle, \phi \rangle\rangle.$$

(45)

Hence by Lemma 9.19 on page 114 of Kuo,[12] together with Eq. (26) and Eq. (32), we have

$$\langle \cdot, \nu_H \rangle \langle : \cdot^{\otimes n} :, f^n \rangle = \langle : \cdot^{\otimes n} : \hat{\otimes} x, \nu_H \hat{\otimes} f^n \rangle$$

$$= n \langle : x^{\otimes(n-1)} : \hat{\otimes} \tau, \nu_H \hat{\otimes} f^n \rangle + \langle : x^{\otimes(n+1)} :, \nu_H \hat{\otimes} f^n \rangle$$

$$= n \langle : x^{\otimes(n-1)} :, \langle \nu_H, f^n \rangle \rangle + \langle : x^{\otimes(n+1)} :, \nu_H \hat{\otimes} f^n \rangle$$

$$= D_\nu^H \langle : \cdot^{\otimes n} :, f^n \rangle + D_\nu^{H*} \langle : \cdot^{\otimes n} :, f^n \rangle.$$

(46)

Hence

$$\langle\langle\langle\cdot,\nu\rangle\circ_H\phi,\langle:\cdot^{\otimes n}:f^n\rangle\rangle\rangle = \langle\langle\langle\cdot,\nu_H\rangle\langle:\cdot^{\otimes n}:,f^n\rangle,\phi\rangle\rangle$$

$$= \langle\langle D_\nu^H\langle:\cdot^{\otimes n}:,f^n\rangle + D_\nu^{H*}\langle:\cdot^{\otimes n}:,f^n\rangle,\phi\rangle\rangle$$

$$= \langle\langle(D_\nu^H + D_\nu^{H*})\phi,\langle:\cdot^{\otimes n}:,f^n\rangle\rangle\rangle$$

(47)

which implies

$$\langle\cdot,\nu\rangle\circ_H\phi = (D_\nu^H + D_\nu^{H*})\phi.$$

(48)

In particular, for

$$\dot{B}^H(t)(x) = \langle x_H,\delta_t\rangle,$$

we have

$$\dot{B}^H(t)(x) = \langle x,\delta_t^H\rangle$$

$$= D_{\delta_t}^H + D_{\delta_t}^{H*}$$

(49)

$$= \partial_t^H + \partial_t^{H*}.$$

That is,

$$\dot{B}^H(t)\circ_H\phi = (\partial_t^H + \partial_t^{H*})\phi.$$

(50)

□

5. A Binomial-Type Finance Model

Consider an economy with only two states: the upstate and the downstate, respectively. The probability operator of the upstate is:

$$p = \frac{\dot{B}_{H,u}^2}{\dot{B}_{H,u}^2 + \dot{B}_{H,d}^2}$$

(51)

and the probability operator of the downstate is:

$$p = \frac{\dot{B}_{H,d}^2}{\dot{B}_{H,u}^2 + \dot{B}_{H,d}^2},$$

(52)

where $\dot{B}_{H,u}$ and $\dot{B}_{H,d}$ are interpreted as fractional white noise operater with parameter H as discussed in the previous sections, and we define the squares $\dot{B}_{H,u}^2$ and $\dot{B}_{H,d}^2$ as:

$$\dot{B}_{H,u}^2 \xi = ((\partial_{H,u,\cdot} + \partial_{H,u,\cdot}^*)\xi)^2; \tag{53}$$

and

$$\dot{B}_{H,d}^2 \xi = ((\partial_{H,d,\cdot} + \partial_{H,d,\cdot}^*)\xi)^2. \tag{54}$$

There are two securities: one risk-free bond with interest rate R and one stock with initial price $S(0)$ and time $t = 1$ price $S(1)$ with u and d being the returns of the stock at the upstate and at the downstate at time $t = 1$.

A trading strategy for a portfolio is a pair (B_0, Δ_0) where B_0 is the dollar amounts in the bond and the Δ_0 the number of shares of the stock at time $t = 0$. Thus the value of the portfolio at times $t = 0$ and $t = 1$ are:

$$V(0) = B_0 + \Delta_0 S(0) \tag{55}$$

and

$$V(1) = B_0(1 + R) + \Delta_0 S(1). \tag{56}$$

The price system of the bond and the stock admits arbitrage if and only if there is a trading strategy $(B_0), \Delta_0)$ with $V(0) = B_0 + \Delta_0 S(0) = 0$ such that $V(1) \geq 0$ and $P(V(1) > 0) > 0$. It can be shown that the price system does not admit arbitrage if and only if $d < 1 + R < u$. Suppose that the condition $d < 1 + R < u$ is violated, say, $1 + R \leq d$. It is easy to verify that $\Delta_0 = 1$ and $B_0 = -S(0)$ is an arbitrage trading strategy. The case $u \leq 1 + R$ is argued similarly. Conversely, if $d < 1 + R < u$, then for any strategy (B_0, Δ_0) with $V(0) = B_0 + \Delta_0 S(0)$,

$$V(1) = B_0(1 + R) + \Delta_0 S(1), \tag{57}$$

and is equal to $\Delta_0 S(0)[-(1 + R) + u]$ if in the upstate, and is equal to $\Delta_0 S(0)[-(1 + R) + d]$ if in the downstate, implying that $V(1)$ cannot be non-negative. In other words, there is no-arbitrage strategy. In the following we assume that the price system does not admit arbitrage and derive option prices using a no-arbitrage argument.

Consider a European option which gives its holder the right (not obligation) to receive or pay a pre-specified amount that is contingent on the
state of the economy on a specific date. For example, a Euorpean call option written on a risky security gives its holder the right (not obligation)
to buy the underlying security at a pre-specified price on a specific date;
while a European put option written on a security gives its holder the right
(not obligation) to sell the underlying security at a pre-specified prie on
a specific date. The pre-specified price is called the strike price and the
specific date is called the expiration or maturity time. The payoff functions
for a European call option and a European put option are $maxS(T) - K, 0$
and $maxK - S(T), 0$, respectively, where K is the strike price and T is the
expiration time.

Consider now that there is an option which pays C_u at the upstate and C_d
at the downstate at time $t = 1$. To determine the price of this option, we
construct a portfolio hoping that the payoff of this portfolio is the same as
that of the option. If we can do so, then the price of the option must be
equal to the initial value of the portfolio in order to avoid arbitrage. For
the option with payoff C_u or C_d, the corresponding portfolio is constructed
as follows. The trading strategy (B_0, Δ_0) should be such that the following
payoff equations are satisfied:

$$B_0(1 + R) + \Delta_0 S(0)u = C_u \tag{58}$$

and

$$B_0(1 + R) + \Delta_0 S(0)d = C_d. \tag{59}$$

Solving these equations yields

$$\Delta_0 S(0) = \frac{C_u - C_d}{u - d}, \tag{60}$$

and

$$B_0 = \frac{uC_d - dC_u}{(1 + R)(u - d)}. \tag{61}$$

Therefore the price of the option is given by

$$B_0 + \Delta_0 S(0) = (1 + R)^{-1}[pC_u + (1 - p)C_d], \tag{62}$$

where

$$p = \frac{\dot{B}_{H,u}^2}{\dot{B}_{H,u}^2 + \dot{B}_{H,d}^2} = \frac{1 + R - d}{u - d}. \tag{63}$$

Therefore we have

$$\dot{B}_{H,u}^2 = \frac{1 + R - d}{u - 1 - R}\dot{B}_{H,d}^2. \tag{64}$$

Note that the right hand side of the above equation is intepreted as an operator defined on a suitable domain of generalized functions.

References

1. E. Alos, O. Mazet and D. Nualart: Stochastic calculus with respect to Gaussian Processes, *Ann. of Probability,* Vol. **29**, No. 2 (2001), p. 766-801.
2. R.J. Barton: Signal Detection in Fractional Gaussian Noise, *IEEE Transactions on Information Theory,* Vol. **34**, No. 5, (1988), p. 943-959.
3. D.M. Chung, U.C. Ji and K. Saitô: Cauchy problems associated with the Lévy Laplacian in white noise analysis. *Infinite dimensional Analysis, Quantum Probability and Related Topics,* Vol. **2**, No.1 (1999). 131-153.
4. T.E. Duncan, Y. Hu and B. Pasik-Duncan: Stochastic calculus for fractional Brownian motion. I. Theory. *SIAM Journal of Control and Optimization* **38**, no.2 (2000), 582- 612.
5. I.M. Gel'fand and G.E. Shilov: Generalized Functions. Vol. I. Academic Press 1964.
6. T. Hida: Canonical representations of Gaussian processes and their applications. Memoirs Coll. Sci., Univ. Kyoto **A33** (1960), 109-155.
7. T. Hida and M. Hitsuda: Gaussian Processes. American Math. Soc. Translation of Math. Monographs Vol. **12**, 1993.
8. T. Hida, H.-H. Kuo, J. Potthoff and L. Streit: White Noise Analysis. Kluwer, 1993.
9. H. Holden, B. Oksendal, J. Uboe and T. Zhang: Stochastic Partial Differential Equations. Birkhauser 1996.
10. Y. Hu: Ito-Wiener chaos expansion with exact residual and correlation, variance inequalities. Journal of Theoretical Probability, **10** (1997), 835- 848.
11. I. Karatzas and S.E. Shreve: Brownian Motion and Stochastic Calculus. Second Edition. Springer-Verlag, New York 1991.
12. H.-H. Kuo: White Noise Distribution Theory. CRC Press 1996.
13. S.J. Lin: Stochastic analysis of fractional Brownian motion. Stochastics and Stochastic Reports **55** (1995), 121- 140.
14. B.B. Mandelbrot and J.W. Van Ness: Fractional Brownian motion, fractional noises and applications. SIAM Rev. **10** (1968), 422- 437.

15. K. Nishi, K. Saitô and A.H. Tsoi: Fractional Brownian motion and the Lévy Laplacian. Quantum Information V. World Scientific (2006), p. 181-191.
16. I. Podlubny: Fractional Differential Equations. Academic Press, 1999.
17. P. Protter: Stochastic Integration and Differential Equations. Springer- Verlag, 1990.
18. L.C.G. Rogers: Arbitrage with fractional Brownian motion. Math. Finance **7** (1997), 95-105.
19. K. Saitô and A.H. Tsoi: The Levy Laplacian as a self- adjoint operator. Quantum Information. Ed. T. Hida and K. Saitô, World Scientific (1999), 159- 171.

Invariance Principle of Regime-Switching Diffusions

C. Zhu

Department of Mathematical Sciences
University of Wisconsin-Milwaukee
Milwaukee, WI 53201, USA
Email: zhu@uwm.edu

G. Yin

Department of Mathematics
Wayne State University
Detroit, MI 48202, USA
Email: gyin@math.wayne.edu

This work is devoted to a two-component process termed switching diffusions. The underlying process has a continuous component with diffusive behavior and a discrete component responsible for the movement from one discrete regime to another. Different from the Markov-modulated switching diffusions, the discrete component depends on the continuous component. Invariance principles of LaSalle are derived under suitable conditions.

Keywords: Switching diffusion; stability; invariance theorem.

1. Introduction

1.1. *Introduction*

This work is concerned with large-time behavior of switching diffusion processes, which are two-component Markov processes. In such processes, because of the two components, continuous dynamics and discrete events co-exist and are intertwined. The models are versatile and much enlarge the domain of applicability of diffusion-type processes. Our motivation of the study stems from a wide variety of applications in financial engineering, production planning, wireless communications, and other related fields, in which the traditional stochastic differential equation formulation is not adequate. In addition to the usual dynamic systems modeled by differential equations, there is another factor in the model that could be used to capture the environmental changes and other random factors. In lieu of one dynamic system represented by a differential equation, we encounter a finite number of such equations for different regimes. Across the different regimes, the behavior of the system could be drastically different. For example, one of the early efforts of using such hybrid models for financial applications can be traced back to Ref. 1, in which both the appreciation rate and the volatility of a stock depend on a continuous-time Markov chain. The introduction of regime-switching models makes it possible to describe stochastic volatility in a relatively simple manner. To illustrate, in the simplest case, a stock market may be considered to have two "modes" or "regimes," up and down, resulted from the state of the underlying economy, the general mood of investors in the market, and so on. The rationale is that in the different modes or regimes, the volatility and return rates are very different.

The rapid progress in natural science, life science, engineering, as well as in social science demands the consideration of such systems. In fact, the advent of switching diffusions is largely because of the practical needs in modeling complex dynamic systems. In recent year, there have been much research activities in studying such systems; see Refs. 2,5,9,12,15,21,23,25, 28,29. Some most recent development on switching diffusion processes can be found in Ref. 24 and references therein. For systems running for a long time, it is crucial to have definitive characterization of such systems' long run behavior; see Refs. 12,15,25 for recent progress on stability of such systems.

1.2. *Our Contributions*

Most of the work to date has been concentrated on Markov-modulated diffusions, in which the Brownian motion and the switching force are independent, whereas less is known for systems with continuous-component (termed x-dependent henceforth for simplicity of presentation) dependent switching processes. As demonstrated in Ref. 29, when x-dependent switching diffusions are encountered, even such properties as continuous and smooth dependence on the initial data are nontrivial and are fairly difficult to establish. Nevertheless, studying such systems are both practically useful and theoretically interesting. In our recent work, basic properties such as recurrence, positive recurrence, and ergodicity are studied in Ref. 28; stability is treated in Ref. 12; stability of randomly switching ordinary differential equations is treated in Ref. 30. Continuing in this direction, this paper takes up the issue of examination of invariance principle akin to the LaSalle's theorem for deterministic systems Refs. 6,7,14. It should be mentioned that study of invariance principles for stochastic systems can also be found in Refs. 13,16 etc.

1.3. *Applications to Mathematical Finance*

This work will be of interest to researchers working in the area of mathematical finance. Early work on regime-switching models for option pricing can be found in Ref. 1. Subsequent work in economic time series Ref. 8 extends much of the traditional ARMA model setup. As this time, there have been resurgent interests in using regime-switching diffusions to depict the financial market, where the switching or jump processes are used to describe stochastic volatility resulting from market modes, interest rates, as well as other economic factors. Some recent work in this direction can be found in Refs. 2,3,18,26,27 and many references therein. In addition to finite-time horizon properties, it is desirable to learn large-time behavior of various models including the well-know geometric Brownian motion models, the regime-switching models, and mean-reversion models. In this regard, various notions of stability including stability in probability as well as almost surely, and stability in distribution are of considerable interests. In related insurance and risk models, one is interested in the ruin probability and possibility of default etc., where regime-switching models have also gained popularity (see, for example, Ref. 20). These consideration include finite time analysis and large-time behavior, which is related to stability as well. In this work, we aim to establish invariance principles for regime-switching diffusions, which will help us further in understanding the underpinning of scenarios arising in financial problems. It will open up new domains for further investigation on issues including stability, mean reversion, ergodic measures, and impact on mathematical finance and related issues.

1.4. *Outline*

The rest of the paper is arranged as follows. Section 2 begins with the formulation of the problem. Section 3 is devoted to the invariance principles. Liapunov function type criteria are obtained first. Then linear (in x) systems are treated. Finally, a few more remarks are made in Section 4 to conclude the paper.

2. Formulation

Throughout the paper, we use z' to denote the transpose of $z \in \mathbb{R}^{\ell_1 \times \ell_2}$ with $\ell_i \geq 1$, whereas $\mathbb{R}^{\ell \times 1}$ is simply written as $\mathbb{R}^{\ell}$; $\mathbb{1} = (1, 1, \ldots, 1)' \in \mathbb{R}^{m_0}$

is a column vector with all entries being 1; the Euclidean norm for a row or a column vector x is denoted by $|x|$. As usual, I denotes the identity matrix. For a matrix A, its trace norm is denoted by $|A| = \sqrt{\text{tr}(A'A)}$. If a matrix A is real and symmetric, we use $\lambda_{\max}(A)$ and $\lambda_{\min}(A)$ to denote the maximal and minimal eigenvalues of A, respectively, and set $\rho(A) := \max\{|\lambda_{\max}(A)|, |\lambda_{\min}(A)|\}$. When B is a set, $I_B(\cdot)$ denotes the indicator function of B.

Let $(\Omega, \mathcal{F}, \mathbf{P})$ be a complete probability space. Suppose that $(X(t), \alpha(t))$ is a two-component Markov process such that $X(\cdot)$ is a continuous component taking values in $\mathbb{R}^r$ and $\alpha(\cdot)$ is a jump component taking values in a finite set $\mathcal{M} = \{1, 2, \ldots, m_0\}$. The process $(X(t), \alpha(t))$ has a generator $\mathcal{L}$ given as follows. For each $i \in \mathcal{M}$ and any twice continuously differentiable function $g(\cdot, i)$,

$$
\begin{aligned}
\mathcal{L}g(x, i) &= \frac{1}{2} \sum_{j,k=1}^{r} a_{jk}(x, i) \frac{\partial^2 g(x, i)}{\partial x_j \partial x_k} + \sum_{j=1}^{r} b_j(x, i) \frac{\partial g(x, i)}{\partial x_j} + Q(x)g(x, \cdot)(i), \\
&= \frac{1}{2} \text{tr}(a(x, i)\nabla^2 g(x, i)) + b'(x, i)\nabla g(x, i) + Q(x)g(x, \cdot)(i),
\end{aligned}
\tag{1}
$$

where $x \in \mathbb{R}^r$, and $Q(x) = (q_{ij}(x))$ is an $m_0 \times m_0$ matrix depending on x satisfying $q_{ij}(x) \geq 0$ for $i \neq j$ and $\sum_{j \in \mathcal{M}} q_{ij}(x) = 0$,

$$
Q(x)g(x, \cdot)(i) = \sum_{j \in \mathcal{M}} q_{ij}(x)g(x, j) = \sum_{j \in \mathcal{M}, j \neq i} q_{ij}(x)(g(x, j) - g(x, i)), \ i \in \mathcal{M},
$$

and $\nabla g(\cdot, i)$ and $\nabla^2 g(\cdot, i)$ denote the gradient and Hessian of $g(\cdot, i)$, respectively.

The process $(X(t), \alpha(t))$ can be described by

$$
dX(t) = b(X(t), \alpha(t))dt + \sigma(X(t), \alpha(t))dw(t), \ X(0) = x, \ \alpha(0) = \alpha, \tag{2}
$$

and for $i \neq j$

$$
\mathbf{P}\{\alpha(t + \Delta t) = j | \alpha(t) = i, (X(s), \alpha(s)), s \leq t\} = q_{ij}(X(t))\Delta t + o(\Delta t), \tag{3}
$$

where $w(t)$ is a d-dimensional standard Brownian motion, $b(\cdot, \cdot) : \mathbb{R}^r \times \mathcal{M} \mapsto \mathbb{R}^r$, and $\sigma(\cdot, \cdot) : \mathbb{R}^r \times \mathcal{M} \mapsto \mathbb{R}^{r \times d}$ satisfies $\sigma(x, i)\sigma'(x, i) = a(x, i)$. Note that the evolution of the discrete component $\alpha(\cdot)$ can be represented as a stochastic integral with respect to a Poisson random measure (see, for example, Refs. 5 and 19). Indeed, for $x \in \mathbb{R}^r$ and $i, j \in \mathcal{M}$ with $j \neq i$, let $\Delta_{ij}(x)$ be consecutive (with respect to the lexicographic ordering on $\mathcal{M} \times \mathcal{M}$), left closed, right open intervals of the real line, each having length $q_{ij}(x)$. Define a function $h : \mathbb{R}^r \times \mathcal{M} \times \mathbb{R} \mapsto \mathbb{R}$ by

$$h(x, i, z) = \sum_{j=1}^{m_0} (j - i) I_{\{z \in \Delta_{ij}(x)\}}. \tag{4}$$

Then (3) is equivalent to

$$d\alpha(t) = \int_{\mathbb{R}} h(X(t), \alpha(t-), z) \mathfrak{p}(dt, dz), \tag{5}$$

where $\mathfrak{p}(dt, dz)$ is a Poisson random measure with intensity $dt \times m(dz)$, and m is the Lebesgue measure on $\mathbb{R}$. The Poisson random measure $\mathfrak{p}(\cdot, \cdot)$ is independent of the Brownian motion $w(\cdot)$. Denote the natural filtration by $\mathcal{F}_t := \sigma\{(X(s), \alpha(s)), s \leq t\}$. Without loss of generality, assume the filtration $\{\mathcal{F}_t\}_{t \geq 0}$ satisfies the usual condition (i.e., it is right continuous with $\mathcal{F}_0$ containing all **P**-null sets).

Throughout the paper, we assume that both $b(\cdot, i)$ and $\sigma(\cdot, i)$ satisfy the usual Lipschitz condition and linear growth condition for each $i \in \mathcal{M}$ and that $Q(\cdot)$ is bounded and continuous. It is well known that under these conditions, the system (2)–(3) (or equivalently, (2)–(5)) has a unique strong solution; see Refs. 10 or 22 for details. From time to time, we often wish to emphasize the initial data $(X(0), \alpha(0) = (x, \alpha)$ dependence of the solution of (2)–(3), which will be denoted by $(X^{x,\alpha}(t), \alpha^{x,\alpha}(t))$ similar to its diffusion counter part.

A few words on notations are needed at this point. We define for any $(x, i), (y, j) \in \mathbb{R}^r \times \mathcal{M}$ that

$$d((x, i), (y, j)) = \begin{cases} |x - y|, & \text{if } i = j \\ |x - y| + 1, & \text{if } i \neq j. \end{cases}$$

It is easy to verify that for any $(x, i), (y, j)$, and (z, l) we have (i) $d((x, i), (y, j)) \geq 0$ and $d((x, i), (y, j)) = 0$ if and only if $(x, i) = (y, j)$, (ii) $d((x, i), (y, j)) = d((y, j), (x, i))$, and (iii) $d((x, i), (y, j)) \leq d((x, i), (z, l)) + d((z, l), (y, j))$. Thus d is a distance function of $\mathbb{R}^r \times \mathcal{M}$. Also if U is a subset of $\mathbb{R}^r \times \mathcal{M}$, we define

$$d((x, i), U) = \inf\{d((x, i), (y, j)) : (y, j) \in U\}.$$

Let $\mathcal{M}$ be endowed with the trivial topology. Then it is obvious that for a fixed $U \in \mathbb{R}^r \times \mathcal{M}$, the function $d(\cdot, U)$ is continuous. As usual, we denote by $d(x, D)$ the distance between $x \in \mathbb{R}^r$ and $D \subset \mathbb{R}^r$, that is

$$d(x, D) = \inf\{|x - y| : y \in D\}.$$

Throughout the paper, we use $\mathbf{P}_{x,\alpha}$ and $\mathbf{E}_{x,\alpha}$ to denote the probability and expectation with $(X(0), \alpha(0)) = (x, \alpha)$, respectively.

3. Main Results

This section is the core of the paper. We begin with the definition of invariant sets. Inspired by the study in Ref. 4, we define the invariant set as follows.

Definition 3.1. A Borel measurable set $U \subset \mathbb{R}^r \times \mathcal{M}$ is said to be *invariant* with respect to the solutions of (2)–(5) or simply, U is *invariant* with respect to the process $(X(t), \alpha(t))$ if

$$\mathbf{P}_{x,i} \{(X(t), \alpha(t)) \in U, \text{ for all } t \geq 0\} = 1, \text{ for any } (x, i) \in U,$$

that is, a process starting from in U will remain in U a.s.

As shown in Refs. 12 and 15, when the coefficients of (2)–(5) satisfy

$$b(0, \alpha) = \sigma(0, \alpha) = 0, \text{ for each } \alpha \in \mathcal{M},$$

then almost every sample path of any solution with initial condition (x, i) satisfying $x \neq 0$ will never reach the origin, in other words, the set $(\mathbb{R}^r - \{0\}) \times \mathcal{M}$ is invariant with respect to the solutions of (2)–(5).

Using the terminologies in Refs. 4 and 11, we present the definitions of stability and asymptotic stability. Then general results in terms of Liapunov function are provided.

Definition 3.2. A closed and bounded set $K \subset \mathbb{R}^r \times \mathcal{M}$ is said to be

(i) *stable in probability* if for any $\varepsilon > 0$ and $\rho > 0$, there is a $\delta > 0$ such that

$$\mathbf{P}_{x,i} \left\{ \sup_{t \geq 0} d((X(t), \alpha(t)), K) < \rho \right\} \geq 1 - \varepsilon, \text{ if } d((x, i), K) < \delta;$$

(ii) *asymptotically stable in probability* is it is stable in probability, and moreover

$$\mathbf{P}_{x,i} \left\{ \lim_{t \to \infty} d((X(t), \alpha(t)), K) = 0 \right\} \to 1, \text{ as } d((x, i), K) \to 0;$$

(iii) *stochastically asymptotically stable in the large* if it is stable in probability, and

$$\mathbf{P}_{x,i} \left\{ \lim_{t \to \infty} d((X(t), \alpha(t)), K) = 0 \right\} = 1, \text{ for any } (x, i) \in \mathbb{R}^r \times \mathcal{M};$$

(iv) *asymptotically stable with probability one* if

$$\lim_{t \to \infty} d((X(t), \alpha(t)), K) = 0, \text{ a.s.}$$

Theorem 3.3. *Assume that there exists a nonnegative function $V : \mathbb{R}^r \times \mathcal{M} \mapsto \mathbb{R}_+$ such that $\mathrm{Ker}(V) := \{(x, i) \in \mathbb{R}^r \times \mathcal{M} : V(x, i) = 0\}$ is nonempty and bounded, and that for each $\alpha \in \mathcal{M}$, $V(\cdot, \alpha)$ is twice continuously differentiable with respect to x, and*

$$\mathcal{L}V(x, i) \leq 0, \text{ for all } (x, i) \in \mathbb{R}^r \times \mathcal{M}. \tag{1}$$

Then

 (i) $\mathrm{Ker}(V)$ *is an invariant set for the process* $(X(t), \alpha(t))$, *and*
 (ii) $\mathrm{Ker}(V)$ *is stable in probability.*

Proof. Let $(x_0, i_0) \in \mathrm{Ker}(V)$. By virtue of the generalized Itô Lemma (Ref. 19), we have for any $t \geq 0$,

$$V(X(t), \alpha(t)) = V(x_0, i_0) + \int_0^t \mathcal{L}V(X(s), \alpha(s))ds + M(t), \tag{2}$$

where $M(t) = M_1(t) + M_2(t)$ is a local martingale with

$$M_1(t) = \int_0^t \langle \nabla V(X(s), \alpha(s)), \sigma(X(s), \alpha(s))dw(s) \rangle,$$

$$M_2(t) = \int_0^t \int_{\mathbb{R}} \big[V(X(s), i_0 + h(X(s), \alpha(s-), z)) \\ - V(X(s), \alpha(s)) \big] \mu(ds, dz),$$

where $\langle \cdot, \cdot \rangle$ denotes the usual inner product, $\mu(ds, dz) = \mathfrak{p}(ds, dz) - ds \times m(dz)$ is a martingale measure ($\mathfrak{p}(dt, dz)$ is the Poisson random measure with intensity $dt \times m(dz)$ as in (5)). Taking expectations on both sides of (2) (as in Ref. 28, use a sequence of stopping times and Fatou's lemma, if necessary), it follows from (1) that

$$\mathbf{E}_{x_0, i_0}[V(X(t), \alpha(t))] \leq V(x_0, i_0) = 0.$$

But V is nonnegative, so we must have $V(X(t), \alpha(t)) = 0$ a.s. for any $t \geq 0$. Then we have

$$\mathbf{P}_{x_0, i_0} \left\{ \sup_{t_n \in \mathbb{Q}^+} V(X(t_n), \alpha(t_n)) = 0 \right\} = 1,$$

where $\mathbb{Q}^+$ denotes the set of nonnegative rational numbers. Now by virtue of Ref. 29, Lemma 3.6, the process $(X(t), \alpha(t))$ is cádlág (sample paths being right continuous and having left limits). Thus we obtain

$$\mathbf{P}_{x_0, i_0} \left\{ \sup_{t \geq 0} V(X(t), \alpha(t)) = 0 \right\} = 1.$$

That is

$$\mathbf{P}_{x_0,i_0}\{(X(t),\alpha(t)) \in \mathrm{Ker}(V), \text{ for all } t \geq 0\} = 1.$$

This proves the first assertion of the theorem.

We proceed to prove the second assertion. For any $\delta > 0$, let U_δ be a neighborhood of $\mathrm{Ker}(V)$ such that

$$U_\delta := \{(x,i) \in \mathbb{R}^r \times \mathcal{M} : d((x,i),\mathrm{Ker}(V)) < \delta\}. \tag{3}$$

Let the initial condition $(x,i) \in U_\delta - \mathrm{Ker}(V)$ and τ be the first exit time of the process from U_δ. That is,

$$\tau = \inf\{t : (X(t),\alpha(t)) \notin U_\delta\}.$$

Then for any $t \geq 0$, we have by virtue of generalized Itô's lemma that

$$V(X(t \wedge \tau),\alpha(t \wedge \tau)) = V(x,i) + \int_0^{t \wedge \tau} \mathcal{L}V(X(s),\alpha(s))ds + M(t \wedge \tau),$$

where

$$\begin{aligned}
M(t \wedge \tau) = & \int_0^{t \wedge \tau} \langle \nabla V(X(s),\alpha(s)),\sigma(X(s),\alpha(s))dw(s)\rangle \\
& + \int_0^{t \wedge \tau} \int_{\mathbb{R}} \big[V(X(s),i + h(X(s),\alpha(s-),z)) \\
& \qquad\qquad - V(X(s),\alpha(s))\big]\mu(ds,dz).
\end{aligned}$$

As argued in the previous paragraph, by virtue of (1), we can use a sequence of stopping times and Fatou's Lemma, if necessary, to obtain

$$\mathbf{E}_{x,i}[V(X(t \wedge \tau),\alpha(t \wedge \tau))] \leq V(x,i) + \mathbf{E}_{x,i}\int_0^{t \wedge \tau} \mathcal{L}V(X(s),\alpha(s))ds \leq V(x,i).$$

Since V is nonnegative, we further have

$$\begin{aligned}
V(x,i) & \geq \mathbf{E}_{x,i}[V(X(\tau),\alpha(\tau))I_{\{\tau < t\}}] + \mathbf{E}_{x,i}[V(X(t),\alpha(t))I_{\{t \leq \tau\}}] \\
& \geq \mathbf{E}_{x,i}[V(X(\tau),\alpha(\tau))I_{\{\tau < t\}}].
\end{aligned} \tag{4}$$

For notational simplicity, denote $(\xi,\ell) = (X(\tau),\alpha(\tau))$. We claim that

$$V(\xi,\ell) > \rho, \text{ for some constant } \rho > 0. \tag{5}$$

To this end, write $\mathrm{Ker}(V) = \bigcup_{l=1}^k (N_{j_l} \times \{j_l\})$, where $k \leq m_0$, $N_{j_l} \subset \mathbb{R}^r$ and $j_l \in \mathcal{M}$, for $j_l = 1,\ldots,k$. Further, we denote $J = \{j_1,\ldots,j_k\} \subset \mathcal{M}$. For example, with $\mathbb{R}^r = \mathbb{R}$, $m_0 = 3$, $V(x,1) = x^2$, $V(x,2) = (x+1)^2$, and $V(x,3) = 1 + x^2$. Then $\mathrm{Ker}(V) = (\{0\} \times \{1\}) \cup (\{-1\} \times \{2\})$ or $N_1 = \{0\}$ and $N_2 = \{-1\}$.

Let us first consider the case when $\ell \notin J$. Note that $\xi \in D$, where D is a bounded neighborhood of $\bigcup_{l=1}^{k} N_l$ (such a neighborhood D exists because $\mathrm{Ker}(V)$ is bounded by assumption). Then we have

$$\inf \{V(x,\ell) : x \in D\} \geq \rho_1 > 0. \tag{6}$$

Suppose (6) were not true. Then there would exist a sequence $\{x_n\} \subset D$ such that $\lim_{n\to\infty} V(x_n, \ell) = 0$. Since $\{x_n\}$ is bounded, there exists a subsequence $\{x_{n_k}\}$ such that $x_{n_k} \to \widetilde{x}$. Thus by the continuity of $V(\cdot, \ell)$, we have

$$V(\widetilde{x}, \ell) = \lim_{k\to\infty} V(x_{n_k}, \ell) = 0.$$

That is, $(\widetilde{x}, \ell) \in \mathrm{Ker}(V)$. This is a contradiction to the assumption that $\ell \notin J$. Thus (6) is true and hence $V(\xi, \ell) \geq \rho_1$.

Now let us consider the case when $\ell \in J$. It follows that $\delta \leq d(\xi, N_\ell) \leq \widetilde{\delta} < \infty$. A similar argument using contradiction as in the previous case shows that

$$\inf \left\{ V(x,\ell) : \delta \leq d(x, N_\ell) \leq \widetilde{\delta} \right\} \geq \rho_2 > 0.$$

Thus it follows that $V(\xi, \ell) \geq \rho_2$. A combination of the two cases gives us $V(\xi, \ell) \geq \rho$, where $\rho = \rho_1 \wedge \rho_2$, Hence the claim follows.

Finally, we have from (4) and (5) that

$$\mathbf{P}_{x,i}\{\tau < t\} \leq \frac{1}{\rho}V(x,i).$$

Letting $t \to \infty$,

$$\mathbf{P}_{x,i}\{\tau < \infty\} \leq \frac{1}{\rho}V(x,i).$$

Note that

$$\{\tau < \infty\} = \left\{ \sup_{0 \leq t < \infty} d((X(t), \alpha(t)), \mathrm{Ker}(V)) \geq \delta \right\}.$$

Therefore it follows that

$$\mathbf{P}_{x,i}\left\{ \sup_{0 \leq t < \infty} d((X(t), \alpha(t)), \mathrm{Ker}(V)) \geq \delta \right\} \leq \frac{1}{\rho}V(x,i) \to 0,$$

as $d((x,i), \mathrm{Ker}(V)) \to 0$. This finishes the proof of the theorem. $\square$

Next we consider asymptotic stability. To this end, we need the following lemma.

Lemma 3.4. *Assume that there exists a nonnegative function* $V : \mathbb{R}^r \times \mathcal{M} \mapsto \mathbb{R}_+$ *with nonempty and bounded* $\mathrm{Ker}(V)$, *such that for each* $\alpha \in \mathcal{M}$,

$V(\cdot, \alpha)$ is twice continuously differentiable with respect to x, and that for any $\varepsilon > 0$

$$\mathcal{L}V(x, i) \leq -\kappa_\varepsilon < 0, \text{ for any } (x, i) \in (\mathbb{R}^r \times \mathcal{M}) - \overline{U}_\varepsilon, \tag{7}$$

where κ_ε is a positive constant depending on ε, U_ε is a neighborhood of $\text{Ker}(V)$ as defined in (3), and $\overline{U}_\varepsilon$ denotes the closure of U_ε. Then for any $0 < \varepsilon < r$, we have

$$\mathbf{P}_{x,i}\{\tau_{\varepsilon,r} < \infty\} = 1, \text{ for any } (x, i) \in U_{\varepsilon,r},$$

where

$$U_{\varepsilon,r} = \{(y, j) \in \mathbb{R}^r \times \mathcal{M} : \varepsilon < d((y, j), \text{Ker}(V)) < r\},$$

and $\tau_{\varepsilon,r}$ is the first exit time from $U_{\varepsilon,r}$, that is

$$\tau_{\varepsilon,r} := \inf\{t \geq 0 : (X(t), \alpha(t)) \notin U_{\varepsilon,r}\}.$$

Proof. Fix any $(x, i) \in U_{\varepsilon,r}$. By virtue of generalized Itô's Lemma, we have that for any $t \geq 0$,

$$V(X(t \wedge \tau_{\varepsilon,r}), \alpha(t \wedge \tau_{\varepsilon,r})) = V(x, i) + \int_0^{t \wedge \tau_{\varepsilon,r}} \mathcal{L}V(X(s), \alpha(s)) ds + M(t \wedge \tau_{\varepsilon,r}),$$

where $M(t \wedge \tau_{\varepsilon,r})$ is a martingale with mean zero. Thus by taking expectations on both sides, and using (7), we obtain

$$\mathbf{E}_{x,i}\left[V(X(t \wedge \tau_{\varepsilon,r}), \alpha(t \wedge \tau_{\varepsilon,r}))\right] \leq V(x, i) - \mathbf{E}_{x,i} \int_0^{t \wedge \tau_{\varepsilon,r}} \kappa_\varepsilon ds$$

$$= V(x, i) - \kappa_\varepsilon \mathbf{E}_{x,i}[t \wedge \tau_{\varepsilon,r}].$$

Note that V is nonnegative, hence we have

$$\mathbf{E}_{x,i}[t \wedge \tau_{\varepsilon,r}] \leq \frac{1}{\kappa_\varepsilon} V(x, i).$$

But

$$\mathbf{E}_{x,i}[t \wedge \tau_{\varepsilon,r}] = t\mathbf{P}_{x,i}\{\tau_{\varepsilon,r} > t\} + \mathbf{E}_{x,i}[\tau_{\varepsilon,r} I_{\{\tau_{\varepsilon,r} \leq t\}}] \geq t\mathbf{P}_{x,i}\{\tau_{\varepsilon,r} > t\}.$$

Thus it follows that

$$t\mathbf{P}_{x,i}\{\tau_{\varepsilon,r} > t\} \leq \frac{1}{\kappa_\varepsilon} V(x, i).$$

Now letting $t \to \infty$, we have

$$\mathbf{P}_{x,i}\{\tau_{\varepsilon,r} = \infty\} = 0 \text{ or } \mathbf{P}_{x,i}\{\tau_{\varepsilon,r} < \infty\} = 1.$$

The assertion thus follows. $\qquad\square$

Theorem 3.5. *Assume that there exists a function V satisfying the conditions of Lemma 3.4. Then $\mathrm{Ker}(V)$ is an invariant set for the process $(X(t), \alpha(t))$ and $\mathrm{Ker}(V)$ is asymptotically stable in probability.*

Proof. Our proof is motivated by Ref. 17, Theorem 5.36; we use similar ideas. By virtue of Theorem 3.3, we know that $\mathrm{Ker}(V)$ is an invariant set for the process $(X(t), \alpha(t))$ and that $\mathrm{Ker}(V)$ is stable in probability. Hence it remains to show that

$$\mathbf{P}_{x,i}\left\{\lim_{t\to\infty} d((X(t), \alpha(t)), \mathrm{Ker}(V)) = 0\right\} \to 1 \text{ as } d((x,i), \mathrm{Ker}(V)) \to 0.$$

Since $\mathrm{Ker}(V)$ is stable in probability, for any $\varepsilon > 0$ and any $\theta > 0$, there exists some $\delta > 0$ (without loss of generality, we may assume that $\delta < \theta$) such that

$$\mathbf{P}_{x,i}\left\{\sup_{t\geq 0} d((X(t), \alpha(t)), \mathrm{Ker}(V)) < \theta\right\} \geq 1 - \frac{\varepsilon}{2}, \tag{8}$$

for any $(x,i) \in U_\delta$, where U_δ is defined in (3). Now fix any $(x, \alpha) \in U_\delta - \mathrm{Ker}(V)$ and $r > d((x,i), \mathrm{Ker}(V))$. let $\rho > 0$ be arbitrary satisfying $0 < \rho < d((x,i), \mathrm{Ker}(V))$ and choose some $\varrho \in (0, \rho)$. Define

$$\tau_\varrho := \inf\{t \geq 0 : d((X(t), \alpha(t)), \mathrm{Ker}(V)) \leq \varrho\},$$
$$\tau_\theta := \inf\{t \geq 0 : d((X(t), \alpha(t)), \mathrm{Ker}(V)) \geq \theta\}.$$

Then it follows from Lemma 3.4 that

$$\mathbf{P}_{x,\alpha}\{\tau_\varrho \wedge \tau_\theta < \infty\} = \mathbf{P}_{x,\alpha}\{\tau_{\varrho,\theta} < \infty\} = 1, \tag{9}$$

where $\tau_{\varrho,\theta}$ is the first exit time from $U_{\varrho,\theta}$, and

$$U_{\varrho,\theta} := \{(y,j) \in \mathbb{R}^r \times \mathcal{M} : \varrho < d((X(t), \alpha(t)), \mathrm{Ker}(V)) < \theta\}.$$

But (8) implies that $\mathbf{P}_{x,\alpha}\{\tau_\theta < \infty\} \leq \frac{\varepsilon}{2}$. Note also

$$\mathbf{P}_{x,\alpha}\{\tau_\varrho \wedge \tau_\theta < \infty\} \leq \mathbf{P}_{x,\alpha}\{\tau_\varrho < \infty\} + \mathbf{P}_{x,\alpha}\{\tau_\theta < \infty\}.$$

Thus it follows that

$$\mathbf{P}_{x,\alpha}\{\tau_\varrho < \infty\} \geq \mathbf{P}_{x,\alpha}\{\tau_\varrho \wedge \tau_\theta < \infty\} - \mathbf{P}_{x,\alpha}\{\tau_\theta < \infty\} \geq 1 - \frac{\varepsilon}{2}. \tag{10}$$

Now let

$$\tau_\rho := \inf\{t \geq \tau_\varrho : d((X(t), \alpha(t)), \mathrm{Ker}(V)) \geq \rho\}.$$

(We use the convention that $\inf \{\emptyset\} = \infty$.) For any $t \geq 0$, we apply generalized Itô's lemma and (7) to obtain

$$\mathbf{E}_{x,\alpha} V(X(\tau_\rho \wedge t), \alpha(\tau_\rho \wedge t))$$

$$\leq \mathbf{E}_{x,\alpha} V(X(\tau_\varrho \wedge t), \alpha(\tau_\varrho \wedge t)) + \mathbf{E}_{x,\alpha} \int_{\tau_\varrho \wedge t}^{\tau_\rho \wedge t} \mathcal{L}V(X(s), \alpha(s))ds \quad (11)$$

$$\leq \mathbf{E}_{x,\alpha} V(X(\tau_\varrho \wedge t), \alpha(\tau_\varrho \wedge t)).$$

Note that $\tau_\varrho \geq t$ implies $\tau_\rho \geq t$ and hence on the set $\{\omega \in \Omega : \tau_\varrho(\omega) \geq t\}$ we have

$$\mathbf{E}_{x,\alpha} V(X(\tau_\rho \wedge t), \alpha(\tau_\rho \wedge t)) = \mathbf{E}_{x,\alpha} V(X(t), \alpha(t))$$
$$= \mathbf{E}_{x,\alpha} V(X(\tau_\varrho \wedge t), \alpha(\tau_\varrho \wedge t)). \quad (12)$$

Hence it follows from (11) and (12) that

$$\mathbf{E}_{x,\alpha} \left[I_{\{\tau_\varrho < t\}} V(X(\tau_\rho \wedge t), \alpha(\tau_\rho \wedge t)) \right]$$
$$\leq \mathbf{E}_{x,\alpha} \left[I_{\{\tau_\varrho < t\}} V(X(\tau_\varrho \wedge t), \alpha(\tau_\varrho \wedge t)) \right]$$
$$= \mathbf{E}_{x,\alpha} \left[I_{\{\tau_\varrho < t\}} V(X(\tau_\varrho), \alpha(\tau_\varrho)) \right] \leq \widehat{V}_\varrho,$$

where $\widehat{V}_\varrho := \sup \{V(y, j) : d((y, j), \mathrm{Ker}(V)) = \varrho\}$. Note that $\tau_\rho < t$ implies $\tau_\varrho < t$. Hence we further have

$$\widehat{V}_\varrho \left[I_{\{\tau_\varrho < t\}} V(X(\tau_\rho \wedge t), \alpha(\tau_\rho \wedge t)) \right]$$
$$\geq \mathbf{E}_{x,\alpha} \left[I_{\{\tau_\varrho < t\}} I_{\{\tau_\rho < t\}} V(X(\tau_\rho \wedge t), \alpha(\tau_\rho \wedge t)) \right]$$
$$= \mathbf{E}_{x,\alpha} \left[I_{\{\tau_\rho < t\}} V(X(\tau_\rho \wedge t), \alpha(\tau_\rho \wedge t)) \right]$$
$$= \mathbf{E}_{x,\alpha} \left[I_{\{\tau_\rho < t\}} V(X(\tau_\rho), \alpha(\tau_\rho)) \right]$$
$$\geq V_\rho \mathbf{P}_{x,\alpha} \{\tau_\rho < t\},$$

where $V_\rho := \inf \{V(y, j) : \rho \leq d((y, j), \mathrm{Ker}(V)) \leq \widetilde{\rho}\}$, with $\widetilde{\rho} > 0$ being some constant. Recall that we showed in the proof of Theorem 3.3 that $V_\rho > 0$. Now since V is continuous, we choose ϱ sufficiently small so that

$$\mathbf{P}_{x,\alpha} \{\tau_\rho < t\} \leq \frac{\widehat{V}_\varrho}{V_\rho} \leq \frac{\varepsilon}{2}.$$

Letting $t \to \infty$, we obtain

$$\mathbf{P}_{x,\alpha} \{\tau_\rho < \infty\} \leq \frac{\varepsilon}{2}. \quad (13)$$

Finally, it follows from (10) and (13) that

$$\mathbf{P}_{x,\alpha} \{\tau_\varrho < \infty, \tau_\rho = \infty\} \geq \mathbf{P}_{x,\alpha} \{\tau_\varrho < \infty\} - \mathbf{P}_{x,\alpha} \{\tau_\rho < \infty\} \geq 1 - \varepsilon.$$

This implies that

$$\mathbf{P}_{x,\alpha} \left\{ \limsup_{t \to \infty} d((X(t), \alpha(t)), \mathrm{Ker}(V)) \leq \rho \right\} \geq 1 - \varepsilon.$$

But $\rho > 0$ can be chosen to be arbitrarily small. Therefore we have

$$\mathbf{P}_{x,\alpha} \left\{ \lim_{t\to\infty} d((X(t),\alpha(t)), \mathrm{Ker}(V)) = 0 \right\} \geq 1 - \varepsilon.$$

This finishes the proof of the theorem. $\qquad\square$

Theorem 3.6. *Assume there exists a function V satisfying the conditions of Lemma 3.4. If V also satisfies*

$$\lim_{|x|\to\infty} \inf_{\alpha\in\mathcal{M}} V(x,\alpha) = \infty. \tag{14}$$

Then $\mathrm{Ker}(V)$ is asymptotically stable in probability in the large, that is, $\mathrm{Ker}(V)$ is stable in probability and

$$\mathbf{P}_{x,\alpha} \left\{ \lim_{t\to\infty} d((X(t),\alpha(t)), \mathrm{Ker}(V)) = 0 \right\} = 1, \tag{15}$$

for any $(x,\alpha) \in \mathbb{R}^r \times \mathcal{M}$.

Proof. By virtue of Theorem 3.3, $\mathrm{Ker}(V)$ is stable in probability. Thus it remains to verify (15). To this end, as in the proof of Theorem 3.3, write $\mathrm{Ker}(V) = \bigcup_{l=1}^{k} (N_{j_l} \times \{j_l\})$, where $k \leq m_0$, $N_{j_l} \subset \mathbb{R}^r$, and $j_l \in \mathcal{M}$. Since $\mathrm{Ker}(V)$ is bounded by assumption, in particular, $\bigcup_{l=1}^{k} N_{j_l}$ is bounded, there exists some $R > 0$ such that

$$\sup\left\{ |y| : y \in \bigcup_{l=1}^{k} N_{j_l} \right\} \leq R. \tag{16}$$

Let $\varepsilon > 0$ and fix any $(x,\alpha) \in \mathbb{R}^r \times \mathcal{M}$. Then (14) implies that there exists some positive constant $\beta > (R + 2) \vee d((x,\alpha), \mathrm{Ker}(V))$ such that

$$\inf\{V(y,j) : |y| \geq \beta, j \in \mathcal{M}\} \geq \frac{2V(x,\alpha)}{\varepsilon}. \tag{17}$$

Define

$$\tau_\beta := \inf\{t \geq 0 : d((X(t),\alpha(t)), \mathrm{Ker}(V)) \geq 2\beta\}.$$

For any $t \geq 0$, we have by virtue of generalized Itô's lemma and (7) that

$$\mathbf{E}_{x,\alpha} V(X(t\wedge\tau_\beta), \alpha(t\wedge\tau_\beta)) \leq V(x,\alpha). \tag{18}$$

We claim that $|X(\tau_\beta)| \geq \beta$. If this were not true, it would follow from (16) that for any $(y,j) \in \mathrm{Ker}(V)$,

$$d((X(\tau_\beta),\alpha(\tau_\beta)),(y,j)) \leq |X(\tau_\beta) - y| + 1 < \beta + R + 1 < 2\beta - 1,$$

where in the last inequality above, we used the fact that $\beta > R + 2$. Then we have $d((X(\tau_\beta),\alpha(\tau_\beta)), \mathrm{Ker}(V)) \leq 2\beta - 1 < 2\beta$. This is a contradiction with the definition of τ_β. Thus we must have $|X(\tau_\beta)| \geq \beta$. Then it follows from (18) that

$$V(x,\alpha) \geq \mathbf{E}_{x,\alpha}\left[V(X(\tau_\beta),\alpha(\tau_\beta))I_{\{\tau_\beta < t\}}\right]$$
$$\geq \inf\left\{V(y,j) : |y| \geq \beta, j \in \mathcal{M}\right\}\mathbf{P}_{x,\alpha}\left\{\tau_\beta < t\right\},$$

and hence (17) implies that

$$\mathbf{P}_{x,\alpha}\left\{\tau_\beta < t\right\} \leq \frac{\varepsilon}{2}.$$

By letting $t \to \infty$, we have

$$\mathbf{P}_{x,\alpha}\left\{\tau_\beta < \infty\right\} \leq \frac{\varepsilon}{2}.$$

Then we can finish the proof using the same argument in the proof of Theorem 3.5. □

The following theorem provides a criterion for asymptotic stability with probability 1.

Theorem 3.7. *If there exists a nonnegative function* $V : U \mapsto \mathbb{R}_+$ *such that for each* $\alpha \in \mathcal{M}$, $V(\cdot,\alpha)$ *is twice continuously differentiable with respect to* x *and that there exists a continuous function* $\widehat{W} : \mathbb{R}^r \times \mathcal{M} \mapsto \mathbb{R}_+$ *satisfying*

$$\mathcal{L}V(x,i) \leq -\widehat{W}(x,i), \text{ for any } (x,i) \in U, \tag{19}$$

where $U \subset \mathbb{R}^r \times \mathcal{M}$ *is an invariance set for the process* $(X(t),\alpha(t))$. *Assume also that either* U *is bounded or*

$$\lim_{|x| \to \infty,\ (x,\alpha) \in U} V(x,\alpha) = \infty. \tag{20}$$

Then for any initial condition $(x,\alpha) \in \mathbb{R}^r \times \mathcal{M}$, *the following assertions are true*

(i) $\limsup_{t \to \infty} V(X^{x,\alpha}(t), \alpha^{x,\alpha}(t)) < \infty$ *a.s.*,

(ii) $\mathrm{Ker}(\widehat{W}) \neq \emptyset$,

(iii) $\lim_{t \to \infty} d((X^{x,\alpha}(t), \alpha^{x,\alpha}(t)), \mathrm{Ker}(\widehat{W})) = 0$ *a.s., and*

(iv) *if moreover,* $\mathrm{Ker}(\widehat{W}) = \{0\} \times \mathcal{M}$, *then*

$$\lim_{t \to \infty} X^{x,\alpha}(t) = 0 \quad a.s. \ .$$

Proof. This theorem can be proved using the arguments in Ref. 16, Theorem 2.1. Some modifications are needed, though. □

We end this section with the following results on linear systems. More specifically, we assume that the evolution (2) is replaced by

$$dX(t) = b(\alpha(t))X(t)dt + \sum_{j=1}^{d} \sigma_j(\alpha(t))X(t)dw_j(t), \qquad (21)$$

where $b(i), \sigma_j(i)$ are $r \times r$ constant matrices and $w_j(t)$ are independent 1-dimensional standard Brownian motions for $i = 1, 2, \ldots, m_0$ and $j = 1, 2, \ldots, d$.

Note that 0 is an equilibrium point for the system given by (21) and (3). As we indicated earlier, it was shown in Refs. 12 and 15 that the set $\mathbb{R}^r \times \mathcal{M}$ is invariant with respect to the process $(X(t), \alpha(t))$.

Theorem 3.8. *Assume that the discrete component $\alpha(\cdot)$ is ergodic with constant generator $Q = (q_{ij})$ and invariant distribution $\pi = (\pi_1, \ldots, \pi_{m_0}) \in \mathbb{R}^{1 \times m_0}$. Then the equilibrium point $x = 0$ of system given by (21) and (3)*

(i) *is asymptotically stable with probability 1 if*

$$\sum_{i=1}^{m_0} \pi_i \lambda_{\max}\left(b(i) + b'(i) + \sum_{j=1}^{d} \sigma_j(i)\sigma_j'(i) \right) < 0, \qquad (22)$$

(ii) *is unstable in probability if*

$$\sum_{i=1}^{m_0} \pi_i \left[\lambda_{\min}\left(b(i) + b'(i) + \sum_{j=1}^{d} \sigma_j(i)\sigma_j'(i) \right) \right.$$
$$\left. -\frac{1}{2} \left(\rho(\sigma_j(i) + \sigma_j'(i)) \right)^2 \right] > 0. \qquad (23)$$

Proof. We need only prove assertion (i), since assertion (ii) was proved in Ref. 12. For notational simplicity, define the column vector $\mu = (\mu_1, \mu_2, \ldots, \mu_{m_0})' \in \mathbb{R}^{m_0}$ with

$$\mu_i = \frac{1}{2}\lambda_{\max}\left(b(i) + b'(i) + \sum_{j=1}^{d} \sigma_j(i)\sigma_j'(i) \right).$$

Also let $\beta := -\pi\mu$. Note that $\beta > 0$ by (22). It follows from the result in Ref. 12 that the equation

$$Qc = \mu + \beta\mathbb{1}$$

has a solution $c = (c_1, c_2, \ldots, c_{m_0})' \in \mathbb{R}^{m_0}$. Thus we have

$$\mu_i - \sum_{j=1}^{m_0} q_{ij}c_j = -\beta, \quad i \in \mathcal{M}. \qquad (24)$$

For each $i \in \mathcal{M}$, consider the Liapunov function $V(x,i) = (1 - \gamma c_i)|x|^\gamma$, where $0 < \gamma < 1$ is sufficiently small so that $1 - \gamma c_i > 0$ for each $i \in \mathcal{M}$. It is readily seen that for each $i \in \mathcal{M}$, $V(\cdot, i)$ is continuous, nonnegative, vanishes only at $x = 0$, and satisfies (20). Detailed calculations as in the proof of Ref. 12, Theorem 4.3, reveal that for $x \neq 0$, we have

$$
\mathcal{L}V(x,i) = \gamma(1 - \gamma c_i)|x|^\gamma \left\{ \frac{x'b(i)x}{|x|^2} - \sum_{j=1}^{m_0} q_{ij} \frac{c_j - c_i}{1 - \gamma c_i} \right.
$$
$$
\left. + \frac{1}{2} \sum_{j=1}^{d} \left(\frac{x'\sigma_j'(i)\sigma_j(i)x}{|x|^2} + (\gamma - 2)\frac{(x'\sigma_j'(i)x)^2}{|x|^4} \right) \right\}. \tag{25}
$$

Note that

$$
\frac{x'b(i)x}{|x|^2} + \frac{1}{2} \sum_{j=1}^{d} \frac{x'\sigma_j'(i)\sigma_j(i)x}{|x|^2}
$$
$$
= \frac{x'(b'(i) + b(i))x}{2|x|^2} + \frac{1}{2} \sum_{j=1}^{d} \frac{x'\sigma_j'(i)\sigma_j(i)x}{|x|^2} \tag{26}
$$
$$
\leq \frac{1}{2}\lambda_{\max}\left(b(i) + b(i)' + \sum_{j=1}^{d} \sigma_j'(i)\sigma_j(i) \right) = \mu_i.
$$

Next, it follows that when γ is sufficiently small,

$$
\sum_{j=1}^{m_0} q_{ij} \frac{c_j - c_i}{1 - \gamma c_i} = \sum_{j=1}^{m_0} q_{ij}c_j + \sum_{j \neq i} q_{ij} \frac{c_i(c_j - c_i)}{1 - \gamma c_i}\gamma = \sum_{j=1}^{m_0} \widehat{q}_{ij}c_j + O(\gamma). \tag{27}
$$

Hence it follows from (24)–(27) that when $0 < \gamma < 1$ sufficiently small, we have

$$
\mathcal{L}V(x,i) \leq \gamma(1 - \gamma c_i)|x|^\gamma \left\{ \mu_i - \sum_{j=1}^{m_0} \widehat{q}_{ij}c_j + O(\gamma) \right\}
$$
$$
= \gamma(1 - \gamma c_i)|x|^\gamma (-\beta + O(\gamma))
$$
$$
\leq -\frac{\beta}{2}\gamma(1 - \gamma c_i)|x|^\gamma := -\widehat{W}(x,i).
$$

Note that $\mathrm{Ker}(\widehat{W}) = \{0\} \times \mathcal{M}$. Therefore we conclude from Theorem 3.7 point $x = 0$ is asymptotically stable with probability 1. □

4. Further Remarks

This paper has focused on invariance principles of regime-switching diffusion processes. The results obtained accommodate the stability results and provide further insight. For subsequent study, we may consider such problems as the ω-limit sets in the deterministic setup. In lieu of memoryless setup, if delays are considered in the system, one gets switching diffusions with delay. Studying invariance for such systems is a worthwhile effort. Another direction is the investigation of associated problems with Poisson type of jumps, in which the sample paths are no longer continuous. It will be interesting to see if any invariance principle can be obtained in that regards.

Acknowledgments

This research was supported in part by the National Science Foundation under grant DMS-0907753 and in part by the Air Force Office of Scientific Research under grant FA9550-10-1-0210.

References

1. G. Barone-Adesi and R. Whaley, Efficient analytic approximation of American option values, *J. Finance*, **42** (1987), 301–320.
2. G.K. Basak, A. Bisi, and M.K. Ghosh, Stability of Degenerate Diffusions with State-Dependent Switching, *J. Math. Anal. Appl.*, **240** (1999), 219-248.
3. R.J. Elliott, L. Chan, and T.K. Siu, Option pricing and Esscher transform under regime switching, *Annals of Finance*, **1** (2005), 423–432.
4. A. Friedman, *Stochastic Differential Equations and Applications*, Vols. I and II, Academic Press, New York, 1975.
5. M.K. Ghosh, A. Arapostathis, and S.I. Marcus, Ergodic control of switching diffusions, *SIAM J. Control Optim.*, **35** (1997), 1952-1988.
6. J.K. Hale, Sufficient conditions for stability and instability of autonomous functional differential equations, *J. Diff. Equations*, **1** (1965), 452–482.
7. J.K. Hale and E.F. Infante, Extended dynamic systems and stability theorem, *Proc. Nat. Acad. Sci.*, **58** (1967), 405-409.
8. J.D. Hamilton and R. Susmel, Autoregressive conditional heteroskedasticity and changes in regime. *J. Econometrics* **64**, (1994), 307-333.
9. A.M. Il'in, R.Z. Khasminskii, and G. Yin, Asymptotic expansions of solutions of integro-differential equations for transition densities of singularly perturbed switching diffusions, *J. Math. Anal. Appl.*, **238** (1999), 516-539.
10. J. Jacod and A.N. Shiryayev, *Limit Theorems for Stochastic Processes*, Springer-Verlag, New York, 1980.
11. R.Z. Khasminskii, *Stochastic Stability of Differential Equations*, Sijthoff and Noordhoff, Alphen aan den Rijn, Netherlands, 1980.
12. R.Z. Khasminskii, C. Zhu, and G. Yin, Stability of regime-switching diffusions, *Stochastic Proc. Appl.*, **117** (2007), 1037-1051.
13. H.J. Kushner, The concept of invariant set for stochastic dynamical systems and applications to stochastic stability, in H.F. Karreman, editor, *Stochastic Optimization and Control*, John Wiley and Sons, New York, 1968, 47-57.

14. J.P. LaSalle, The extent of asymptotic stability, *Proc. Nat. Acad. Sci.*, **46** (1960), 363–365.

15. X. Mao, Stability of stochastic differential equations with Markovian switching, *Stochastic Proc. Appl.*, **79** (1999), 45-67.

16. X. Mao, A note on the LaSalle-Type theorems for stochastic differential delay equations, *J. Math. Anal. Appl.*, **268** (2002), 125–142.

17. X. Mao and C. Yuan, *Stochastic Differential Equations with Markovian Switching*, Impreial College Press, London, 2006.

18. M. Pemy, Q. Zhang, and G. Yin, Liquidation of a large block of stock with regime switching, *Math. Finance*, **18** (2008), 629-648.

19. A.V. Skorohod, *Asymptotic Methods in the Theory of Stochastic Differential Equations*, Amer. Math. Soc., Providence, RI, 1989.

20. H.L. Yang and G. Yin, Ruin probability for a model under Markovian switching regime, in *Probability, Finance and Insurance*, World Sci., 206-217, T.L. Lai, H. Yang, and S.P. Yung Eds., 2004.

21. G. Yin, V. Krishnamurthy, and C. Ion, Regime switching stochastic approximation algorithms with application to adaptive discrete stochastic optimization, *SIAM J. Optim.*, **14** (2004), 1187-1215.

22. G. Yin, X. Mao, C. Yuan, and D. Cao, Approximatio methods for hybrid diffusion systems with state-dependent switching diffusion processes: Numerical alogorithms and existence and uniqueness of solutions, *SIAM Journal on Mathematical Analysis*, **41** (2010), 2335–2352.

23. G. Yin and C. Zhu, Regularity and recurrence of switching diffusions, *J. Syst. Sci. Complexity*, **20** (2007), 273-283.

24. G. Yin and C. Zhu, *Hybrid Switching Diffusions: Properties and Applications*, Springer, New York, 2010.

25. C. Yuan and X. Mao, Asymptotic stability in distribution of stochastic differential equations with Markovian switching, *Stochastic Proc. Appl.*, **103** (2003), 277-291.

26. Q. Zhang, Stock trading: An optimal selling rule, *SIAM J. Control Optim.*, **40** (2001), 64–87.

27. X.Y. Zhou and G. Yin, Markowitz's mean-variance portfolio selection with regime switching: a continuous-time model, *SIAM J. Control Optim.*, **42** (2003), 1466–1482

28. C. Zhu and G. Yin, Asymptotic properties of hybrid diffusion systems, *SIAM J. Control Optim.*, **46** (2007), 1155-1179.

29. C. Zhu and G. Yin, On strong Feller, recurrence, and weak stabilization of regime-switching diffusions, *SIAM J. Control Optim.*, **48** (2009), 2003–2031.

30. C. Zhu, G. Yin, and Q.S. Song, Stability of random-switching systems of differential equations, *Quarterly Appl. Math.*, **67** (2009), 201–220.

PART II

FINANCE AND STOCHASTICS

Real Options and Competition

Alain Bensoussan

School of Management
University of Texas at Dallas
The Hong Kong Polytechnic University
Richardson, TX 75083-0688, Hong Kong
Email: alain.bensoussan@utdallas.edu

J. David Diltz

Department of Finance and Real Estate
The University of Texas at Arlington
Arlington, TX 76019, USA
Email: diltz@uta.edu

SingRu (Celine) Hoe

School of Management
University of Texas at Dallas
Richardson, TX 75083-0688, USA
Email: celinehoe@utdallas.edu

This paper provides a review of our research, Bensoussan et al. Refs. 3,4, which explore optimal investment policies for irreversible capital investment projects under uncertainty in monopoly and Stackelberg leader-follower frameworks. Optimal policies derived for the Stackelberg duopoly case in an incomplete market are, to our knowledge, heretofore undocumented. We show that competition and market incompleteness have vital impacts on investment decisions. In our survey, we present models in the case of complete markets and in the case of incomplete markets with different types of investment payoffs and market competition frameworks. We discuss model setup, explaining solution technique and procedure. We provide major results and theorem, and refer proofs and details to our original papers Refs. 3,4.

Keywords: Real options; optimal stopping; variational inequality; utility based pricing; differential game.

1. Introduction

This paper provides a review of our research, Bensoussan et al. Refs. 3,4, which explore optimal investment policies for irreversible capital investment projects under uncertainty in monopoly and Stackelberg leader-follower frameworks. We present models in the case of complete markets and in the case of incomplete markets with different types of investment payoffs and market competition frameworks. We discuss model setup, explaining solution technique and procedure. We provide major results and theorem.

Among the most significant developments in corporate finance over the last twenty-five years are arbitrage-free pricing models to valuation and management of the firm's fixed assets. Dubbed "real options" by Myers (1977), research in this area has since proliferated, and landmark works are numerous. An essential work in this domain is the book of Dixit and Pindyck (1994) Ref. 9. Additional references are Constantinides (1978) Ref. 8 applying stochastic optimal control theory to investment decisions, Brennan and Schwartz (1982a, 1982b) Refs. 5,6 who apply stochastic control to the problem of a regulated public utility, Brennan and Schwartz (1985) Ref. 7, McDonald and Siegel (1986) Ref. 18, among others.

Two important issues distinguish real options from their financial counterparts. First, competitive strategy is seldom relevant in finance because financial markets are assumed to be efficient and liquid. No single trader exerts a measurable impact on the market. However, competitive strategy is common to real options because capital investment decisions made by one firm frequently have a significant impact on the cash flows of its competitors (see Dixit and Pindyck (1994, Ch9) Ref. 9 and Grenadier (1996, 1999, 2002) Refs. 10–12). Second, market incompleteness is not a major problem in financial markets because most financial assets are heavily traded. It is a critical issue in real options because assets underlying real options are rarely traded in markets. Financial mathematicians address the market incompleteness issue by employing strategies such as: (1) minimizing tracking error, (2) selecting a martingale measure for pricing using minimal martingale or minimal entropy methods, and (3) employing utility functions to estimate an indifference price. The existing real options literature is quite limited in addressing competition and incompleteness simultaneously due to the mathematical complexity.

Ignoring these issues by applying classical real options models may yield a non-optimal capital investment policy. In our research, Bensoussan et al. Refs. 3,4, we extend the current financial economics research by simultaneously incorporating competitive strategy and market incompleteness. Our primary results concerning optimal investment policies when competitive strategy and market incompleteness appear simultaneously are, to our knowledge, new.

We consider an irreversible capital investment project under uncertainty in a Stackelberg leader-follower game, with predetermined leader and follower. While decisions are made over an infinite horizon, the follower is forbidden to undertake the investment project until the leader has already done so. We model uncertainties with respect to two types of investment payoffs: lump-sum investment payoffs and investment payoffs as a series of cash flows. The lump-sum payoff is characterized by an externally defined project value process. The cash flow payoff is generated by either: (1) an arithmetic Brownian motion cash flow process, or (2) a geometric Brownian motion cash flow process, where the former captures the possibility of losses from investment operation.

The firm's problem is to evaluate the investment project and select the optimal (stopping) time to invest. We use variational inequalities (V.I.s) to solve the optimal stopping problems corresponding to investment decisions. V.I. theory was introduced by Stampacchia (1964) and Lions and Stampacchia (1967) Ref. 17 with applications drawn principally from mechanics. Bensoussan and Lions (1982) Ref. 1 were the first to apply V.I.s to solve stochastic optimal control problems.

The issue of completeness and incompleteness comes in for the valuation of payoffs. In the case of complete markets, there exists a unique martingale measure, so the value of future payoffs are simply discounted mathematical expectations with respect to this unique measure. In the case of incomplete markets, we adopt the traditional utility maximization approach. The investor/ entrepreneur maximizes his/her expected discounted exponential (CARA) utility function with respect to a stopping time for capital investment, an investment strategy, and/or a consumption rule.

For the lump-sum payoff in the case of incomplete markets, we overcome the leader's problem of comparing gains and losses at different times using certainty equivalence. In this case, we can only formulate and solve leader's V.I. in the weak sense. For the case of investment payoffs as a series of cash flows in incomplete markets, we solve the utility maximization problem in two steps: (1) we assume that the investment is already in place (i.e., the "post-investment" period), and solve the control problem for a portfolio strategy and a consumption rate; (2) In the second step, we solve the complete utility maximization problem, choosing an optimal stopping time, an optimal consumption rule and an optimal portfolio investment strategy (i.e., the "pre-investment" period). The idea behind this two-step procedure is that we need the solution of the post-investment period to define the obstacle function of the V.I. for the pre-investment period. This V.I.

has a nice interpretation as a stochastic differential game, different from the Stackelberg game.

The follower's (and monopolist's) V.I. has a continuously differentiable (C^1) "obstacle" function (i.e., payoff received at the stopping time). In this case, the optimal stopping time decision can be obtained by a threshold approach. The Stackelberg leader's V.I. contains a non-smooth obstacle function $(C^0,$ not $C^1)$, and thus we must formulate the V.I. in a *weak* sense because the solution of the V.I. has the regularity of the obstacle. Nevertheless, in most cases, we are able to show a strong solution of the V.I. exists since the obstacle is not continuously differentiable only at a single point. Due to the lack of smoothness of the obstacle, the leader's optimal strategy (the solution of the V.I.) is a *two-interval solution*, characterized by three thresholds.

Except for the case of lump-sum investment payoffs in case of incomplete markets, we are able to characterize the leader's optimal investment policies as three thresholds. To enjoy monopoly rents for a longer period of time, the leader prefers choosing the lower level of threshold. This understandable but non-intuitive situation cannot be predicted without the mathematical theory.

The outline of the paper is as follows. First, we review the valuation and general features of our research problems and models. We present models and summarize our research results in the case of complete markets and in the case of incomplete markets in Sections 3 and 4 respectively. We end with a conclusion in Section 5.

2. Valuation and General Features of Problems and Models

2.1. *Asset Valuation in Complete and Incomplete Markets*

The essential property of complete markets is that there exists a uniquely defined risk-neutral probability measure. In the case of complete markets we will measure the value of losses and gains by risk-neutral expectation. For our future reference, the framework is as follows. Consider a probability space $(\Omega, \mathcal{F}, \mathbb{Q})$ with a Wiener process $W(t)$. We assume that the market is characterized by a single asset, $S(t)$, evolving as:

$$dS(t) = rS(t)dt + \sigma S(t)(\lambda dt + dW(t)), \tag{1}$$

where r is a risk free interest rate, σ is the volatility and λ is the Sharpe ratio; they are all constants. The market is complete and there exists a unique risk-neutral probability measure, obtained as follows. Define:

$$\widehat{W}(t) = W(t) + \lambda t$$

and the process $Z(t)$ by:

$$dZ(t) = -\lambda Z(t)dW(t), \quad Z(0) = 1.$$

The Radon-Nikodym density of the risk neutral probability measure $\widehat{\mathbb{Q}}$ with respect to $\mathbb{Q}$ is:

$$\frac{d\widehat{\mathbb{Q}}}{d\mathbb{Q}}\Big|_{\mathcal{F}_t} = Z(t)$$

with $\mathcal{F}_t = \sigma(W(s), s \leq t)$. Under $\widehat{\mathbb{Q}}$, $\widehat{W}(t)$ is a standard Wiener process. We denote $\widehat{E}[\cdot]$ as the expectation with respect to the risk neutral probability measure $\widehat{\mathbb{Q}}$.

If capital markets are not complete, the analysis changes. The risk-neutral probability is not uniquely defined. The classical risk-neutral valuation approach is no longer appropriate. One proposed solution is to introduce a rational utility-maximizing investor who evaluates unhedgeable risk based on the investor's risk preferences. Utility-based valuation in stochastic dynamic market environments derives from the famous work of Merton (1969) Ref. 19. In our incomplete model, we employ the utility-based valuation, assuming that a hypothetical investor/entrepreneur's risk preferences may be specified by an exponential (CARA) utility function.

2.2. *Market Frameworks*

We consider two different market frameworks: a single market player (monopolist) or two market players (duopolists). Market players' decisions are times to enter into the market. In the monopolist's case, by paying an investment cost K at the market entry time τ, the firm/entrepreneur expects to receive the whole operation income.

In the duopolists' case, we consider a Stackelberg leader-follower game, with predetermined leader and follower. The follower is forbidden to invest until the leader has already done so. Each player pays an investment cost K upon entry. The leader receives the whole operation income prior to the follower's entry. Upon follower's entry, the leader must share market with follower. In our framework, the portion of share is given.

2.3. Types of Investment Payoffs

We model two types of operating income from the capital investment undertaken. One is in the form of a lump-sum, i.e., the investment project yields a one time payoff at the time of investment for a given initial investment K. We model this situation in terms of the investment project value, governed by an externally determined geometric Brownian motion process. In this model, the leader has to incorporate into his/her objective function the surrender value to the follower upon the follower's optimal entry, so one has to compare values of events taking place at different times. In the case of incomplete markets, we work with investor's utility and there is no identity relationship for values at different times. We circumvent the problem by employing equivalence (indifference) considerations.

Alternatively, we consider operating income characterized by cash flows, i.e., upon paying an initial investment K, the entrepreneur receives a series of cash flows thereafter. The cash flow evolves as either an arithmetic Brownian motion process or a geometric Brownian motion process. The cash flow valuation model circumvents the problem of comparing gains and losses at different times.

3. The Complete Market Case

3.1. Single Player

3.1.1. Lump-sum Payoffs - External Value Process

The project value defining lump-sum investment payoffs evolves as:

$$dV(t) = rV(t)dt + \eta V(t)(\xi dt + dW(t)) \tag{2}$$
$$= (r + \eta(\xi - \lambda))V(t)dt + \eta V(t)d\widehat{W}(t), \quad V(0) = v,$$

where η and ξ are constants.

By paying K at τ, the firm expects to receive the corresponding payoff $(1 - a)V_v(\tau)$, where $a \in [0, 1)$ represents market share by other active participants. For the single decision maker case, a is zero. In the Stackelberg games that follow, a represents the leader's market share after the follower's entry into the market.[a]

---

[a]In the context of the leader-follower model, we assume $a > \frac{1}{2}$. Otherwise, the leader will surrender a majority interest in the project, providing a strong disincentive to enter at all.

The manager's expected discounted payoff undertaken at time τ is:

$$J_v(\tau) = \widehat{E}\big[e^{-r\tau}\big((1-a)V_v(\tau) - K\big)\mathbb{1}_{\tau<\infty}\big]. \tag{3}$$

The manager chooses an optimal stopping time to maximize the expected discounted project value:

$$F(v) = \sup_{\tau \geq 0} J_v(\tau). \tag{4}$$

Solving (4) by V.I., we obtain:

$$F(v) = \begin{cases} \dfrac{K}{\beta - 1}\big(\dfrac{v}{\hat{v}}\big)^{\beta} & v \leq \hat{v} \\ (1-a)v - K & v \geq \hat{v} \end{cases}, \tag{5}$$

where

$$\beta = \frac{1}{2} - \frac{r + \eta(\xi - \lambda)}{\eta^2} + \sqrt{\big(\frac{1}{2} - \frac{r + \eta(\xi - \lambda)}{\eta^2}\big)^2 + \frac{2r}{\eta^2}} > 1, \tag{6}$$

$$\hat{v} = \frac{\beta K}{(1-a)(\beta - 1)} \tag{7}$$

and we assume $\xi < \lambda$.

The optimal stopping rule $\hat{\tau}(v)$ which achieves the supremum in (4) is:

$$\hat{\tau}(v) = \inf\{t | V_v(t) \geq \hat{v}\} \tag{8}$$

with $\hat{(v)}$ defined in (7).

3.1.2. *Cash Flow Payoffs - Geometric Brownian Motion Cash Flow Process*

The cash flow process $Y(t)$ from investment operation evolves as:

$$dY(t) = Y(t)(\alpha dt + \varsigma dW(t)) \tag{9}$$
$$= Y(t)\big((\alpha - \lambda\varsigma)dt + \varsigma d\widehat{W}(t)\big), \quad Y(0) = y,$$

where α and ς are constants.

If the firm exploits the investment opportunity by paying cost K, at time t, the firm expects to receive a continuous cash flow $\delta Y(t)$ per unit time. The project value at time t, $V(t)$, is:

$$V(t) = \delta \widehat{E}\Big[\int_t^{\infty} e^{-r(s-t)}Y_y(s)ds\big|\mathcal{F}_t\Big] = \frac{\delta Y_y(t)}{r - (\alpha - \lambda\varsigma)}, \tag{10}$$

provided $r - (\alpha - \lambda\varsigma) > 0$.

From (10), the manager's expected discounted payoff from the capital investment project taken at time τ is:

$$J_y(\tau) = \widehat{E}\Big[e^{-r\tau}\Big(\frac{\delta Y_y(\tau)}{r - (\alpha - \lambda\varsigma)} - K\Big)\mathbb{1}_{\tau<\infty}\Big]. \tag{11}$$

The manager's value function is:

$$F(y) = \sup_{\tau \geq 0} J_y(\tau). \tag{12}$$

Solving (12) by V.I., we obtain:

$$F(y) = \begin{cases} \Big(\dfrac{\delta}{\beta\big(r - (\alpha - \lambda\varsigma)\big)}\Big)^{\beta}\Big(\dfrac{\beta-1}{K}\Big)^{\beta-1} y^{\beta}, & y \leq \hat{y} \\ \dfrac{\delta y}{r - (\alpha - \lambda\varsigma)} - K, & y \geq \hat{y} \end{cases} \tag{13}$$

where

$$\beta = \frac{1}{2} - \frac{\alpha - \lambda\varsigma}{\varsigma^2} + \sqrt{\Big(\frac{1}{2} - \frac{\alpha - \lambda\varsigma}{\varsigma^2}\Big)^2 + \frac{2r}{\varsigma^2}} > 1, \tag{14}$$

$$\hat{y} = \frac{\beta K}{\beta - 1}\frac{r - (\alpha - \lambda\varsigma)}{\delta} \tag{15}$$

and we assume $r - (\alpha - \lambda\varsigma) > 0$.

The optimal stopping rule $\hat{\tau}(y)$ that achieves the supremum in (12) is:

$$\hat{\tau}(y) = \inf\{t | Y_y(t) \geq \hat{y}\} \tag{16}$$

with $\hat{y}$ defined in (15).

3.1.3. Cash Flow Payoffs - Arithmetic Brownian Motion Cash Flow Process

The cash flow process $Y(t)$ from investment operation evolves as:

$$dY(t) = \alpha dt + \varsigma dW(t) \tag{17}$$

$$= (\alpha - \lambda\varsigma)dt + \varsigma d\widehat{W}(t), \quad Y(0) = y,$$

where α and ς are constants.

If the firm exploits the investment opportunity by paying cost K, at time t, the firm expects to receive a continuous cash flow $\delta Y(t)$ per unit time. The project value at time t, $V(t)$, is:

$$V(t) = \delta\widehat{E}\Big[\int_t^{\infty} e^{-r(s-t)}Y_y(s)ds | \mathcal{F}_t\Big] = \delta\Big(\frac{Y_y(t)}{r} + \frac{\alpha - \lambda\varsigma}{r^2}\Big). \tag{18}$$

From (18), the manager's expected discounted payoff from the capital investment project taken at time τ is:

$$J_y(\tau) = \widehat{E}\big[e^{-r\tau}\big(\delta(\frac{Y_y(\tau)}{r} + \frac{\alpha - \lambda_\varsigma}{r^2}) - K\big)\mathbb{1}_{\tau < \infty}\big]. \tag{19}$$

The manager's value function is:

$$F(y) = \sup_{\tau \geq 0} J_y(\tau). \tag{20}$$

Solving (20) by V.I., we obtain:

$$F(y) = \begin{cases} \dfrac{\delta}{r\beta}\exp\{-\beta(\hat{y} - y)\}, & \text{if } y \leq \hat{y} \\[2ex] \delta(\dfrac{y}{r} + \dfrac{\alpha - \lambda_\varsigma}{r^2}) - K, & \text{if } y \geq \hat{y} \end{cases} \tag{21}$$

where

$$\beta = -\frac{\alpha - \lambda_\varsigma}{\varsigma^2} + \sqrt{(\frac{\alpha - \lambda_\varsigma}{\varsigma^2})^2 + \frac{2r}{\varsigma^2}} > 0, \tag{22}$$

and

$$\hat{y} = \frac{1}{\beta} + \frac{r}{\delta}K - \frac{\alpha - \lambda_\varsigma}{r} > 0. \tag{23}$$

The optimal stopping rule $\hat{\tau}(y)$ that achieves the supremum in (20) is:

$$\hat{\tau}(y) = \inf\{t | Y_y(t) \geq \hat{y}\} \tag{24}$$

with $\hat{y}$ defined in (23).

3.2. *Two Players: A Stackelberg Leader-Follower Game*

There are two players with predetermined leader and follower. The follower is forbidden to invest until the leader has already done so. The leader enters at time θ and the follower at time $\tau \geq \theta$ where θ and τ are stopping times of the filtration $\mathcal{F}_t$. Each player invests K upon entry. Upon follower's optimal entry, the leader must share market (i.e., project value) with follower. In our framework, the portion of share is given.

3.2.1. *Lump-sum Payoffs - External Value Process*

Upon follower's optimal entry, leader surrenders $(1 - a)$, $a \in (0, 1)$ portion of project value to the follower.

A. Follower

The follower's problem is similar to the single player's (the monopolist's) since, after the leader enters, the follower makes an optimal stopping decision as if he/she is a single player. The follower's solution is the same as that defined in (8). However, the stopping time $\hat{\tau}(v)$ is the optimal entry if the follower can enter in the market at time zero. Since the follower enters after the leader (who starts at θ), for finite θ, the follower will enter at time: $\hat{\tau}_\theta = \theta + \hat{\tau}(V_v(\theta))$.

B. Leader

When the leader enters at time $\theta < \infty$, by paying cost K, he/she receives:

$$aV_v(\theta)\mathbb{1}_{V_v(\theta) \geq \hat{v}} + (V_v(\theta) - \beta F(V_v(\theta)))\mathbb{1}_{V_v(\theta) < \hat{v}} - K,$$

where $F(v)$, β and $\hat{v}$ are defined in (5), (6) and (7) respectively. The leader's value function is:

$$L(v) = \sup_{\theta \geq 0} \widehat{E}\big[e^{-r\theta}\Psi(V_v(\theta))\mathbb{1}_{\theta < \infty}\big], \tag{25}$$

where $\Psi(v) = av\mathbb{1}_{v \geq \hat{v}} + (v - \beta F(v))\mathbb{1}_{v < \hat{v}} - K$. $L(v)$ must satisfy:

$$L(v) \geq 0; \quad L(v) \geq \Psi(v). \tag{26}$$

Setting $U(v) = L(v) - av + K$, (25) can be re-expressed as:

$$U(v) = \sup_{\theta \geq 0} \widehat{E}\bigg\{e^{-r\theta}\chi(V_v(\theta))\mathbb{1}_{\theta < \infty} + \int_0^\theta e^{-rs}f(V_v(s))ds\bigg\}. \tag{27}$$

This formulation leads to an optimal stopping time problem with an obstacle $\chi(v) = \Psi(v) - av + K$ and a running profit $f(v) = a\eta(\xi - \lambda)v + rK$. Both $U(v)$ and $\chi(v)$ are bounded; we have:

$$0 \leq \chi(v) \leq K, \text{ and } 0 \leq U(v) \leq K. \tag{28}$$

The obstacle $\chi(v)$ is not $C^1(0, \infty)$ with a single non-differentiable point at $\hat{v}$ and we have to formulate the V.I. in the weak sense.[b]

Through studying the regularity of the solution $U(v)$ by partial differential equation techniques, it turns out that the solution $U(v)$ will be smoother than the obstacle. Therefore, the V.I. will have a strong formulation, and we can in fact obtain an explicit *two-interval solution*.

---

[b]We refer the formulation and proof to the original work Bensoussan et al. (2009) Ref. 3.

Theorem 3.1. *There exist three points $0 < v_1 < v_2 < \hat{v} < v_3$ such that:*

$$
\begin{cases}
-\frac{1}{2}U''(v)\eta^2 v^2 - (r + \eta(\xi - \lambda))vU'(v) + rU(v) = a\eta(\xi - \lambda)v + rK \\
\qquad\qquad\qquad for \ 0 < v < v_1 \ and \ v_2 < v < v_3 \\
U(v) = \chi(v) \ \ for \ v_1 \le v \le v_2 \\
U(v) = 0 \ \ for \ v > v_3
\end{cases}
$$

$$(29)$$

with matching conditions: $U'(v_1) = \chi'(v_1)$; $U'(v_2) = \chi'(v_2)$; $U'(v_3) = 0$, *where* $\chi(v) = \big((1 - a)v - \beta F(v)\big)\mathbb{1}_{v < \hat{v}}$ *satisfying*

$$
\begin{cases}
-\frac{1}{2}\chi''(v)\eta^2 v^2 - (r + \eta(\xi - \lambda))v\chi'(v) + r\chi(v) = -(1 - a)\eta(\xi - \lambda)v \\
\qquad\qquad\qquad if \ v < \hat{v} \\
\chi(0) = \chi(\hat{v}) = (1 - a)\hat{v} - \frac{\beta K}{\beta - 1} = 0
\end{cases}
$$

$$(30)$$

with $F(v)$, β and $\hat{v}$ defined in (5), (6), and (7) respectively.

We can now define the leader's optimal stopping rule as:

$$
\hat{\theta}(v) = \begin{cases}
\inf\{t | V_v(t) \ge v_1\} & \text{if } 0 \le v < v_1 \\
0 & \text{if } v_1 \le v \le v_2 \\
\inf\{t | V_v(t) \le v_2 \text{ or } V_v(t) \ge v_3\} & \text{if } v_2 < v < v_3 \\
0 & \text{if } v \ge v_3
\end{cases}
$$

$$(31)$$

3.2.2. *Cash Flow Payoffs - Geometric Brownian Motion Cash Flow Process*

The leader receives a continuous cash flow $\delta_1 Y(t)$ per unit time prior to follower's entry. Upon follower's optimal entry, each gets a continuous cash flow $\delta_2 Y(t)$ per unit time.

A. Follower

The follower's solution is the same as that defined in (16) but replacing δ by δ_2. Since the follower enters after the leader (who starts at $\theta < \infty$), the follower will enter at time: $\hat{\tau}_\theta = \theta + \hat{\tau}(Y_y(\theta))$.

B. Leader

When the leader enters at time $\theta < \infty$, by paying cost K, he/she receives:

$$-K + \frac{\delta_2 Y_y(\theta)}{r - \alpha + \lambda\varsigma} + (\delta_1 - \delta_2)\left(\frac{Y_y(\theta)}{r - \alpha + \lambda\varsigma} - \frac{\beta}{\delta_2}F(Y_y(\theta))\right)\mathbb{1}_{Y_y(\theta)<\hat{y}}$$

where $F(y)$, β and $\hat{y}$ are is defined in (13), (14), and (15) respectively. The leader's value function is:

$$L(y) = \sup_{\theta \geq 0} \widehat{E}\{\exp(-r\theta)\Psi\big(Y_y(\theta)\big)\mathbb{1}_{\theta<\infty}\} \tag{32}$$

where $\Psi(y) = -K + \frac{\delta_2 y}{r-\alpha+\lambda\varsigma} + (\delta_1 - \delta_2)\left(\frac{y}{r-\alpha+\lambda\varsigma} - \frac{\beta}{\delta_2}F(y)\right)\mathbb{1}_{y<\hat{y}}$. $L(y)$ must satisfy:

$$L(y) \geq 0; \quad L(y) \geq \Psi(y). \tag{33}$$

Setting $U(y) = L(y) - \frac{\delta_2 y}{r-\alpha+\lambda\varsigma} + K$, (32) can be re-expressed as:

$$U(y) = \sup_{\theta \geq 0} \widehat{E}\left\{e^{-r\theta}\chi\big(Y_y(\theta)\big)\mathbb{1}_{\theta<\infty} + \int_0^\theta e^{-rs}f\big(Y_y(s)\big)ds\right\}. \tag{34}$$

This formulation leads to an optimal stopping time problem with an obstacle $\chi(y) = \Psi(y) - \frac{\delta_2 y}{r - \alpha + \lambda\varsigma} + K$ and a running profit $f(y) = -\delta_2 y + rK$. Both $U(y)$ and $\chi(y)$ are bounded; we have:

$$0 \leq U(y) \leq K\max(1, \frac{\delta_1 - \delta_2}{\delta_2}),$$

$$\text{and } 0 \leq \chi(y) \leq K\frac{\delta_1 - \delta_2}{\delta_2} \leq K\max(1, \frac{\delta_1 - \delta_2}{\delta_2}). \tag{35}$$

The obstacle $\chi(y)$ is not $C^1(0, \infty)$ with a single non-differentiable point at $\hat{y}$ and we thus have to formulate the V.I. in the weak sense.[c]

Applying the regularity study as in Section 3.2.1, it turns out that the solution $U(y)$ will be smoother than the obstacle. Therefore, the V.I. will have a strong formulation and we can obtain an explicit *two-interval solution*.

Theorem 3.2. *There exist three points* $0 < y_1 < y_2 < \hat{y} < y_3 = \bar{y}$ *such*

---

[c]We refer the formulation and proofs to the original work Bensoussan et al. (2009) Ref. 4.

that:

$$\begin{cases} -\frac{1}{2}U''(y)\varsigma^2 y^2 - (\alpha - \varsigma\lambda)yU'(y) + rU = -\delta_2 y + rk \\ \qquad for\, 0 < y < y_1\, and\, y_2 < y < y_3 \\ U(y) = \chi(y)\, for\, y_1 \leq y \leq y_2 \\ U(y) = 0\, for\, y \leq y_3 \end{cases}$$

with matching conditions: $U'(y_1) = \chi'(y_1);\ U'(y_2) = \chi'(y_2);\ U'(y_3) = 0,$ *where* $\chi(y) = (\delta_1 - \delta_2)\left(\frac{y}{r - \alpha + \lambda\varsigma} - \frac{\beta}{\delta_2}F(y)\right)\mathbb{1}_{y < \hat{y}}$ *satisfying*

$$\begin{cases} -\frac{1}{2}\chi''(y)\varsigma^2 y^2 - (\alpha - \lambda\varsigma)y\chi'(y) + r\chi(y) = (\delta_1 - \delta_2)y,\, for\, y < \hat{y} \\ \chi(0) = \chi(\hat{y}) = 0 \end{cases} \tag{36}$$

with $F(y),\ \beta$ *and* $\hat{y}$ *defined in (13), (14), and (15) respectively. The function* U *vanishes for* y *sufficiently large.*

The leader's optimal stopping rule is defined in the the same way as the lump-sum payoff case (cf.(31)).

3.2.3. *Cash Flow Payoffs - Arithmetic Brownian Motion Cash Flow Process*

The leader receives a continuous cash flow $\delta_1 Y(t)$ per unit time prior to follower's entry. Upon follower's optimal entry, each gets a continuous cash flow $\delta_2 Y(t)$ per unit time.

A. Follower

The follower's solution is the same as that defined in (24) but replacing δ by δ_2. Since the follower enters after the leader (who starts at $\theta < \infty$), the follower will enter at time: $\hat{\tau}_\theta = \theta + \hat{\tau}(Y_y(\theta))$.

B. Leader

When the leader enters at time $\theta < \infty$, by paying cost K, he/she receives:

$$-K + \delta_2\left(\frac{Y_y(\theta)}{r} + \frac{\alpha - \lambda\varsigma}{r^2}\right) + (\delta_1 - \delta_2)\mathbb{1}_{Y_y(\theta) < \hat{y}}\left[\frac{Y_y(\theta)}{r} + \frac{\alpha - \lambda\varsigma}{r^2}\right.$$
$$\left. - \left(\frac{\hat{y}}{r} + \frac{\alpha - \lambda\varsigma}{r^2}\right)e^{-\beta(\hat{y} - Y_y(\theta))}\right],$$

where β and $\hat{y}$ are defined in (22) and (23) respectively. The leader's value function is:

$$L(y) = \sup_{\theta \geq 0} \widehat{E}\big[e^{-r\theta}\Psi(Y_y(\theta))\mathbb{1}_{\theta<\infty}\big] \tag{37}$$

where $\Psi(y) = -K + \delta_2\big(\frac{y}{r} + \frac{\alpha-\lambda\varsigma}{r^2}\big) + (\delta_1 - \delta_2)\mathbb{1}_{y<\hat{y}}\big[\frac{y}{r} + \frac{\alpha-\lambda\varsigma}{r^2} - \big(\frac{1}{r\beta} + \frac{K}{\delta_2}\big)e^{-\beta(\hat{y}-y)}\big]$. $L(y)$ must satisfy:

$$L(y) \geq 0; \quad L(y) \geq \Psi(y). \tag{38}$$

$\Psi(y)$ is unbounded as $y \to \pm\infty$. The obstacle, $\Psi(y)$ is not $C^1(-\infty,\infty)$ with the only non-differentiable point at $\hat{y}$, so we have to formulate the V.I. in the weak sense.[d]

Again, applying the regularity study, it turns out that the solution $L(y)$ will be smoother than the obstacle. Therefore, the V.I. will have a strong formulation and we can obtain an explicit *two-interval solution.*

Theorem 3.3. *Define:*

$$u(y) = L(y) - (\frac{\delta_2}{r}y + \frac{\delta_2}{r^2}(\alpha - \varsigma\lambda)) + K$$

and

$$m(y) = \Psi(y) - (\frac{\delta_2}{r}y + \frac{\delta_2}{r^2}(\alpha - \varsigma\lambda)) + K$$

with $\Psi(y)$ *satisfying:*

$$\begin{cases} -\frac{1}{2}\varsigma^2\Psi''(y) - (\alpha - \varsigma\lambda)\Psi'(y) + r\Psi = \delta_1 y - rK, & y < \hat{y} \\ \Psi(\hat{y}) = \frac{\delta_2}{r}\hat{y} + \frac{\delta^2}{r^2}(\alpha - \varsigma\lambda) - K = \frac{\delta_2}{r\beta} \end{cases} \tag{39}$$

and $m(y)$ *satisfying*

$$-\frac{1}{2}\varsigma^2 m''(y) - (\alpha - \varsigma\lambda)m(y) + rm(y) = (\delta_1 - \delta_2)y, \quad y < \hat{y}. \tag{40}$$

There exists a triple $y_1 < y_2 < \hat{y} < y_3$ *such that:*

$$\begin{cases} -\frac{1}{2}\varsigma^2 u''(y) - (\alpha - \varsigma\lambda)u'(y) + ru = -\delta_2 y + rK, & y < y_1 \text{ and } y_2 < y < y_3 \\ u(y) = m(y), & y_1 \leq y \leq y_2 \\ u(y) = 0, & y \geq y_3 \end{cases}$$

with matching conditions: $u'(y_1) = m'(y_1)$; $u'(y_2) = m'(y_2)$; $u'(y_3) = 0$. *The function u vanishes for y sufficiently large.*

---

[d]We refer the formulation and proofs to the original work Bensoussan et al. (2009) Ref. 3.

The leader's optimal stopping rule is defined in the the same way as the lump-sum payoff case (cf.(31)).

4. The Incomplete Market Case

4.1. *Utility-Based Pricing*

The asset S representing the market still evolves as (1). However, the project value defining lump-sum investment payoffs now evolves according to:

$$dV(t) = rV(t)dt + \eta V(t)\big(\xi dt + \rho dW(t) + \sqrt{1 - \rho^2}dW^0(t)\big), \qquad (41)$$

the cash flow investment payoffs by geometric Brownian motion process evolve as:

$$dY(t) = Y(t)\big(\alpha dt + \varsigma(\rho dW(t) + \sqrt{1 - \rho^2}dW^0(t))\big), \quad Y(0) = y, \qquad (42)$$

and the cash flow investment payoffs by arithmetic Brownian motion process evolve as:

$$dY(t) = \alpha dt + \varsigma\big(\rho dW(t) + \sqrt{1 - \rho^2}dW^0(t)\big), \quad Y(0) = y \qquad (43)$$

where $W(t)$ and $W^0(t)$ are independent Wiener processes. The parameter $|\rho| < 1$ is the correlation coefficient between market uncertainty and investment payoff uncertainty. The market asset S can only span a portion of the investment payoff risk driven by the Wiener process $W(t)$, leaving the remaining risk driven by $W^0(t)$ unhedgeable. The unique equivalent martingale measure is undefined, so the risk-neutral pricing is no longer appropriate, and an alternative must be developed in this framework.

We adopt the utility-based pricing approach where the risk averse investor's preferences are characterized by an exponential utility function given by:

$$U(x) = -\frac{1}{\gamma}e^{-\gamma x} \qquad (44)$$

where γ is his/her risk aversion parameter, $\gamma > 0$.

The valuation formulation is: given initial wealth, the investor considers utility maximization by choosing an optimal stopping time to invest the project together with a portfolio investment-consumption strategy. We will formulate the investor's specific utility maximization problem in each corresponding scenarios.

4.2. Single Player

4.2.1. Lump-sum Payoffs - External Value Process

The rational investor maximizes his/her expected utility of wealth. Given initial wealth, x, the risk averse investor optimizes his/her portfolio by dynamically choosing allocations in the market asset S and the riskless bond. The investor's wealth X, evolves as:

$$dX(t) = rX(t)dt + X(t)\pi(t)\sigma(\lambda dt + dW(t)), \tag{45}$$

where $\pi(t)$ is the portion of wealth invested in asset S. We use the discounted value $\widetilde{V}(t) = V(t)e^{-rt}$ and $\widetilde{X}(t) = X(t)e^{-rt}$ where:

$$d\widetilde{V}(t) = \widetilde{V}(t)\eta\big(\xi dt + \rho dW(t) + \sqrt{1-\rho^2}dW^0(t)\big), \ \ \widetilde{V}(0) = v \tag{46}$$

$$d\widetilde{X}(t) = \widetilde{X}(t)\pi(t)\sigma\big(\lambda dt + dW(t)\big), \ \ \widetilde{X}(0) = x. \tag{47}$$

Let $\mathcal{F}_t = \sigma(W(s), W^0(s); s \leq t)$. We consider stopping times τ with respect to $\mathcal{F}_t$. At τ, the individual invests and receives $\widetilde{X}_x(\tau) + (\widetilde{V}_v(\tau) - K)^{+}$[e]. To a pair $(\pi(\cdot), \tau)$, we associate with the objective function:

$$J_{x,v}(\pi(\cdot), \tau) = E\left[e^{\frac{\lambda^2}{2}\tau}U\left(\widetilde{X}_x(\tau) + \big(\widetilde{V}_v(\tau) - K\big)^{+}\right)\right]. \tag{48}$$

We assume τ is finite a.s.. The function (48) is well defined but the value $-\infty$ is possible.

We define the associated value function as:

$$F(x, v) = \sup_{\pi(\cdot), \tau} J_{x,v}(\pi(\cdot), \tau). \tag{49}$$

Solving (49) by V.I., we obtain:

$$F(x, v) = \begin{cases} U(x)\big(\chi(v)\big)^{\frac{1}{1-\rho^2}}, & \text{if } 0 < v < \hat{v} \\ U(x)e^{-\gamma\big((1-a)v-K\big)}, & \text{if } v \geq \hat{v} \end{cases} \tag{50}$$

where

$$\chi(v) = 1 - (1 - e^{-\varpi})\Big(\frac{v}{\hat{v}}\Big)^{\beta}, \tag{51}$$

$$\beta = 1 - \frac{2(\xi - \rho\lambda)}{\eta} > 1, \tag{52}$$

---

[e]This formulation is similar to the specification that the investment cost grows at the risk-free rate.

$\hat{v}$ solves

$$(1-a)\hat{v} = K + \frac{\varpi}{\gamma(1-\rho^2)} \tag{53}$$

with ϖ the unique positive solution of

$$e^{\varpi} - \frac{\varpi}{\beta} = 1 + \frac{K\gamma(1-\rho^2)}{\beta}, \tag{54}$$

and we assume $\xi - \rho\lambda < 0$.

We next define the optimal stopping rule which achieves supremum in (49):

$$\hat{\tau}(x,v) = \inf\left\{t \geq 0 \middle| F\big(\widetilde{X}_x(t), \widetilde{V}_v(t)\big) = U\big(\widetilde{X}_x(t) + \big((1-a)\widetilde{V}_v(t) - K\big)^+\big)\right\}$$

$$= \inf\{t \geq 0 | \widetilde{V}_v(t) \text{ is outside } (0, \hat{v})\} = \hat{\tau}(v) \tag{55}$$

and $\hat{\tau}(v) < \infty$ a.s..

4.2.2. *Cash Flow Payoffs - Geometric Brownian Motion Cash Flow Process*

The rational investor maximizes his/her expected utility of consumption.[f] Given the initial wealth, x, the risk averse investor optimizes his/her portfolio by dynamically choosing allocations in the market asset S, the riskless bond, and the consumption rate, C. The investor's wealth X, evolves as:

$$\begin{cases} dX(t) = \pi(t)X(t)\sigma(\lambda dt + dW(t)) + rX(t)dt - C(t)dt, \ t < \tau \\ X(\tau) = X(\tau - 0) - K \\ dX(t) = \pi(t)X(t)\sigma(\lambda dt + dW(t)) + rX(t)dt - C(t)dt + \delta Y(t)dt, \ t > \tau, \\ dY(t) = Y(t)\big(\alpha dt + \varsigma(\rho dW(t) + \sqrt{1-\rho^2}dW^0(t))\big) \\ X(0) = x, \ Y(0) = y \end{cases} \tag{56}$$

where $\pi(t)$ is the proportion of wealth invested in asset S, $C(t)$ is the consumption rate, and τ is a stopping time to undertake the investment, chosen optimally by the investor. The wealth process is discontinuous at τ.

From (56), we observe that the wealth process has two possible evolution regimes, so we introduce:

$$dX^0(t) = \pi(t)X^0(t)\sigma(\lambda dt + dW(t)) + rX^0(t)dt - C(t)dt, \tag{57}$$

---

[f]We allow for negative consumption.

and

$$dX^1(t) = \pi(t)X^1(t)\sigma(\lambda dt + dW(t)) + rX^1(t)dt - C(t)dt + \delta Y(t)dt, \quad (58)$$

where X^0 is the wealth process X before the stopping time τ and X^1 is the wealth process X after the stopping time τ.

A. Post-Investment Utility Maximization

We begin with a control problem relative to the process $X^1(t)$. Introduce the set of admissible controls, $\mathcal{U}^1_{x,y} = \{C(\cdot), \pi(\cdot)\}$, satisfying:

$$\begin{cases} E \int_0^T \left(\pi(t)X^1(t)\right)^2 dt < \infty \ \forall\, T \\ E \int_0^T \left(C(t)\right)^2 dt < \infty \ \forall\, T \end{cases}, \quad (59)$$

$$\tau_N \uparrow \infty \text{ a.s. as } N \uparrow \infty, \quad \text{where } \tau_N = \inf\{t | X^1(t) \le -N\}, \quad (60)$$

and a transversality condition:

$$e^{-\mu T} E[e^{-r\gamma\left(X^1(T) + f(Y(T))\right)}] \to 0, \text{ as } T \to \infty \quad (61)$$

where μ is the discount rate, $f(y)$ is a positive function which will be made precise later (cf.(69),(68)) and $f(y)$ has linear growth with $f(0) = 0$ (cf.(67)). To a pair of $\left(C(\cdot), \pi(\cdot)\right)$, we introduce the objective function:

$$J\left(C(\cdot)\right) = E[\int_0^\infty e^{-\mu t} U\left(C(t)\right) dt]. \quad (62)$$

This function is well defined, but it may take the value $-\infty$. We define the value function:

$$F^1(x,y) = \sup_{\{\pi(\cdot), C(\cdot)\} \in \mathcal{U}^1_{x,y}} J\left(C(\cdot)\right). \quad (63)$$

First, we note that if $y = 0$, then $Y(t) = 0$, $\forall\, t$. The problem reduces to the classical investment-consumption problem with the solution given by:

$$F(x) = -\frac{1}{r\gamma} \exp\{-r\gamma x + 1 - \frac{\mu + \frac{\lambda^2}{2}}{r}\}. \quad (64)$$

We have:

$$F^1(x,0) = F(x). \quad (65)$$

We look for a solution of (63) as follows:

$$F^1(x,y) = -\frac{1}{r\gamma} \exp\{-r\gamma(x + f(y)) + 1 - \frac{\mu + \frac{\lambda}{2}}{r}\} \quad (66)$$

in which, by (65), implies:

$$f(0) = 0. \qquad (67)$$

By (66), the Bellman equation that the value function $F^1(x,y)$ (63) must satisfy reduces to:

$$\frac{1}{2}y^2\varsigma^2 f'' + (\alpha - \lambda\varsigma\rho)yf' - \frac{1}{2}r\gamma y^2\varsigma^2(1-\rho^2)f'^2 - rf + \delta y = 0. \qquad (68)$$

Proposition 4.1. *The function,*

$$f(y) = \inf_{\{v(\cdot)\in\mathcal{U}_y\}} E\Big[\int_0^\infty e^{-rt}\big(\delta Y_y(t) + \frac{1}{2}v^2(t)\big)dt\Big] \qquad (69)$$

with

$$\begin{cases} dY(t) = Y(t)\big[\alpha - \lambda\varsigma\rho + \varsigma\sqrt{r\gamma(1-\rho^2)}v(t)\big]dt + Y(t)\varsigma dW(t), \ \ Y(0) = y \\ \mathcal{U}_y = \{v(\cdot)|E[\int_0^\infty e^{-rt}v(t)^2dt] < \infty, \ e^{-rT}E[Y_y(T)] \to 0 \ as \ T \to \infty\} \end{cases}$$

is the unique function in $C^2(0,\infty)$, solution of (68), (67) in the interval $[0, \epsilon y + M_\epsilon]^g$, such that $f(y) \uparrow \infty$ as $y \uparrow \infty$. Moreover, the function $f'(y)$ is bounded.

Note that the function $f(y)$ defined in (69) is the function entering into the transversality condition in (61).

Theorem 4.1. *The function $F^1(x,y)$ given by (66) coincides with the value function given by (63).*

B. Pre-Investment Utility Maximization

We now turn to the problem of optimal stopping with the obstacle defined by $F^1(x,y)$. Before the stopping time τ, the wealth process X is governed by $X^0(t)$ (57) and the cash flow process evolves as (42).

Introduce the set of admissible controls, $\mathcal{U}^0_{x,y} = \{\pi(.), C(.), \tau\}$, satisfying:

$$\begin{cases} E[\int_0^T \big(\pi(t)X^0(t)\big)^2 dt] < \infty, \ \forall\, T \\ E[\int_0^T C(t)^2 dt] < \infty, \ \forall\, T \end{cases} , \qquad (70)$$

$$\tau \wedge \theta^0 < \infty \ \text{a.s. where } \theta^0 = \inf\{t|Y(t) = 0\}, \qquad (71)$$

$^g M_\epsilon$ is defined as: $M_\epsilon = \dfrac{\big(\frac{\delta}{\epsilon} - r + \alpha - \lambda\varsigma\rho\big)^2}{2\varsigma^2 r^2\gamma(1-\rho^2)}$. Note that ϵ can be arbitrarily small.

and $\tau^* = \lim \uparrow \tau_N \geq \tau \wedge \theta^0$ a.s. where $\tau_N = \inf\{t|X^0(t) \leq -N\}$.   (72)

If $\theta^0 \leq \tau$, the investment never takes place and the investor receives $F\big(X^0(\theta^0)\big)$(cf.(64)). If $\tau < \theta^0$, the investor will receive $F^1(X^0(\tau) - K, Y(\tau))$(cf.(66)) at the stopping time τ. Therefore, the objective function is:

$$J_{x,y}\big(C(\cdot), \pi(\cdot), \tau\big) = E\bigg[\int_0^{\tau \wedge \theta^0} e^{-\mu t} U\big(C(t)\big) dt + F^1\big(X^0(\tau) - K, Y(\tau)\big)$$
$$\times e^{-\mu\tau} \mathbb{1}_{\tau < \theta^0} + F\big(X^0(\theta^0)\big) e^{-\mu\theta^0} \mathbb{1}_{\theta^0 \leq \tau} \bigg]$$
(73)

and we define the associated value function:

$$F(x,y) = \sup_{\{\pi(\cdot), C(\cdot), \tau\} \in \mathcal{U}_{xy}^0} J_{x,y}\big(C(\cdot), \pi(\cdot), \tau\big).$$
(74)

We look for a solution of (74) in the form:

$$F(x,y) = -\frac{1}{r\gamma} \exp\Big[-r\gamma\big(x + g(y)\big) + 1 - \frac{\mu + \frac{\lambda^2}{2}}{r} \Big].$$
(75)

The V.I. that the value function $F(x,y)$ (74) must satisfy reduces to:

$$\begin{cases} \frac{1}{2}y^2\varsigma^2 g'' + g'y(\alpha - \lambda\varsigma\rho) - \frac{1}{2}y^2\varsigma^2 r\gamma(1-\rho^2)g'^2 - rg \leq 0 \\ g(y) \geq f(y) - K \\ \big(g(y) - f(y) + K\big)\big[\frac{1}{2}y^2\varsigma^2 g'' + g'y(\alpha - \lambda\varsigma\rho) - \frac{1}{2}y^2\varsigma^2 r\gamma(1-\rho^2)g'^2 - rg\big] \\ \qquad\qquad = 0 \\ g(0) = 0. \end{cases}$$
(76)

In (76), the nonlinear operator is connected to a minimization problem and the inequalities are connected to a maximization problem for a stopping time. $g(y)$ can be interpreted as the value function of a differential game. Define: $u(y) = g(y) - f(y) + K$, yielding:

$$\begin{cases} -\frac{1}{2}y^2\varsigma^2 u'' - yu'\big(\alpha - \lambda\varsigma\rho - yf'\varsigma^2 r\gamma(1-\rho^2)\big) + \frac{1}{2}y^2\varsigma^2 r\gamma(1-\rho^2)u'^2 + ru \\ \qquad\qquad\qquad\qquad\qquad\qquad\qquad\qquad \geq -\delta y + rK \\ u \geq 0 \\ u\big[-\frac{1}{2}y^2\varsigma^2 u'' - yu'\big(\alpha - \lambda\varsigma\rho - yf'\varsigma^2 r\gamma(1-\rho^2)\big) + \frac{1}{2}y^2\varsigma^2 r\gamma(1-\rho^2)u'^2 \\ \qquad\qquad\qquad\qquad\qquad\qquad\qquad\qquad + ru + \delta y - rK \big] = 0 \\ u(0) = K. \end{cases}$$
(77)

We study (77) by the threshold approach. For fixed $\hat{y}$, we solve the Dirichlet problem:

$$\begin{cases} -\frac{1}{2}y^2\varsigma^2 u'' - yu'\left(\alpha - \lambda\varsigma\rho - yf'\varsigma^2 r\gamma(1-\rho^2)\right) + \frac{1}{2}y^2\varsigma^2 r\gamma(1-\rho^2)u'^2 + ru \\ \qquad\qquad = -\delta y + rK, \quad 0 < y < \hat{y} \\ u(0) = K, \quad u(\hat{y}) = 0. \end{cases}$$

(78)

Equation (78) is a Bellman equation of the following control problem with the controlled diffusion:

$$\begin{cases} dY = Y\left[\alpha - \lambda\varsigma\rho - Yf'(Y)\varsigma^2 r\gamma(1-\rho^2) + \varsigma\sqrt{r\gamma(1-\rho^2)}v\right]dt + Y\varsigma dW \\ Y(0) = y, \quad 0 < y < \hat{y} \end{cases}$$

(79)

and the value function

$$u(y) = \inf_{v(\cdot)} E\left[\int_0^{\theta_y\left(v(\cdot)\right)} e^{-rt}\left[-\delta Y_y(t) + rK + \frac{1}{2}v^2(t)\right]dt + K \times e^{-r\theta_y\left(v(\cdot)\right)}\right.$$

$$\left. \times \mathbb{1}_{Y_y\left(\theta_y(v(\cdot))\right)=0}\right]$$

(80)

where $\theta_y\left(v(\cdot)\right) = \inf\{t | Y(t) \text{ is outside } (0,\hat{y})\}$, and it is a.s. finite.

Theorem 4.2. *There exists a unique value $\hat{y}$ such that*

$$u'(\hat{y}) = 0, \quad \hat{y} \geq \frac{Kr}{\delta}.$$

(81)

The value function $u(y)$ (cf.80) extended by 0 beyond $\hat{y}$ is the unique solution of the V.I. (77). It is C^1 and piecewise C^2.

Hence, there exists a unique $\hat{y}$ such that:

$$\begin{cases} -\frac{1}{2}y^2\varsigma^2 g'' - g'y(\alpha - \lambda\varsigma\rho) + \frac{1}{2}y^2\varsigma^2 r\gamma(1-\rho^2)g'^2 + rg = 0, \quad y < \hat{y} \\ g(y) = f(y) - K, \quad y \geq \hat{y} \\ g'(\hat{y}) = f'(\hat{y}) \\ g(0) = 0 \end{cases}$$

(82)

Note that $g(y) \geq 0$ since $u(y) \geq -f(y) + K$.

Theorem 4.3. *The function $F(x,y)$ defined by (75) coincides with the value function given by (74).*

We next define the optimal stopping rule as: $\hat{\tau}(y) = \inf\{t | Y_y(t) \geq \hat{y}\}$, where $\hat{y}$ is the unique value defined by the V.I. (77) and (81) (the smooth matching point).

4.2.3. *Cash Flow Payoffs - Arithmetic Brownian Motion Cash Flow Process*

As in Section 4.2.2, the rational investor maximizes his/her expected utility of consumption by dynamically choosing allocations in the market asset S, the riskless bond, and the consumption rate, C. Given the initial wealth, x, his/her wealth X evolves the same as (56).[h] And again, before the stopping time τ, his/her wealth process X corresponds to X^0 (57), and after the stopping time τ, it corresponds to X^1 (58).

A. Post-Investment Utility Maximization

As in Section 4.2.2, we begin with a control problem relative to the process $X^1(t)$. Introduce the set of admissible controls, $\mathcal{U}^1_{x,y} = \{C(\cdot), \pi(\cdot)\}$, satisfying:

$$\begin{cases} E \int_0^T \left(\pi(t)X^1(t)\right)^2 dt < \infty \ \forall\, T \\ E \int_0^T \left(C(t)\right)^2 dt < \infty \ \forall\, T \end{cases}, \qquad (83)$$

$$\tau_N \uparrow \infty \text{ a.s. as } N \uparrow \infty, \text{ where } \tau_N = \inf\{t | rX^1(t) + \delta Y(t) \le -N\}, \quad (84)$$

and a transversality condition:

$$e^{-\mu T} E[e^{-\gamma\left(rX^1(T) + \delta Y(T)\right)}] \to 0, \text{ as } T \to \infty. \qquad (85)$$

To a pair of $(C(\cdot), \pi(\cdot))$, we introduce the objective function:

$$J\big(C(\cdot)\big) = E[\int_0^\infty e^{-\mu t} U\big(C(t)\big) dt]. \qquad (86)$$

This function may take the value $-\infty$. We define the associated value function as:

$$F^1(x, y) = \sup_{\{\pi(\cdot), C(\cdot)\} \in \mathcal{U}^1_{x,y}} J\big(C(\cdot)\big). \qquad (87)$$

Theorem 4.4.

$$F^1(x, y) = -\frac{1}{r\gamma} \exp\Big\{ -r\gamma(x + \frac{\delta}{r}y) + 1 - \frac{\mu + \frac{\lambda^2}{2}}{r}$$
$$- \frac{\delta\gamma}{r}\big(\alpha - \lambda\varsigma\rho - \frac{1}{2}\varsigma^2\delta\gamma(1 - \rho^2)\big)\Big\} \qquad (88)$$

is the solution of the value function (87).

---

[h]The cash flow process now evolves as: $dY(t) = \alpha dt + \varsigma(\rho dW(t) + \sqrt{1 - \rho^2} dW^0(t))$.

B. Pre-Investment Utility Maximization

We now solve the optimal stopping problem with the obstacle defined by the value function $F^1(x, y)$. Before the stopping time τ, the wealth process $X^0(t)$ evolves as (57) and the cash flow process $Y(t)$ evolves as (43). Introduce the set of admissible controls $\mathcal{U}^0_{x,y} = \{\pi(\cdot), C(\cdot), \tau\}$ satisfying:

$$\begin{cases} E[\int_0^T \left(\pi(t)X^0(t)\right)^2 dt] < \infty, \ \forall \, T \\ E[\int_0^T C(t)^2 dt] < \infty, \ \forall \, T \end{cases}, \tag{89}$$

$$\tau < \infty \text{ a.s., and } \tau^* = \lim \uparrow \tau_N \geq \tau \text{ a.s. where} \tau_N = \inf\{t | X^0(t) \leq -N\}. \tag{90}$$

At time τ, the investor undertakes the investment and receives $F^1(X^0(\tau) - K, Y(\tau))$(cf.(88)). Therefore, the objective function is:

$$J_{x,y}\left(C(\cdot), \pi(\cdot), \tau\right) = E\left[\int_0^\tau e^{-\mu t} U\left(C(t)\right) dt + F^1\left(X^0(\tau) - K, Y(\tau)\right) e^{-\mu \tau} \right] \tag{91}$$

and we define the value function:

$$F(x, y) = \sup_{\{\pi(\cdot), C(\cdot), \tau\} \in \mathcal{U}^0_{xy}} J_{x,y}\left(C(\cdot), \pi(\cdot), \tau\right). \tag{92}$$

We look for a solution of the form:

$$F(x, y) = -\frac{1}{r\gamma} \exp\left[-r\gamma\left(x + g(y)\right) + 1 - \frac{\mu + \frac{\lambda^2}{2}}{r} \right]. \tag{93}$$

Recall $F^1(x, y) = -\frac{1}{r\gamma} \exp\left[-r\gamma\left(x + f(y)\right) + 1 - \frac{\mu + \frac{\lambda^2}{2}}{r} \right]$.[i] The V.I. that the value function $F(x, y)$ (92) must satisfy reduces to:

$$\begin{cases} \frac{1}{2}\varsigma^2 g'' + g'(\alpha - \lambda\varsigma\rho) - \frac{1}{2}\varsigma^2 r\gamma(1 - \rho^2)g'^2 - rg \leq 0 \\ g(y) \geq f(y) - K \\ \left(g(y) - f(y) + K\right)\left[\frac{1}{2}\varsigma^2 g'' + g'(\alpha - \lambda\varsigma\rho) - \frac{1}{2}\varsigma^2 r\gamma(1 - \rho^2)g'^2 - rg\right] = 0 \end{cases} \tag{94}$$

Considering $u(y) = g(y) - f(y) + K$, the V.I. (94) becomes:

$$\begin{cases} -\frac{1}{2}\varsigma^2 u'' - u'\left(\alpha - \lambda\varsigma\rho - \varsigma^2\delta\gamma(1 - \rho^2)\right) + \frac{1}{2}\varsigma^2 r\gamma(1 - \rho^2)u'^2 + ru \\ \qquad\qquad\qquad\qquad\qquad\qquad\qquad\qquad \geq -\delta y + rK \\ u \geq 0 \\ u\left[-\frac{1}{2}\varsigma^2 u'' - u'\left(\alpha - \lambda\varsigma\rho - \varsigma^2\delta\gamma(1 - \rho^2)\right) + \frac{1}{2}\varsigma^2 r\gamma(1 - \rho^2)u'^2 \right. \\ \qquad\qquad\qquad\qquad\qquad\qquad\qquad\qquad \left. + ru + \delta y - rK\right] = 0 \end{cases} \tag{95}$$

---

[i] $f(y) = \frac{\delta y}{r} + \frac{\delta}{r^2}\left(\alpha - \lambda\varsigma\rho - \frac{1}{2}\varsigma^2\rho\gamma(1 - \rho^2)\right)$.

We study (95) by the threshold approach. For $\hat{y}$ fixed, we solve the Dirichlet problem:

$$\begin{cases} -\frac{1}{2}\varsigma^2 u'' - u'\big(\alpha - \lambda\varsigma\rho - \varsigma^2\delta\gamma(1-\rho^2)\big) + \frac{1}{2}\varsigma^2 r\gamma(1-\rho^2)u'^2 + ru \\ \qquad\qquad\qquad\qquad\qquad\qquad\qquad = -\delta y + rK, \quad y < \hat{y} \\ u(\hat{y}) = 0 \end{cases}$$

(96)

and we require a linear growth for $y \to -\infty$.

Theorem 4.5. *For each $\hat{y}$, there exists a unique solution of (96) with the estimate:*

$$-\frac{(-\delta\hat{y} + rK)^-}{r} \le u(y) \le -\frac{\delta}{r}(y - \hat{y}) + \frac{1}{r}\bigg(-\delta\hat{y} + rK$$
$$-\frac{\delta}{r}(\alpha - \lambda\varsigma\rho - \gamma\delta\varsigma^2(1-\rho^2))\bigg)^+,$$

for $y < \hat{y}$.

(97)

The solution is C^2 in $(-\infty, \hat{y})$. There exists a unique $\hat{y}$ such that:

$$u'(\hat{y}) = 0, \qquad \hat{y} \ge \frac{rK}{\delta}.$$

(98)

The corresponding solution of (96) extended by 0 beyond $\hat{y}$ is the unique solution of the V.I. (95). It is C^1 and piecewise C^2.

Hence, there exists a unique $\hat{y}$ such that:

$$\begin{cases} -\frac{1}{2}\varsigma^2 g'' - g'(\alpha - \lambda\varsigma\rho) + \frac{1}{2}\varsigma^2 r\gamma(1-\rho^2)g'^2 + rg = 0, & y < \hat{y} \\ g(y) = \frac{\delta}{r}y - K + \frac{\delta}{r^2}(\alpha - \lambda\varsigma\rho - \frac{1}{2}\varsigma^2\delta\gamma(1-\rho^2)), & y \ge \hat{y} \\ g'(\hat{y}) = \frac{\delta}{r} \end{cases}$$

(99)

We have also: $g(y) \to 0$ as $y \to -\infty$, and $g(y) \ge 0$.

Theorem 4.6. *The function $F(x, y)$ given by (93) coincides with the value function (92).*

We next define the optimal stopping rule, which achieves supremum in (92), as: $\hat{\tau}(y) = \inf\{t | Y_y(t) \ge \hat{y}\}$, where $\hat{y}$ is the unique value defined by the V.I. (95) and (98).

4.3. *Two Players: A Stackelberg Leader-Follower Game*

The setting is the same as Section 3.2.

4.3.1. *Lump-sum Payoffs - External Value Process*

A. Follower

The follower's strategy is identical to that described in the Section 4.2.1. We have the follower's value function as defined in (50). The optimal stopping strategy for the follower is:

$$\hat{\tau}(v) = \inf\{t \geq 0 | \widetilde{V}_v(t) \text{ is outside } (0, \hat{v})\},$$

where $\hat{v}$ is defined in (53). The stopping time $\hat{\tau}(v)$ is the optimal entry if the follower can enter in the market at time zero. Since the follower enters after the leader (who starts at θ), for finite θ, the follower will enter at time: $\hat{\tau}_\theta = \theta + \hat{\tau}(\widetilde{V}_v(\theta))$.

B. Leader

As the case of a single player, the leader dynamically optimizes his/her investment portfolio and solves his/her utility maximization problem with joint decisions of stopping times and portfolio investment strategies. When the leader enters at time $\theta < \infty$,

(1) If $\widetilde{V}(\theta) > \hat{v}$, the follower enters immediately and the leader gets $\widetilde{X}(\theta) + a\widetilde{V}(\theta) - K$.
(2) If $\widetilde{V}(\theta) < \hat{v}$, then $\hat{\tau}(\widetilde{V}(\theta)) > 0$, the leader gets immediately $\widetilde{X}(\theta) + (\widetilde{V}(\theta) - K)$, but he/she surrenders at time $\theta + \hat{\tau}(\widetilde{V}(\theta))$, a percentage $(1 - a)$ of project value to the follower.

A difficulty occurs in the latter scenario. At time θ, the leader must determine the value surrendered to the follower, taking into account the follower's optimal entry at $\hat{\tau}_\theta$. However, we work with investor's utility and there is no identity relationship for values at different times. We will circumvent this problem converting the surrender value by employing equivalence (indifference) considerations.

Let $\widetilde{X}(t)$, $\widetilde{V}(t)$ be the wealth of the follower and the value process governed by (47) and (46). Suppose $\theta = 0$ and $v > \hat{v}$, then the follower receives $(1 - a)v$ at time 0 and he/she values this operation by $U(x + (1 - a)v)$. If $\theta = 0$ and $0 < v < \hat{v}$, then the follower receives $(1 - a)\widetilde{V}(\hat{\tau}(v))$ at time $\hat{\tau}(v)$; the corresponding value is: $e^{\frac{\lambda^2}{2}\hat{\tau}(v)} U(\widetilde{X}_x(\hat{\tau}(v)) + (1-a)\widetilde{V}_v(\hat{\tau}(v)))$. Since the follower can dynamically optimize his/her portfolio between time

0 and $\hat{\tau}(v)$, the value is in fact:

$$H(x,v) = \max_{\pi(\cdot)} E\big[e^{\frac{\lambda^2}{2}\hat{\tau}(v)} U\big(\widetilde{X}_x(\hat{\tau}(v)) + (1-a)\widetilde{V}_v(\hat{\tau}(v))\big)\big] \qquad (100)$$

and we have the boundary conditions: $H(x,\hat{v}) = U\big(x + (1-a)\hat{v}\big)$ and $H(x,0) = U(x)$.

Proposition 4.2. *The solution of (100) is:*

$$H(x,v) = U(x)\Lambda(v) \qquad (101)$$

where: $\Lambda(v) = \big(1 + Bv^\beta\big)^{\frac{1}{1-\rho^2}}$ *and* $B = -\dfrac{1 - e^{-\left(\varpi + K\gamma(1-\rho^2)\right)}}{\hat{v}^\beta}$, *with* β, $\hat{v}$, ϖ *defined in (52), (53) and (54) respectively.*

We next define the follower's indifference value $H^e(v)$ by:

$$H(x + H^e(v), 0) = H(x,v) = U(x)e^{-\gamma H^e(v)}; \qquad (102)$$

then:

$$\begin{cases} e^{-\gamma H^e(v)} = \Lambda(v), \ 0 < v < \hat{v} \\ H^e(v) \leq (1-a)v, \text{ if } v \leq \hat{v} \\ H^e(v) = (1-a)v, \ v \geq \hat{v} \end{cases} \qquad (103)$$

$H^e(v)$ is the value that makes the follower indifferent between receiving it at time 0 and losing the right to undertake investment at time $\hat{\tau}(v)$, or to exert the right to undertake the investment. The leader must deduct this amount in addition to K at time 0 if he/she decides to invest at time 0.

We now formulate completely the leader's problem. For each pair of $(\pi(\cdot), \theta)$, the leader's objective function is :

$$J_{x,v}(\pi(\cdot), \theta) = E\left[e^{\frac{\lambda^2}{2}\theta} U\left(\widetilde{X}(\theta) + \big(\widetilde{V}(\theta) - K - H^e(\widetilde{V}(\theta))\big)^+\right)\right] \qquad (104)$$

and we define associated value function:

$$L(x,v) = \sup_{\pi(\cdot),\theta} J_{x,v}(\pi(\cdot), \theta). \qquad (105)$$

We impose $\theta < \infty$ a.s. We look for a solution of the form:

$$L(x,v) = U(x)L(v) = U(x)\Sigma(v)^{\frac{1}{1-\rho^2}} \quad \text{with } 0 \leq L(v) \leq 1. \qquad (106)$$

We obtain the V.I.:

$$
\begin{cases}
\Sigma'(\xi - \lambda\rho) + \frac{1}{2}v\eta\Sigma'' \geq 0 \\
\Sigma(v) \leq e^{-\gamma(1-\rho^2)\left(v - K - H^e(v)\right)^+} \\
\left[\Sigma'(\xi - \lambda\rho) + \frac{1}{2}v\eta\Sigma''\right]\left[\Sigma(v) - e^{-\gamma(1-\rho^2)\left(v - K - H^e(v)\right)^+}\right] = 0 \\
0 \leq \Sigma(v) \leq 1 \\
\Sigma(0) = 1
\end{cases}
\quad (107)
$$

The obstacle, $\psi(v) = e^{-\gamma(1-\rho^2)\left(v - K - H^e(v)\right)^+}$ is continuous but not C^1, and its derivative is discontinuous at v^o ($K < v^o < \hat{v}$) and $\hat{v}$. Therefore, we must consider (107) in a weak sense. Nevertheless, (107) is the V.I. of the following optimal stopping problem with the state equation:

$$
dV(t) = V(t)\eta\big((\xi - \lambda\rho)dt + dW(t)\big), \quad V(0) = v \qquad (108)
$$

and the value function:

$$
\Sigma(v) = \inf_{\tau_\Sigma} J_v(\tau_\Sigma) \text{ with } J_v(\tau_\Sigma) = E\big[\psi\big(V_v(\tau_\Sigma)\big)\big] \qquad (109)
$$

and τ_Σ must be a stopping time which is a.s. finite. To formulate (109) in a weak form sense, we introduce the Sobolev space $H_\varrho^1(0, \infty)$ with the scalar product:$((\phi, \tilde\phi))_\varrho = (\phi, \tilde\phi)_\varrho + \int_0^\infty v^2 \frac{\phi'(v)\tilde\phi'(v)}{(1+v^2)\varrho} dv$. We define the bilinear form:

$$
b(\phi, \tilde\phi) = \int_0^\infty \phi'(v)\Big\{ -\eta(\xi - \lambda\rho) + \eta^2 \frac{1 - v^2(\varrho - 1)}{1 - v^2}\Big\} \frac{\tilde\phi(v)}{(1 + v^2)\varrho} dv
$$
$$
+ \frac{1}{2}\int_0^\infty \frac{\phi'(v)\tilde\phi'(v)v^2\eta^2}{(1 + v^2)\varrho} dv.
$$

The V.I. corresponding to (109) is:

$$
b(\Sigma, \tilde\Sigma - \Sigma) \geq 0, \qquad \forall \tilde\Sigma \in \mathcal{K}, \ \Sigma \in \mathcal{K} \qquad (110)
$$

where $\mathcal{K} = \{\phi \in H_\varrho^1(0, \infty) | 0 \leq \phi(v) \leq \psi(v), \ \forall v\}$ is the convex subset of $H_\varrho^1(0, \infty)$.

Theorem 4.7. *Assume $\xi - \lambda\rho < 0$. Then there exists one and only one solution of (110). It coincides with the value function (109).*

Because of the presence of the term $H^e(v)$, we cannot have a general result as it is the case of complete markets.

4.3.2. *Cash Flow Payoffs - Geometric Brownian Motion Cash Flow Process*

A. Follower

We have the follower's value function as defined by (75) with $g(y)$ satisfying (76) and δ replaced by δ_2. The optimal stopping strategy for the follower is:

$$\hat{\tau}(y) = \inf\{t|Y(t) \geq \hat{y}\}$$

where $\hat{y}$ is a fixed number defined by the V.I. (77),and (81)(the smooth matching point). We must take $\delta = \delta_2$ and $\hat{y} > \frac{rK}{\delta_2}$. Since the follower can enter only after the leader (who starts at $\theta < \infty$), the follower will enter at time: $\hat{\tau}_\theta = \theta + \hat{\tau}\big(Y_y(\theta)\big)$.

B. Leader

By investing K, the leader expects to receive a continuous cash flow $\delta_1 Y(t)$ per unit time prior to the follower's entry, and $\delta_2 Y(t)$ per unit time afterwards.

B.1 Post-Investment Utility Maximization

As in Section 4.2.2, we solve the leader's utility maximization problem in two steps, beginning with the control problem assuming the capital investment project has been undertaken.

Suppose $\theta = 0$, the leader's wealth is x, and the cash flow $y > 0$; then the leader's wealth becomes immediately $x - K$ since he/she must pay the fixed cost of entry, K. The leader must share the market upon follower's entry at $\hat{\tau}(y)$. Thus, for a generic initial wealth x, the leader's wealth evolves as:

$$\begin{cases} dX^1(t) = \pi(t)X^1(t)\sigma\big(\lambda dt + dW(t)\big) + rX^1(t)dt + \delta_1 Y(t)dt - C(t)dt, \\ \qquad\qquad\qquad\qquad\qquad\qquad\qquad\qquad\qquad \text{for } 0 < t < \hat{\tau}(y) \\ X^1(0) = x \\ dX^2(t) = \pi(t)X^2(t)\sigma\big(\lambda dt + dW(t)\big) + rX^2(t)dt + \delta_2 Y(t)dt - C(t)dt, \\ \qquad\qquad\qquad\qquad\qquad\qquad\qquad\qquad\qquad \text{for } t > \hat{\tau}(y) \\ X^2\big(\hat{\tau}(y)\big) = X^1\big(\hat{\tau}(y)\big). \end{cases}$$

$$(111)$$

If $\theta = 0$ and $y > \hat{y}$, the follower enters immediately, and the leader's problem is identical to the follower's, i.e., (63) with $\delta = \delta_2$. So, we consider the function:

$$L^2(x, y) = -\frac{1}{r\gamma} e^{-r\gamma\left(x + f(y)\right) + 1 - \frac{\mu + \frac{\lambda^2}{2}}{r}}, \tag{112}$$

where f is the solution of (68), (67) in the interval $[0, \epsilon y + M_\epsilon]^{\text{j}}$ with δ being replaced by δ_2.

If $\theta = 0$ and $y < \hat{y}$, the leader's problem is described as follows. The wealth process is described by X^1 in (111) and the cash flow process follows (42). Introduce the set of admissible controls, $\mathcal{U}^1_{x,y} = \{C(\cdot), \pi(\cdot)\}$, satisfying:

$$\begin{cases} E\left[\int_0^T \left(\pi(t)X^1(t)\right)^2 dt\right] < \infty, & \forall\, T < \infty \\ E\left[\int_0^T C(t)^2 dt\right] < \infty, & \forall\, T < \infty \end{cases}, \tag{113}$$

and

$$\tau^* = \lim \uparrow \tau_N \geq \hat{\tau}(y) \wedge \theta^0 \text{ a.s. where}$$

$$\tau_N = \inf\{t \mid X^1(t) \leq -N\} \text{ and } \hat{\tau}(y) = \inf\{t \mid Y_y(t) \geq \hat{y}\}. \tag{114}$$

Recall $\theta^0 = \inf\{t \mid Y_y(t) = 0\}$. If $\theta^0 < \hat{\tau}(y)$, the follower never invests, and the leader's value function at time θ^0 corresponds to $F(X^1(\theta^0))(cf.(64))$. If $\hat{\tau}(y) < \theta^0$, the leader's value function corresponds to $L^2(X^1(\hat{\tau}(y)), Y(\hat{\tau}(y)))$, at the follower's entry time, $\hat{\tau}(y)$. Thus, to a pair of $(C(\cdot), \pi(\cdot))$, we associate the objective function:

$$J\left(C(\cdot), \pi(\cdot)\right) = E\Big[\int_0^{\hat{\tau}(y) \wedge \theta^0} e^{-\mu t} U(C(t)) dt + F\left(X^1(\theta^0)\right) e^{-\mu\theta^0} \mathbb{1}_{\theta^0 \leq \hat{\tau}(y)}$$
$$+ L^2\left(X^1(\hat{\tau}(y)), Y(\hat{\tau}(y))\right) e^{-\mu\hat{\tau}(y)} \mathbb{1}_{\hat{\tau}(y) < \theta^0}\Big]. \tag{115}$$

We consider the value function:

$$L^1(x, y) = \sup_{\{\pi(\cdot), C(\cdot)\} \in \mathcal{U}^1_{x,y}} J\left(C(\cdot), \pi(\cdot)\right). \tag{116}$$

We study the Bellman equation associated with (116) for $0 < y < \hat{y}$. We look for a solution of the form:

$$L^1(x, y) = -\frac{1}{r\gamma} e^{-r\gamma\left(x + g(y)\right) + 1 - \frac{\mu + \frac{\lambda^2}{2}}{r}}, \tag{117}$$

---

[j]See footnote 4.1 for the definition of M_ϵ.

and we define:

$$L^1(x,y) = L^2(x,y), \quad \text{if } y > \hat{y},$$

where $L^2(x,y)$ is defined in (112). The Bellman equation that the value function $L^1(x,y)$ (116) must satisfy reduces to:

$$\begin{cases} \frac{1}{2}y^2\varsigma^2 g'' + (\alpha - \lambda\varsigma\rho)yg' - \frac{1}{2}r^2\gamma^2 y^2\varsigma^2(1-\rho^2)g'^2 - rg + \delta_1 y = 0, \ 0 < y < \hat{y} \\ g(0) = 0, \qquad g(\hat{y}) = f(\hat{y}). \end{cases}$$

$$(118)$$

Where $f(y)$ is the solution of (68), (67) in the interval $[0, \epsilon y + M_\epsilon]^k$ with δ being replaced by δ_2. We extend $g(y)$ by $f(y)$ for $y > \hat{y}$.
Similar to the study of (68), $g(y)$ may be interpreted as a function of a control problem, and there exists a unique solution which is $C^2(0, \hat{y})$.

Theorem 4.8. *The function $L^1(x,y)$ defined by (117) coincides with the value function given in (116).*

B.2 Pre-Investment Utility Maximization

We now turn to the leader's optimal stopping problem (i.e., choice of θ) with obstacle defined by $L^1(x,y)$. Before the stopping time θ, the leader's wealth evolves according to (57) and the cash flow evolves as (42). Introduce the set of admissible controls, $\mathcal{U}^0_{x,y} = \{C(\cdot), \pi(\cdot), \theta\}$, satisfying

$$\begin{cases} E\left[\int_0^T \left(\pi(t)X^0(t)\right)^2 dt\right] < \infty, \quad \forall T < \infty \\ E\left[\int_0^T C(t)^2 dt\right] < \infty, \quad \forall T < \infty \end{cases}, \quad (119)$$

$$\theta \wedge \theta^0 < \infty \text{ a.s. where } \theta^0 = \inf\{t | Y_y(t) = 0\}, \quad (120)$$

and $\tau^* = \lim \uparrow \tau_N \geq \theta \wedge \theta^0$ a.s. where $\tau_N = \inf\{t | X^0(t) \leq -N\}$. (121)

If $\theta^0 \leq \theta$, the leader never takes the investment and receives $F(X^0(\theta^0))$(cf.(64)). If $\theta < \theta^0$, the leader receives $L^1(X^0(\theta) - K, Y(\theta))$ (cf.(117)) at θ.
Therefore, the objective function is:

$$J_{x,y}(C(\cdot), \pi(\cdot), \theta) = E\Big[\int_0^{\theta \wedge \theta^0} U(C(t))e^{-\mu t}dt + L^1(X^0(\theta) - K, Y(\theta)) \times$$

$$e^{-\mu\theta}\mathbb{1}_{\theta < \theta^0} + F(X^0(\theta^0))e^{-\mu\theta^0}\mathbb{1}_{\theta^0 \leq \theta}\Big]$$

$$(122)$$

---

^{k}See footnote 4.1 for the definition of M_ϵ.

and we define the value function:

$$L(x,y) = \sup_{\{\pi(\cdot),C(\cdot),\theta\}\in\mathcal{U}^0_{x,y}} J_{x,y}\big(C(\cdot),\pi(\cdot),\theta\big). \tag{123}$$

We look for a solution as follows:

$$L(x,y) = -\frac{1}{r\gamma}e^{-r\gamma\big(x+h(y)\big)+1-\frac{\mu+\frac{\lambda^2}{2}}{r}}. \tag{124}$$

The V.I. that the value function $L(x,y)$ (123) must satisfy reduces to:

$$\begin{cases} \frac{1}{2}y^2\varsigma^2 h'' + h'y(\alpha - \lambda\varsigma\rho) - \frac{1}{2}y^2\varsigma^2 r\gamma(1-\rho^2)h'^2 - rh \leq 0 \\ h(y) \geq g(y) - K \\ \big(h(y) - g(y) + K\big)\big[\frac{1}{2}y^2\varsigma^2 h'' + h'y(\alpha-\lambda\varsigma\rho) - \frac{1}{2}y^2\varsigma^2 r\gamma(1-\rho^2)h'^2 - rh\big] \\ \hspace{8cm} = 0 \\ h(0) = 0. \end{cases} \tag{125}$$

The obstacle $g(y) - K$ is C^0 but not C^1. Thus the V.I.(125) must be interpreted in a weak sense. Considering: $u(y) = h(y) - f(y) + K$, we yield:

$$\begin{cases} -\frac{1}{2}y^2\varsigma^2 u'' - y(\alpha - \lambda\varsigma\rho - y\varsigma^2 r\gamma(1-\rho^2)f')u' + \frac{1}{2}\varsigma^2 r\gamma y^2(1-\rho^2)u'^2 + ru \\ \hspace{9cm} \geq -\delta_2 y + rK \\ u \geq m \\ (u-m)\big[-\frac{1}{2}y^2\varsigma^2 u'' - y(\alpha - \lambda\varsigma\rho - y\varsigma^2 r\gamma(1-\rho^2)f')u' \\ \hspace{4cm} + \frac{1}{2}\varsigma^2 r\gamma y^2(1-\rho^2)u'^2 + ru + \delta_2 y - rK\big] = 0 \\ u(0) = K \end{cases} \tag{126}$$

where $m = g(y) - f(y)$ is the solution of:

$$\begin{cases} -\frac{1}{2}y^2\varsigma^2 m'' - y(\alpha - \lambda\varsigma\rho - y\varsigma^2 r\gamma(1-\rho^2)f')m' + \frac{1}{2}\varsigma^2 r\gamma y^2(1-\rho^2)m'^2 \\ \hspace{4cm} + ry = (\delta_1 - \delta_2)y, \quad 0 < y < \hat{y} \\ m(0) = m(\hat{y}) = 0 \\ m(y) = 0, \quad y > \hat{y}. \end{cases} \tag{127}$$

We next show that $u(y)$ is more appropriately the value function of a stochastic differential game.

Theorem 4.9. *We assume* $\frac{r-\alpha+\lambda\varsigma\rho}{\gamma\varsigma^2(1-\rho^2)} > \delta_1\hat{y}$. *There exists a unique* $u(y) \in C^1(0,\infty)$, *piecewise* C^2, *solution of (126). This function vanishes for y* sufficiently large. Moreover, it is the value function given

$$u(y) = \inf_{v(\cdot)} \sup_{\theta} J_y\big(v(\cdot), \theta\big) = \sup_{\theta} \inf_{v(\cdot)} J_y\big(v(\cdot), \theta\big) \tag{128}$$

with

$$
\begin{cases}
dY(t) = Y(t)\big(\alpha - \lambda\varsigma\rho - \varsigma^2 r\gamma(1-\rho^2)Y(t)f'(Y(t)) + v(t)\varsigma\sqrt{r\gamma(1-\rho^2)}\big)\,dt \\
\qquad + \varsigma Y(t)dW(t) \\
Y(0) = y \\
J_y\big(v(\cdot), \theta\big) = E\Big[\int_0^{\theta^0 \wedge \theta} \big(-\delta_2 Y_y(t) + rK + \tfrac{1}{2}v^2(t)\big)e^{-rt}dt + Ke^{-r\theta^0}\mathbb{1}_{\theta^0 < \theta} \\
\qquad\qquad\qquad\qquad\qquad\qquad + m\big(Y_y(\theta)\big)e^{-r\theta}\mathbb{1}_{\theta < \theta^0}\Big]
\end{cases}
$$

where $\theta^0 = \inf\{t | Y_y(t) = 0\}$, and m is defined previously, the solution of (127) and extended by 0 for $y > \hat{y}$.
There exists a saddle point $\hat{v}(\cdot)$, $\hat{\theta}$ such that:

$$J_y\big(\hat{v}(\cdot), \hat{\theta}\big) = \inf_{v(\cdot)} \sup_{\theta} J_y\big(v(\cdot), \theta\big) = \sup_{\theta} \inf_{v(\cdot)} J_y\big(v(\cdot), \theta\big)$$

and we have $0 \leq u(y) \leq K + \bar{m}$, where $\bar{m} = \sup m(y) \leq \frac{\delta_1 - \delta_2}{r}\hat{y}$ with $m(y) = g(y) - f(y)$, solution of (127) and extended by 0 for $y > \hat{y}$.

The solution of (126) is characterized by two intervals.

Theorem 4.10. There exists a unique triple y_1, y_2, y_3 with $0 < y_1 < y_2 < \hat{y} < y_3$ such that

$$-\frac{1}{2}y^2\varsigma^2 u'' - y\big(\alpha - \lambda\varsigma\rho - y\varsigma^2 r\gamma(1-\rho^2)f'\big)u' + \frac{1}{2}\varsigma^2 r\gamma y^2(1-\rho^2)u'^2 + ru$$

$$= -\delta_2 y + rK,$$

$$0 < y < y_1 \quad \text{and} \quad y_2 < y < y_3 \tag{129}$$

with the smooth pasting conditions:

$$
\begin{cases}
u(y_1) = m(y_1),\ u'(y_1) = m'(y_1) \\
u(y_2) = m(y_2),\ u'(y_2) = m'(y_2)\ , \\
u(y_3) = 0, \qquad u'(y_3) = 0
\end{cases}
\tag{130}
$$

where $m(y) = g(y) - f(y)$, the solution of (127) and extended by 0 for $y > \hat{y}$.

Theorem 4.11. The function $L(x, y)$ defined by (124) coincides with the value function (122).

The leader's optimal stopping rule is defined in the the same way as the lump-sum payoff case in case of complete markets (cf.(31)).

4.3.3. *Cash Flow Payoffs - Arithmetic Brownian Motion Cash Flow Process*

A. Follower

The follower's value function as defined by (93) with $g(y)$ satisfying (94), where $\hat{y}$ is the unique value defined by the V.I. (95) and (98). We take $\delta = \delta_2$. The optimal stopping stopping strategy for the follower is:

$$\hat{\tau}(y) = \inf\{t | Y_y(t) \geq \hat{y}\},$$

where $\hat{y}$ is the unique value defined by the V.I. (95) and (98). Note again that we must take $\delta = \delta_2$ and thus $\hat{y} \geq \frac{rK}{\delta_2}$. Since the follower enters after the leader (who starts at $\theta < \infty$), the follower will enter at time: $\hat{\tau}_\theta = \theta + \hat{\tau}(Y_y(\theta))$.

B. Leader

By paying cost K, the leader expects to receive a continuous cash flow $\delta_1 Y(t)$ per unit time prior to the follower's entry, and $\delta_2 Y(t)$ per unit time after the follower's entry.

B.1 Post-Investment Utility Maximization

Suppose $\theta = 0$, his/her wealth is x, and the cash flow $y > 0$. The wealth becomes immediately $x - K$ since he/she has to pay the fixed cost of entry. The follower will enter at $\hat{\tau}(y)$. The leader's wealth evolves as (111).

If $\theta = 0$, and $y \geq \hat{y}$, the follower enters immediately and the leader's problem is exactly the same as the follower's, i.e., (87) with $\delta = \delta_2$. So we consider the function:

$$L^2(x, y) = -\frac{1}{r\gamma} e^{-r\gamma\left(x + f(y)\right) + 1 - \frac{\mu + \frac{\lambda^2}{2}}{r}} \tag{131}$$

$$\text{with:} \quad f(y) = \frac{\delta_2 y}{r} + \frac{\delta_2}{r^2}\left(\alpha - \lambda\varsigma\rho - \frac{1}{2}\varsigma^2\delta_2\gamma(1 - \rho^2)\right). \tag{132}$$

If $\theta = 0$ and $y < \hat{y}$, the leader's problem is described as follows. The leader's wealth evolves according to (58) with δ being replaced by δ_1 and the cash flow process evolves as (43). Introduce the set of admissible controls

$\mathcal{U}^1_{x,y} = \{C(\cdot), \pi(\cdot)\}$ satisfying:

$$\begin{cases} E\big[\int_0^T \big(\pi(t)X^1(t)\big)^2 dt\big] < \infty, & \forall T < \infty \\ E\big[\int_0^T C(t)^2 dt\big] < \infty, & \forall T < \infty \end{cases}, \tag{133}$$

and $\tau^* = \lim \uparrow \tau_N \geq \hat{\tau}(y)$ a.s. where $\tau_N = \inf\{t|X^1(t) \leq -N\}$. (134)

At $\hat{\tau}(y)$, the leader gets $L^2\big(X^1(\hat{\tau}(y)), Y(\hat{\tau}(y))\big)$ (cf.(131)). To a pair $(C(\cdot), \pi(\cdot))$, we associate the objective function:

$$J\big(C(\cdot), \pi(\cdot)\big) = E\Big[\int_0^{\hat{\tau}(y)} e^{-\mu t}C(t)dt + L^2\big(X^1(\hat{\tau}(y)), Y(\hat{\tau}(y))\big)e^{-\mu\hat{\tau}(y)}\Big] \tag{135}$$

where $\hat{\tau}(y) < \infty$ a.s. We consider the value function:

$$L^1(x,y) = \sup_{\{\pi(\cdot), C(\cdot)\} \in \mathcal{U}^1_{x,y}} J\big(C(\cdot), \pi(\cdot)\big). \tag{136}$$

We look for a solution:

$$L^1(x,y) = -\frac{1}{r\gamma} e^{-r\gamma\big(x+g(y)\big)+1-\frac{\mu+\frac{\lambda^2}{2}}{r}}. \tag{137}$$

The V.I. that the value function $L^1(x,y)$ (136) must satisfy reduces to:

$$\begin{cases} -\frac{1}{2}\varsigma^2 g'' - (\alpha - \varsigma\rho)g' + \frac{1}{2}r^2\gamma^2\varsigma^2(1-\rho^2)g'^2 + rg = \delta_1 y, & y < \hat{y} \\ g(\hat{y}) = f(\hat{y}), & g \text{ has linear growth at } -\infty. \end{cases} \tag{138}$$

Considering the difference $m = g - f$, (138) becomes:

$$\begin{cases} -\frac{1}{2}\varsigma^2 m'' - m'\big(\alpha - \lambda\varsigma\rho - \varsigma^2\delta_2\gamma(1-\rho^2)\big) + \frac{1}{2}\varsigma^2 r\gamma(1-\rho^2)m'^2 + rm \\ \hspace{6cm} = (\delta_1 - \delta_2)y \\ m(\hat{y}) = 0, \quad m \text{ has linear growth at } y \to -\infty. \end{cases} \tag{139}$$

Proposition 4.3. *There exists one and only one solution of (139) in the interval:*

$$\frac{\delta_1 - \delta_2}{r}(y - y^*) \leq m(y) \leq \frac{\delta_1 - \delta_2}{r}\big[y - y_0 + (y_0 - \hat{y})e^{\beta(y-\hat{y})}\big], \quad \text{for } y < \hat{y}, \tag{140}$$

where $\beta > 0$ is the the solution of: $-\frac{1}{2}\varsigma^2\beta^2 - \beta\big(\alpha - \lambda\varsigma\rho - \gamma\delta\varsigma^2(1-\rho^2)\big) + r = 0$, and $y_0 = -\frac{\alpha - \lambda\varsigma\rho - \delta_2\gamma\varsigma^2(1-\rho^2)}{r}$ with $f\big(y_0 + \frac{\delta_2^2\gamma\varsigma^2(1-\rho^2)}{r^2}\big) = 0$, and we take: $y^* = \max\big(\hat{y}, y_0 + \frac{\gamma\varsigma^2(1-\rho^2)(\delta_1-\delta_2)}{2r}\big)$.

Theorem 4.12. *The function $L^1(x,y)$ defined by (137) coincides with the value function given in (136).*

B.2 Pre-Investment Utility Maximization

We now turn to the leader's optimal stopping problem (i.e., choice of θ). Before the stopping time θ, the leader's wealth evolves according to (57) and cash flow evolve as (43). Introduce the set of admissible controls, $\mathcal{U}^0_{x,y} = \{C(\cdot), \pi(\cdot), \theta\}$, satisfying:

$$\begin{cases} E\big[\int_0^T \big(\pi(t)X^0(t)\big)^2 dt\big] < \infty, & \forall\, T < \infty \\ E\big[\int_0^T C(t)^2 dt\big] < \infty, & \forall\, T < \infty \end{cases}, \tag{141}$$

$$\theta < \infty \text{ a.s., and } \tau^* = \lim \uparrow \tau_N \geq \theta \quad \text{a.s. where } \tau_N = \inf\{t | X^0(t) \leq -N\}. \tag{142}$$

At time θ, the leader invests and receives $L^1(X^0(\theta) - K, Y(\theta))$(cf.((137)); the leader's objective function is:

$$J_{x,y}\big(C(\cdot), \pi(\cdot), \theta\big) = E\Big[\int_0^\theta U\big(C(t)\big)e^{-\mu t}dt + L^1\big(X^0(\theta) - K, Y(\theta)\big)e^{-\mu\theta}\Big] \tag{143}$$

and we define the value function:

$$L(x,y) = \sup_{\{\pi(\cdot), C(\cdot), \theta\} \in \mathcal{U}^0_{x,y}} J_{x,y}\big(C(\cdot), \pi(\cdot), \theta\big). \tag{144}$$

We look for a solution of the form:

$$L(x,y) = -\frac{1}{r\gamma}e^{-r\gamma\big(x+h(y)\big)+1-\frac{\mu+\frac{\lambda^2}{2}}{r}}. \tag{145}$$

The V.I. that the value function $L(x,y)$ (144) must satisfy reduces to:

$$\begin{cases} \frac{1}{2}\varsigma^2 h'' + (\alpha - \lambda\varsigma\rho)h' - \frac{1}{2}\varsigma^2 r\gamma(1 - \rho^2)h'^2 - rh \leq 0 \\ h(y) \geq g(y) - K \\ \big(h(y) - g(y) + K\big)\big[\frac{1}{2}\varsigma^2 h'' + (\alpha - \lambda\varsigma\rho)h' - \frac{1}{2}\varsigma^2 r\gamma(1 - \rho^2)h'^2 - rh\big] = 0 \end{cases}. \tag{146}$$

The obstacle $g(y) - K$ is C^0 but not C^1, so that the V.I.(146) must be interpreted in a weak sense. Considering: $u(y) = h(y) - f(y) + K$, we yield:

$$\begin{cases} -\frac{1}{2}\varsigma^2 u'' - \left(\alpha - \lambda\varsigma\rho - \varsigma^2\delta_2\gamma(1-\rho^2)\right)u' + \frac{1}{2}\varsigma^2 r\gamma(1-\rho^2)u'^2 + ru \\ \qquad\qquad\qquad\qquad\qquad\qquad\qquad\qquad \geq -\delta_2 y + rK \\ u \geq m \\ (u-m)\left[-\frac{1}{2}\varsigma^2 u'' - \left(\alpha - \lambda\varsigma\rho - \varsigma^2\delta_2\gamma(1-\rho^2)\right)u' + \frac{1}{2}\varsigma^2 r\gamma(1-\rho^2)u'^2 \right. \\ \left. \qquad\qquad\qquad\qquad\qquad\qquad\qquad + ru + \delta_2 y - rK\right] = 0 \end{cases}$$
$$\tag{147}$$

The function $m(y) = g(y) - f(y)$ is defined by (139). To exclude the case that the leader and the follower have the same strategy, we will consider:

$$m \text{ is not always negative.} \tag{148}$$

As in the geometric Brownain motion cash flow case, we next show that $u(y)$ is more appropriately the value function of a stochastic differential game.

Theorem 4.13. *Assume (148). There exists a unique $u \in C^1(-\infty, \infty)$, piecewise C^2 solution of (147). This function vanishes for y sufficiently large. It is the value function:*

$$u(y) = \inf_{v(\cdot)} \sup_{\theta} J_y\big(v(\cdot), \theta\big) = \sup_{\theta} \inf_{v(\cdot)} J_y\big(v(\cdot), \theta\big) = J_y\big(\hat{v}(\cdot), \hat{\theta}\big), \tag{149}$$

where

$$\begin{cases} J_y\big(v(\cdot), \theta\big) = E\Big[\int_0^\theta \big(-\delta_2 Y_y(t) + rK + \frac{1}{2}v^2(t)\big)e^{-rt}dt + m\big(Y_y(\theta)\big) \\ \qquad\qquad\qquad\qquad\qquad\qquad\qquad\qquad\qquad\qquad \times e^{-r\theta}\mathbb{1}_{\theta<\infty}\Big]. \\ v(\cdot) \in \mathcal{U}_y = \{\limsup_{T\to\infty} \big(-EY_y(T)e^{-rT}\big) = 0\} \end{cases}$$
$$\tag{150}$$

Moreover, $u(y) + \frac{\delta_2}{r}y$ is bounded for $y \to -\infty$, and $u \geq 0$.

The solution to (147) is characterized by two intervals.

Theorem 4.14. *Assume (148). There exists a unique triple y_1, y_2, y_3*

with $y_1 < y_2 < \hat{y} < y_3$ such that

$$
\begin{cases}
-\frac{1}{2}\varsigma^2 u''(y) - \left(\alpha - \lambda\varsigma\rho - \varsigma^2\delta_2\gamma(1-\rho^2)\right)u'(y) + \frac{1}{2}\varsigma^2 r\gamma(1-\rho^2)u'^2(y) \\
\qquad + ru(y) = -\delta_2 y + rK \qquad \text{for } y < y_1 \text{ and } y_2 < y < y_3 \\
u(y) = m(y) \quad \text{for } y_1 \le y \le y_2 \\
u'(y_1) = m'(y_1) \\
u'(y_2) = m'(y_2) \\
u(y) = 0 \quad \text{for } y \ge y_3
\end{cases}
$$

(151)

We can take $y_3 = \bar{y} = k\hat{y}$ with k sufficiently large.

Theorem 4.15. *Assume (148). Then the function $L(x, y)$ defined by (145) is the value function (144).*

The leader's optimal stopping rule can be defined in the same way as the lump-sum payoff case in case of complete markets (cf.(31)).

5. Conclusion

From our research results, we caution that, for managers, naively assuming market completeness and ignoring the potential strategic interactions from competitors will likely lead to non-optimal investment decisions. The leader's three threshold solution, which is understandable but non-intuitive, cannot be predicted without the mathematical theory.

References

1. Bensoussan, A. and J.L. Lions, *Applications of Variational Inequalities in Stochastic Control*, Elsevier North-Holland, 1978.
2. Bensoussan, A., 2008, "Real Options", Handbook of Mathematical Modelling and Numerical Methods in Finance, A. Bensoussan and Q. Zhang(Eds.), 15(1), Elsevier(December), 531-572.
3. Bensoussan, A., J. D. Diltz and S. Hoe, 2009, "Real Options Games in Complete and Incomplete Markets with Several Decision Makers," forthcoming, *SIAM Journal of Financial Mathematics*.
4. Bensoussan, A., J. D. Diltz and S. Hoe, 2009, "Real Options in a Stackelberg Game with a Stochaastic Demand Process," to be published.
5. Brennan, M. and E. Schwartz, 1982, Regulation and Corporate Investment Policy, *Journal of Finance*, **37(2)**, 289-300.
6. Brennan, M. and E. Schwartz, 1982, Consistent Regulatory Policy Under Uncertainty, *Bell Journal of Economics*, **13(2)**, 506-524.
7. Brennan, M. and E. Schwartz, 1985, Evaluating Natural Resource Investments, *Journal of Business*, **58(2)**, 135-157.

8. Constantinides, George M., 1978, "Market Risk Adjustment in Project Valuation", *Journal of Finance*, **33(2)**, 603-616.
9. Dixit, A. and R.S. Pindyck, *Investment Under Uncertainty*, Princeton University Press, 1994.
10. Grenadier, S., 1996, "The Strategic Exercise Of Options: Development Cascades And Overbuilding In Real Estate Markets," *Journal of Finance*, **51(5)**, 1653-1679.
11. Grenadier, S., 1999, "Information Revelation Through Option Exercise," *Review of Financial Studies*, **12(1)**, 95-129.
12. Grenadier, S., 2002, "Option Exercise Games: An Application To The Equilibrium Investment Strategies Of Firms," *Review of Financial Studies*, **15(3)**, 691-721.
13. Henderson, V., 2008, "Valuing the Option to Invest in an Incomplete Market," *Mathematics and Financial Economics*, **1(2)**, 103-128.
14. Henderson, V. and D. Hobson, 2002, Real Options with Constant Relative Risk Aversion, *Journal of Economic Dynamics and Control*, **27(2)**, 329-355.
15. Huisman, K.J.M., *Technology Investment: A Game Theoretical Real Options Approach*, Kluwer Academic Pbulishing, 2001.
16. Karatzas, I. and S. Shreve, *Methods of Mathematical Finance*, Springer, 1998.
17. Lions, J.L., and G. Stampacchia, 1967, Variational Inequalities, *Comm. P. Appl. Math.*, **XX**, 493-519.
18. McDonald, R. and D. Siegel, 1986, The Value of Waiting to Invest, *Quarterly Journal of Economics*, **101(4)**, 708-728.
19. Merton, R.C., 1969, "Lifetime Portfolio Selection Under Uncertainty: The Continuous Time Case", *Review of Economics and Statistics*, **51**, 247-257.
20. Miao, J. and N. Wang, 2007, "Investment, Consumption and Hedging under Incomplete Markets", *Journal of Financial Economics*, **86**, 608-642.
21. Musiela, M. and T. Zariphopoulou, 2001, "Pricing and Risk Management of Derivatives Written on Non-Traded Assets," Working Paper.
22. Oberman, A. and T. Zariphopoulou, 2003, "Pricing Early Exercise Contracts in Incomplete Markets," *Computational Management Science*, 75-107.

Finding Expectations of Monotone Functions of Binary Random Variables by Simulation, with Applications to Reliability, Finance, and Round Robin Tournaments

Mark Brown

Department of Mathematics
City College, CUNY, New York, NY, USA
Email: cybergarf@aol.com

Erol A. Peköz

School of Management
Boston University
595 Commonwealth Avenue
Boston, MA 02215, USA
Email: pekoz@bu.edu

Sheldon M. Ross

Department of Industrial and Systems Engineering
University of Southern California
Los Angeles, CA 90089, USA
Email: smross@usc.edu

We study the quantity $E[\phi(X_1, \ldots, X_n)]$ when $X_1, \ldots, X_n$ are independent Bernoulli random variables, and ϕ is a nondecreasing function. With $T = \sum_{i=1}^{n} X_i$, we note that $E[\phi(X_1, \ldots, X_n)|T]$ is a nondecreasing function of T, and show how it can be efficiently estimated by a simulation study that stratifies on T. Our results are applied to static and dynamic reliability systems, the pricing of derivatives related to basket default swaps, and to round robin tournaments.

1. Introduction

Let $X_1, \ldots, X_n$ be independent Bernoulli random variables, and let $T = \sum_{i=1}^{n} X_i$ be their sum. Also, suppose that ϕ is a nondecreasing function of the vector $\mathbf{X} = (X_1, \ldots, X_n)$. In Section 2 we show how we can efficiently estimate $E[\phi(\mathbf{X})]$ by a simulation approach that stratifies on T. In Section 3, we note that $E[\phi(\mathbf{X})|T]$ is a nondecreasing function of T, and show how this result can sometimes be used to modify our simulation

estimates. In Section 4 we apply our results to the classical reliability problem of finding the probability that a system composed of n independent
binary components will function. In Section 5 we consider a dynamic version of the preceding model, in which each component works for a random
length of time and then fails. Depending on the structure of the system,
the component failures eventually cause the system to fail. Supposing that
there is a cost incurred if the system fails before some specified time t^*,
with the cost depending on the time of failure, we present a stratification
approach to estimate the expected cost incurred. In Section 6 we apply our
results to the pricing of derivatives related to basket default swaps, where
our methods are similar to, but an improvement on, a recently proposed
simulation approach (see[2]). In Section 7 we apply our results to estimating
win probabilities in round robin tournaments.

2. Estimating $E[\phi(X_1, \ldots, X_n)]$ by Simulation

Suppose that $\mathbf{X} = (X_1, \ldots, X_n)$ is a vector of independent Bernoulli random variables with $E[X_i] = p_i$, and that we are interested in using simulation to estimate $E[\phi(\mathbf{X})]$. Our approach is to stratify on $T = \sum_{i=1}^{n} X_i$. To
do so we need to first determine the probability mass function of T, and
then show how to simulate $\mathbf{X}$ conditional on $T = i$, for $i = 0, \ldots, n$.

To start, we note that if

$$P_j(i) = P(X_j + \ldots + X_n = i)$$

then, analogous to Example 3.22 of,[6] $P_j(i)$ satisfy

$$P_j(i) = p_j P_{j+1}(i - 1) + (1 - p_j)P_{j+1}(i) \tag{1}$$

$$P_n(1) = p_n, \quad P_n(0) = 1 - p_n, \quad P_n(i) = 0, \ i \neq 0, 1.$$

Starting with the preceding expression for $P_n(i)$, the equations (1) are easily
solved by a recursion that first solves for P_{n-1}, then P_{n-2}, and so on.

To generate $(X_1, \ldots, X_n)$ conditional on $T = i$, generate in sequence

- X_1 given $T = i$
- X_2 given $T = i, X_1$
- X_3 given $T = i, X_1, X_2$

and so on. To generate X_j, given both that $T = i$ and the values of

$X_1, \ldots, X_{j-1}$, use that

$$P(X_j = 1 | T = i, X_1, \ldots, X_{j-1}) = P(X_j = 1 | \sum_{k=j}^{n} X_k = i - \sum_{k=1}^{j-1} X_k)$$

$$= \frac{p_j P_{j+1}(i - 1 - \sum_{k=1}^{j-1} X_k)}{P_j(i - \sum_{k=1}^{j-1} X_k)}. \qquad (2)$$

Suppose one is planning on doing r simulation runs. One way to employ the stratification approach is to do $rP_1(i)$ runs conditional on $T = i$ for each $i = 0, \ldots, n$. Call the preceding the "standard proportional stratification" approach, Because T and $\phi(\mathbf{X})$ are both increasing functions of the vector of independent random variables $\mathbf{X}$ it follows that they are positively correlated and thus the standard proportional stratification will result in an estimator with smaller variance than the estimator based on r non-stratified runs (see[7]).

However, a more efficient approach than using standard proportional stratification is to first do a preliminary study by simulating $\phi(\mathbf{X})$ conditional on $T = i$ enough times so as to get a rough estimate of

$$s_i^2 \equiv \mathrm{Var}(\phi(\mathbf{X}) | T = i), \quad i = 0, \ldots, n.$$

Then, if you are planning to do r simulation runs, do $r \frac{s_i P_1(i)}{\sum_k s_k P_1(k)}$ of these runs conditional on $T = i$. (That this results in the minimal variance of the final estimator see.[7]) Letting ϕ_i be the average of the runs done conditional on $T = i$, estimate $E[\phi(\mathbf{X})] = \sum_{i=0}^{n} E[\phi(\mathbf{X}) | T = i] P_1(i)$ by $\sum_{i=0}^{n} \phi_i P_1(i)$.

Remark. Whether using the standard or the more efficient procedure, further variance reduction can be obtained by using antithetic variables. That is, in generating $(X_1, \ldots, X_n)$ conditional on $T = i$, first generate random numbers $U_1, \ldots, U_n$, and then generate the value of X_j, given both that $T = i$ and the values of $X_1, \ldots, X_{j-1}$, by letting it equal 1 if U_j is less than the right side of (2) In the next generation of $(X_1, \ldots, X_n)$ conditional on $T = i$, we can utilize the same set of random numbers, but this time subtracting each from 1. (That is, we use the random numbers $1 - U_1, \ldots, 1 - U_n$ in the next run.) Because ϕ is a monotone function the reuse of the random number set will lead to a variance reduction when compared with using a new independent set of random numbers (see[7] for a proof).

Example 1. Let $X_i, i = 1, \ldots, 20$, be independent Bernoulli random variables with common mean $P(X_i = 1) = 1/2$, and let

$$\phi(x_1, \ldots, x_{20}) = (\sum_{i=1}^{20} ix_i)^2.$$

The standard deviation of the standard Monte Carlo (commonly referred to as the *raw*) simulation estimator of $E[\phi(X_1, \ldots, X_n)]$ based on 10^4 runs is 57, whereas the standard deviation of the stratified estimator based on the same number of runs is 26, a variance reduction by a factor of approximately 4. However, noting that the lower indexed X_i have a much smaller effect on the value of $\phi(X_1, \ldots, X_n)]$ than do the higher ones, it seems reasonable that rather than stratifying on $\sum_{i=1}^{20} X_i$ it would be better to stratify on, say, $\sum_{i=10}^{20} X_i$. A further simulation indicated that stratifying on $\sum_{i=10}^{20} X_i$ reduced the standard deviation (again based on 10^4 runs) down to 8, a variance reduction over the standard Monte Carlo estimator by a factor of about 50.

3. Monotonicity of the Conditional Expectation Given T

Not only, as noted in the preceding section, are T and $\phi(X_1, \ldots, X_n)$ positively correlated but, even stronger, is that the conditional distribution of $\phi(\mathbf{X})$ given that $T = k$ is stochastically increasing in k. That is, we have the following result.

Theorem 3.1. *If $X_1, \ldots, X_n$ are independent Bernoulli random variables, and ϕ a nondecreasing function, then $E[\phi(X_1, \ldots, X_n)|T = k]$ is a nondecreasing function of k.*

Theorem 3.1 is a special case of a general result of[3] which states that the preceding is true whenever the X_i are independent and have logconcave densities or mass functions. As the proof of[3] is rather involved, we now present a proof that is not only elementary but also in the spirit of this paper. We being with a lemma.

Lemma 3.1. *The conditional distribution of X_n given T is stochastically increasing in T. That is, $P(X_n = 1|T = k)$ is a nondecreasing function of k.*

Proof. The proof makes use of the log concavity result that $P(T = k)/P(T = k-1)$ is nonincreasing in k. (For a proof, see.[5]) Let $p_n =$

$P(X_n = 1) = 1 - q_n$. Also, let $b_r = P(\sum_{i=1}^{n-1} X_i = r)$. Then, we have

$$P(X_n = 1 | T = k) = \frac{p_n b_{k-1}}{P(T = k)}$$

$$= \frac{p_n b_{k-1}}{p_n b_{k-1} + q_n b_k}$$

$$= \frac{p_n}{p_n + q_n b_k / b_{k-1}}$$

and the result follows since b_k / b_{k-1} is nonincreasing by log concavity. $\square$

We now prove the theorem.

Proof of Theorem 3.1. Let $S_r = \sum_{i=1}^{r} X_i$. Also, for fixed $k > 0$, let $Y_1, ..., Y_n$ be distributed as $X_1, ..., X_n$ conditional on $S_n = k$; and let $Z_1, ..., Z_n$ be distributed as $X_1, ..., X_n$ conditional on $S_n = k - 1$. We now show how we can generate random vectors $Y_1, ..., Y_n$ and $Z_1, ..., Z_n$ that are distributed according to the preceding and are such that $Y_i \geq Z_i, i = 1, ..., n$. To do so, first note that by the Lemma 3.1

$$P(Y_n = 1) \geq P(Z_n = 1).$$

We generate the random vectors as follows:

KEY STEP   Generate a random number U. Then,

if $U \leq P(Z_n = 1)$, set $Z_n = 1$, else set it equal to 0,

if $U \leq P(Y_n = 1)$, set $Y_n = 1$, else set it equal to 0.

There are now 3 cases:

Case 1: $Z_n = Y_n = 1$
Given the scenario of this case, $Y_1, ..., Y_{n-1}$ is distributed as $X_1, .., , X_{n-1}$ conditional on $S_{n-1} = k - 1$; and $Z_1, ..., Z_{n-1}$ is distributed as $X_1, ..., X_{n-1}$ conditional on $S_{n-1} = k - 2$. Consequently, by Lemma 3.1

$$P(Y_{n-1} = 1 | Y_n = 1) \geq P(Z_{n-1} = 1 | Z_n = 1).$$

Thus, we can repeat KEY STEP to generate Y_{n-1} and Z_{n-1} so that $Y_{n-1} \geq Z_{n-1}$.

Case 2: $Z_n = Y_n = 0$
Given the scenario of this case, $Y_1, ..., Y_{n-1}$ is distributed as $X_1, .., , X_{n-1}$

conditional on $S_{n-1} = k$; and $Z_1, ..., Z_{n-1}$ is distributed as $X_1, .., , X_{n-1}$ conditional on $S_{n-1} = k - 1$. Consequently, by Lemma 3.1

$$P(Y_{n-1} = 1 | Y_n = 0) \geq P(Z_{n-1} = 1 | Z_n = 0).$$

Thus, we can repeat KEY STEP to generate Y_{n-1} and Z_{n-1} so that $Y_{n-1} \geq Z_{n-1}$.

Case 3. $Z_n = 0, Y_n = 1$

Given the scenario of this case, the random vectors $Y_1, ..., Y_{n-1}$ and $Z_1, ..., Z_{n-1}$ have the same joint distribution. Thus, we can generate them so that $Y_i = Z_i, i = 1, ..., n - 1$.

The preceding shows how to generate the vectors so that $Y_i \geq Z_i, i = 1, ..., n$. By the monotonicity of ϕ this implies that $\phi(Y_1, ..., Y_n) \geq \phi(Z_1, ..., Z_n)$. Consequently,

$$\begin{aligned} E[\phi(X_1, ..., X_n) | T = k] &= E[\phi(Y_1, ..., Y_n)] \\ &\geq E[\phi(Z_1, ..., Z_n)] \\ &= E[\phi(X_1, ..., X_n) | T = k - 1]. \qquad \square \end{aligned}$$

Remark. While Theorem 1 might seem quite intuitive, it does depend on the X_i being Bernoulli random variables. For instance, suppose that X_1 and X_2 both put all their mass on the values $0, 2, 3$. Then, with

$$\phi(x_1, x_2) = x_1^2 + x_2^2$$

we would have that

$$E[\phi(X_1, X_2) | T = 3] > E[\phi(X_1, X_2) | T = 4]. \qquad \square$$

Suppose now that our simulation of the preceding section resulted in the estimate ϕ_i of $E[\phi(\mathbf{X}) | T = i]$ for $i = 0, ..., n$. If it results that ϕ_i is not nondecreasing in i, then we can modify these estimates by using the ideas of isotonic regression. Isotonic regression takes preliminary estimates $e_1, ..., e_n$ of unknown quantities that are known to be nondecreasing, and obtains final estimates $a_1, ..., a_n$ by solving the minimization problem

$$\min_{a_1 \leq ... \leq a_n} \sum_{i=1}^{n} (e_i - a_i)^2,$$

which can generally be solved in time linear in n (see[1] for details).

The following corollary will be used in the sequel.

Corollary 3.1. *Suppose that the random vectors* $(X_i, W_i), i = 1, \ldots, n$ *are independent, where* $X_i, i = 1, \ldots, n$ *are Bernoulli random variables, and where*

$$W_i | X_i = 1 \geq_{st} W_i | X_i = 0$$

where by the preceding we mean that $P(W_i \geq y | X_i = 1) \geq P(W_i \geq y | X_i = 0)$, *for all* y. *Let* $T = \sum_{i=1}^{n} X_i$. *Then, for any nondecreasing function* h, $E[h(W_1, \ldots, W_n) | T = k]$ *is a nondecreasing function of* k.

Proof. Let $g(\mathbf{X}) = E[h(\mathbf{W}) | \mathbf{X}]$ where $\mathbf{W} = (W_1, \ldots, W_n)$ and $\mathbf{X} = (X_1, \ldots, X_n)$. Because the random vectors are independent and W_i given $X_i = 1$ is stochastically larger than W_i given $X_i = 0$, it follows that $g(\mathbf{X})$ is a nondecreasing function of $\mathbf{X}$. Hence, by Theorem 3.1, it follows that $E[g(\mathbf{X}) | T = k]$ is a nondecreasing function of k. The result now follows because

$$\begin{aligned}
E[h(\mathbf{W}) | T = k] &= E[E[h(\mathbf{W}) | T = k, \mathbf{X}] | T = k] \\
&= E[E[h(\mathbf{W}) | \mathbf{X}] | T = k] \\
&= E[g(\mathbf{X}) | T = k]. \qquad \square
\end{aligned}$$

4. The Classical Reliability Model

Consider an n component system in which each component is either working or failed, and suppose that there exists a nondecreasing binary function ϕ such $\phi(x_1, \ldots, x_n)$ is 1 if the system works when x_i is the indicator variable for whether component i is working, $i = 1, \ldots, n$. The function ϕ is called the structure function, and to rule out trivialities we assume that $\phi(0, 0, \ldots, 0) = 0$ and $\phi(1, 1, \ldots, 1) = 1$. Now consider the problem of determining $E[\phi(X_1, \ldots, X_n)]$ when the X_i are independent Bernoulli random variables with $p_i = E[X_i]$, $i = 1, \ldots, n$. With $T = \sum_{i=1}^{n} X_i$, we can use the approach of the preceding sections to estimate $E[\phi(X_1, \ldots, X_n)]$ by doing a simulation that stratifies on T.

Call any set of components having the property that the system necessarily works when all of these components are working a *path* set. If no proper subset of a path set is itself a path set, call the path set a *minimal path set*. Let m_1 denote the size of the smallest minimal path set, and let

m_2 be the size of the largest minimal path set. Because the system works whenever $T > m_2$ and does not work when $T < m_1$, we have that

$$E[\phi(X_1, \ldots, X_n)] = \sum_{i=m_1}^{m_2} E[\phi(X_1, \ldots, X_n)|T = i]P(T = i) + P(T > m_2).$$

Consequently, we need only estimate the quantities $E[\phi(X_1, \ldots, X_n)|T = i]$ for $m_1 \le i \le m_2$, and this can be done using the approach of Section 2. If r simulation runs are planned then one can either do $\frac{rP(T=i)}{P(m_1 \le T \le m_2)}$ runs conditional on $T = i$, $m_1 \le i \le m_2$ or, better, do an initial small size simulation to estimate the conditional variances $\mathrm{Var}(\phi(X_1, \ldots, X_n)|T = i)$, $m_1 \le i \le m_2$ and then do a larger simulation in which the number of runs done conditional on $T = i$ is proportional to $P(T = i)$ times the square root of the estimate of the $\mathrm{Var}(\phi(X_1, \ldots, X_n)|T = i)$. If the resulting estimates of $E[\phi(X_1, \ldots, X_n)|T = i]$ are not monotone in i, then the estimates can be modified by an isotonic regression.

5. A Dynamic Reliability Model

Again suppose that ϕ is a structure function for an n component system. Suppose that each of the n components is initially working, and that component i works for random time W_i, $i = 1, \ldots, n$. In addition, suppose that $W_1, \ldots, W_n$ are independent, with W_i having distribution function F_i. If L is the amount of time that the system itself works, then

$$L = \max_{i=1,\ldots,s} \min_{j \in M_i} W_j$$

where $M_1, \ldots, M_s$ are the minimal path sets for the structure function ϕ.

Suppose that a cost $C(t)$ is incurred if the lifetime of the system is t, where, for some specified time t^*,

$$C(t) = h((t^* - t)^+)$$

where h is a nondecreasing function having $h(0) = 0$. In other works, there is no cost if system life exceeds t^* and a cost $h(s)$ if the system fails at time $t^* - s$, $s < t^*$. We are interested in using simulation to estimate $E[C(L)]$.

Let X_i equal 1 if component i is still working at time t^* and let it equal 0 otherwise. Then $E[\phi(\mathbf{X})]$ is the probability that the system life exceeds t^*. Let $T = \sum_{i=1}^{n} X_i$. With $p_i = 1 - F_i(t^*)$, $i = 1, \ldots, n$, let $P_j(i), i, j = 1, \ldots, n$ be the solution of (1).

We propose to estimate $E[C(L)]$ by stratifying on T. To begin, let m be the size of the largest minimal path set. Because the system cannot be failed if there are more than m working components, we have

$$E[C(L)] = \sum_{i=0}^{m} E[C(L)|T = i]P_1(i).$$

To simulate $C(L)$ conditional on $T = i$, $i = 0, \ldots, m$, first use the method given in Section 2 to generate $X_1, \ldots, X_n$ conditional on $T = i$. If $\phi(X_1, \ldots, X_n) = 1$ then take 0 as the estimate of $E[C(L)|T = i]$ from that run. If $\phi(X_1, \ldots, X_n) = 0$, then for any j for which $X_j = 0$, generate W_j according to the distribution

$$F_j^*(t) = P(W_j \leq t | W_j \leq t^*) = F_j(t)/F_j(t^*), \quad 0 \leq t \leq t*.$$

For j such that $X_j = 1$, set $W_j = t^*$. With L being the lifetime of the system according to the preceding values of W_j, take $C(L)$ as the estimate of $E[C(L)|T = i]$ from that run. Of course the number of runs to do conditional on $T = i$ should be determined by a preliminary small simulation study to estimate the quantities $\mathrm{Var}(C(L)|T = i)$, $i = 0, \ldots, m$.

Because the random vectors (X_i, W_i) satisfy the conditions of Corollary 2 and $C(L)$ is a nonincreasing function of $(W_1, \ldots, W_n)$ it follows from that Corollary that $E[C(L)|T = i]$ is a nonincreasing function of i. Hence, if the resulting estimates of $E[C(L)|T = i]$ are not monotone, then an isotonic regression can be employed to modify them.

A set of components is said to be a *cut* set if the system is necessarily failed when all components in this set are failed. A cut set is said to be a *minimal cut set* if none of its proper subsets are cut sets. A second simulation approach, which can be used when the number of minimal cut sets is not too large, uses an identity for Bernoulli sums. Suppose there are r minimal cut sets, $C_1, \ldots, C_r$, and for the set C_i define an indicator variable Z_i equal to 1 if all the components in C_i fail before time t^* and equal to 0 otherwise. Let $Z = \sum_{i=1}^{r} Z_i$, and let

$$\lambda = E[Z] = \sum_{i=1}^{r} \prod_{j \in C_i} (1 - p_j).$$

Now, it can be shown (see Section 11.3 of[7] for a proof) that for any random variable R

$$E[ZR] = \lambda E[R|Z_I = 1] \tag{3}$$

where I is independent of Z, R and is equally likely to be any of the values $1, \ldots, r$. Now, let

$$R = \begin{cases} 0 & \text{if } Z = 0 \\ \frac{C(L)}{Z} & \text{if } Z > 0 \end{cases}.$$

Because $Z = 0$ if and only if $C(L) = 0$, the identity (3) yields that

$$E[C(L)] = \lambda E[\frac{C(L)}{Z} | Z_I = 1]. \tag{4}$$

Using that

$$P(I = j | Z_I = 1) = \frac{P(Z_j = 1)}{\sum_{i=1}^{r} P(Z_i = 1)} = \frac{a_j}{\sum_{i=1}^{r} a_i}$$

where

$$a_i = \prod_{j \in C_i} (1 - p_j)$$

we can use the preceding to obtain a simulation estimate of $E[C(L)]$ by performing each simulation run as follows:

(1) Generate I such that $P(I = j) = \frac{a_j}{\sum_{i=1}^{r} a_i}$, $j = 1, \ldots, r$. Suppose $I = j$.
(2) For $i \in C_j$, generate W_i according to the distribution
$$F_i^*(t) = P(W_i \le t | W_i \le t^*) = F_i(t)/F_i(t^*), \quad 0 \le t \le t^*.$$
(3) For $i \notin C_j$, generate W_i according to the distribution F_i.
(4) Determine Z_i, $i = 1, \ldots, r$.
(5) Determine $Z = \sum_{i=1}^{r} Z_i$.
(6) Determine L.
(7) Determine $C(L)$.
(8) Return the estimator $\frac{\lambda C(L)}{Z}$.

In contrast to our first estimator, the preceding estimator need not have a smaller variance than the raw simulation estimator. However, it should be very efficient when λ is small.

6. Modeling Basket Default Costs

A model for basket default swaps in which a portfolio consists of n assets, the ith of which defaults at a random time having distribution F_i, was considered in[4] and.[2] It was supposed in these papers that if at least r

assets default by a fixed time t^* then a cost depending on L, the time of the rth default, was incurred. Thus, the model of[4] is a special case of the model of the preceding section, in which the system structure is an $n-r+1$ of n system which works if and only if at least $n-r+1$ of the n components fail. Letting T be the number of assets that do not fail (i.e., do not default) by time t^*, it was suggested in,[2] as an improvement on the method of,[4] that $E[C(L)]$ be estimated by continually simulating the system conditional on the event that $T \leq n-r$, and then use the average of the values obtained for $C(L)$ multiplied by $P(T \leq n-r)$. That is,[2] notes that

$$E[C(L)] = E[C(L)|T \leq n-r]P(T \leq n-r)$$

and then uses simulation to estimate $E[C(L)|T \leq n-r]$. However, because our approach estimates $E[C(L)|T \leq n-r]$ by estimating all the quantities $E[C(L)|T = j]$, $j \leq n-r$, while using the known probabilities $P(T = j)$, it is a stratified version of the estimator of[2] and thus necessarily has a smaller variance. (Intuitively, there will be a lot more variance in the conditional distribution of $C(L)$ given that $T \leq n-r$, than there will be in the conditional distribution of $C(L)$ given that $T = j$, for any $j \leq n-r$.)

In cases where $P(T \geq r)$ is small, and $\binom{n}{r}$ is not too large, the second simulation method of the previous section should also be considered.

Example 2. The paper[2] gave an example in which there are 10 independent exponential random variables $W_i, i = 1, \ldots, 10$, with respective rates $0.03, 0.01, 0.02, 0.01, 0.005, 0.001,$
$0.002, 0.002, 0.017, 0.003,$
and with respective additive costs $0.3, 0.1, 0.2, 0.1, 0.3, 0.1, 0.2, 0.2, 0.1, 0.3$ incurred if $W_i < t, i = 1, \ldots, 10$. The method of[2] performed slightly better than raw simulation. The following indicates the variance of the raw estimator and of our estimator for different values of t.

Table 1.

t	raw estimator variance	variance of our estimator
5	0.021	0.003
10	0.034	0.006
15	0.045	0.008
20	0.052	0.009
25	0.057	0.010
30	0.062	0.011

As can be seen our estimator is far superior to the raw simulation estimator, and thus to the estimator of.[2]

7. A Round Robin Tournament

In a round robin tournament of $n + 1$ players, each of the $\binom{n+1}{2}$ pairs play a match. The players who win the greatest number of matches are the winners of the tournament. Suppose that the results from all matches are independent, and that $P(i, j) = 1 - P(j, i)$ is the probability that i beats j in their match. Let I be the indicator of the event that player $n + 1$ is the sole tournament winner, and suppose further that we want to use simulation to estimate $E[I] = P(I = 1)$. Letting X_i be the indicator of the event that player $n + 1$ beats player i in their match, $i = 1, \ldots, n$, an efficient way to estimate $E[I]$ would be to let $T = \sum_{i=1}^{n} X_i$ and then do the simulation stratified on T. That is, compute $P(T = j), j = 0, 1, \ldots, n$, and use

$$E[I] = \sum_{j=[n/2]+1}^{n-1} E[I|T = j]P(T = j) + P(T = n).$$

To estimate $E[I|T = j]$, we would generate $X_1, \ldots, X_n$ conditional on $T = j$ and then generate the outcomes of the $\binom{n}{2}$ games that do not involve player $n + 1$, and then take I as the estimator of $E[I|T = j]$ from that run. Antithetic variables should be effective, so when doing the next run we should use (in the same manner) the same uniforms (subtracted from 1) that were just used. As always it is advised to do a small simulation preparatory study to estimate the conditional variances, so as to set the number of runs done conditional in each strata in the final study.

Let W_i be the event that player i is the sole tournament winner, and suppose now that, rather than just wanting to estimate $P(W_{n+1})$, we want to estimate $P(W_i)$ for all $i = 1, \ldots, n + 1$. In this situation, we suggest a post-stratification technique. Start by solving the $n + 1$ sets of linear equations so as to determine the quantities

$$P(T_i = j), \quad j = 1, \ldots, n, \quad i = 1, \ldots, n + 1$$

where T_i is the number of matches that player i wins. Now perform a fixed number of raw simulations of the tournament. Based on the results, let $N(i, j)$ be the number of simulation runs in which player i wins exactly j matches, and let $W(i, j)$ denote the number of simulation runs in which player i both wins exactly j matches and, in addition, is the sole winner of the tournament. Now, take $W(i,j)/N(i,j)$ as an estimate of $P(W_i|T_i = j)$. Since

$$P(W_i) = \sum_{j=1}^{n} P(W_i|T_i = j)P(T_i = j)$$

this yields our estimate of $P(W_i)$; namely,

$$\sum_{j=1}^{n} \frac{W(i,j)}{N(i,j)} P(T_i = j).$$

Acknowledgments

The research of Mark Brown was supported by the National Security Agency, under Grant H98230-06-01-0149.

References

1. Barlow, R. E., Bartholomew, D. J., Bremner, J. M., and Brunk, H. D., (1972), Statistical Inference under Order Restrictions: Isotonic Regression, Wiley, New York.
2. Chen, Z. and Glasserman, P., (2008), Fast Pricing of Basket Default Swaps, *Operations Research*, **56**, 2, 286-303.
3. Efron, B., (1965) Increasing Properties of Polya Frequency Functions. *Annals of Mathematical Statistics*, **36**, 272-279.
4. Joshi, M. and Kainth, D., (2004), Rapid and Accurate development of Prices and Greeks for nth to Default Credit Swaps in the Li Model, *Quantitative Finance*, Vol. 4, Institute of Physics Publishing, London, UK, 266-275.
5. Keilson, J., and Gerber, H., (1971), Some Results for Discrete Unimodality, *Jour. Amer. Statist. Assc.* **66**, 386-389.
6. Ross, S. M., (2007), Introduction to Probability Models, Nineth ed., *Academic Press*, Burlington, MA.
7. Ross, S. M., (2006), Simulation, fourth ed., *Academic Press*, Burlington, MA.

Filtering with Counting Process Observations and Other Factors: Applications to Bond Price Tick Data

Xing Hu

Department of Economics
Princeton University
Princeton, 08544, USA
Email: xinghu@princeton.edu

David R. Kuipers

Department of Finance
Henry W. Bloch School of Business and Public Administration
University of Missouri at Kansas City
Kansas City, MO 64110, USA
Email: kuipersd@umkc.edu

Yong Zeng

Department of Mathematics and Statistics
University of Missouri at Kansas City
Kansas City, MO 64110, USA
Email: zengy@umkc.edu

In this paper, we propose an extended filtering micromovement model. The model captures the two main stylized facts of the bond price tick data: random trading times and trading noises. In the intrinsic value process for the transaction price of 5-year U.S. Treasury note, we extend the volatility part by adding the buyer-seller initiation dummy. For the extended model, we present the normalized and un-normalized filtering equations, a robustness theorem and the consistency of Bayes estimates. Based on the robustness theorem, we employ the Markov chain approximation method to construct a robust recursive algorithm for computing the posteriors and Bayes estimates. We present a Monte Carlo example to demonstrate that the computed Bayes estimates converge to their true values. The algorithm is applied to one and an half month of intraday transaction prices of 5-year Treasury notes. Bayes estimates are obtained. Especially, the sign of the buyer-seller initiation dummy is significantly negative, supporting that the inventory theory dominates in the bond trading.

Keywords: Ultra high frequency data; filtering; counting process; Markov chain approximation method; Bayes parameter estimation; price discreteness; and price clustering.

1. Introduction

Asset pricing models in financial markets can broadly be classified into two types based on the observed data frequency: macromovement and micromovement models. Macromovement models generally refer to data observed at daily or less frequent horizons. For example, Figure 1 depicts a time series plot of daily closing prices for trading in a particular 5-year U.S. Treasury note in the bond market during the six-week period November 15, 2000 through December 29, 2000. The macromovement for this security consists of 31 data points for the 31 business days during this time period. Despite the brevity, the stochastic nature of the time series is evident in the figure, and is often modelled in the econometrics and mathematical finance literature by continuous-time diffusion models such as geometric Brownian motion (GBM), stochastic volatility (SV), jump diffusion, or more elaborate Markov-class stochastic price processes.

Alternatively, transactional price data on an intraday basis is the domain for micromovement asset pricing models. Engle[5] refers to data of this type as ultra high frequency (UHF), because it contains the trading times and prices for all market activity at the maximum level of disaggregation. Because traders choose to transact at random times during the trading day, due to both information-based and liquidity-based motives (O'Hara[21]), econometricians model the duration between trades as a stochastic phenomena resulting in irregularly-spaced time series. Engle[5] develops a general framework for modelling such time series with many recent developments.

Figure 2 depicts the micromovement data for the same 5-year U.S. Treasury note graphed in Figure 1, during the same time period. There are over 2,200 transactions shown in the figure. The general movement in Figure 2 can be seen as its subsample macromovement from Figure 1, with an overlay of an additional noise process that results from the UHF data. The source of this noise process can fundamentally be related to the asynchronicity of trading in security markets; actual price processes are not continuous-time realizations, but are observed with discrete sampling, at the irregular intervals where trades take place. Further, markets set minimum price variation conventions for trading–the so-called "tick size" in the market–which results in price discreteness noise. In the 5-year Treasury note market, the minimum tick size is 1/128th percent of par value, so that realized price processes exhibit some level of noise as prices move from discrete tick to tick, or level by level.

Finally, even after consideration of asynchronous trading and price discreteness, additional noise is observed in actual security market price pro-

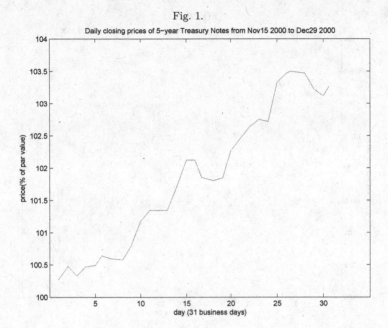

Fig. 1.

cesses due to the tendency for trades and price changes to cluster on certain ticks and in particular increments compared to others. This price clustering noise, extensively modelled in the finance literature by Harris[15] and others, can be seen for the 5-year U.S. Treasury note in Figure 3.

Given this background, we reach the simple intuition, prevalent in the finance and econometrics modelling literature, that micromovement models should be built upon fundamental intrinsic value price processes, with an overlay of market microstructure noise derived from any of the several sources noted above. Zeng[25] proposes one such model for asset prices in a trading market, a filtering micromovement (hereafter, FM) model, that incorporates UHF data. Corresponding to the macromovement, an unobservable intrinsic value process for an asset is assumed and it is the permanent component and has a long-term impact on price. Trading times are assumed to be driven by a conditional Poisson process. Prices are corrupted observations of the intrinsic value process at the trading times by market microstructure noise, which is explicitly and flexibly modeled by a random transformation. Comparing with intrinsic value, noise is the transient component and only has a short-term impact (when a trade happens) on price. The FM model can be framed as a filtering problem with counting

Fig. 2.

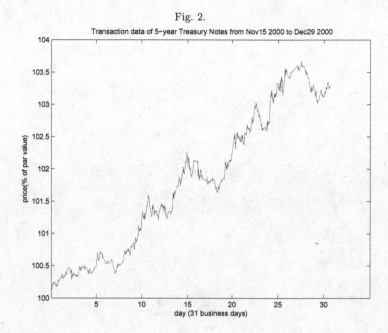

Transaction data of 5-year Treasury Notes from Nov15 2000 to Dec29 2000

process observations. This connection introduces the powerful stochastic filtering theory, which has found great success in engineering and networking.
Then, the unnormalized Duncan-Mortensen-Zakai-like filtering equation
and the normalized, Kushner-Stratonovich (KS) (or Fujisaki-Kallianpur-
Kunita)-like filtering equations are derived. They uniquely characterize the
evolutions of the continuous-time likelihoods and posteriors, respectively.
Moreover, the Bayes estimation via filtering for the intrinsic value process
and the related parameters are developed by employing the Markov chain
approximation method to numerically solve the filtering equation. Furthermore, the Bayesian hypothesis testing or model selection via filtering for
the FM model is developed in Kouritzin and Zeng.[17]

In the paper, we study and perform Bayes estimation via filtering for an
extended FM model by allowing an observable economic variable to affect
the volatility of the intrinsic value process. The variable is the buyer-seller
initiation dummy, which is observable in the bond market. This is a special
feature of bond market. Such variable is not observable in stock market, but
it can be inferred by some tests and a widely-used test is the Lee and Ready
test (see[18]). By testing whether this dummy is statistically significant, we
can infer whether the information theory or the inventory theory better

Fig. 3.

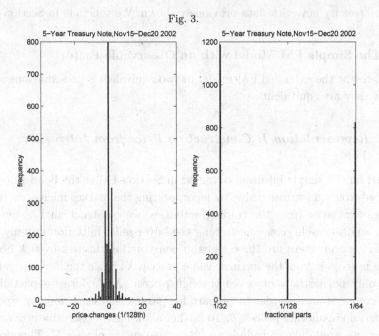

fit the bond market. Moreover, the trading times are assumed to follow an *Exponential Autoregressive Duration* (EACD) model proposed by Engle and Russel[6] instead an inhomogeneous Poisson process. The EACD model is estimated and tested. We find that the EACD model fits the trading times pretty well.

To the best of our knowledge, Frey[11] and Frey and Runggaldier[12] are the first papers that employ the non-linear filtering technique to model UHF data. Their viewpoint is to model the unobserved volatility process, which is crucial for option pricing. Their model is able to capture the Poisson random arrival times in UHF data. Cvitnic, Lipster and Rozovskii[4] extends the previous model to a more general framework. However, market microstructure noise is missing in these models and they did not consider the impact of other observable variable on volatility.

In Section 2, we present the extended FM model with an observable variable in two equivalent fashions. In Section 3, we present the continuous-time Bayes estimation via filtering for the extended FM model including the filtering equations, a robustness theorem, a recursive algorithm, and a theorem on the consistency of the Bayes estimates. In Section 4, we present a Monte Carlo example. In Section 5, we provide empirical results for the 5 year Treasury note tick data previously shown. We conclude in Section 6.

2. The Simple FM Model with an Observable Factor

We present the extended FM model in two equivalent representations and show they are equivalent.

2.1. Representation I: Constructing Price from Intrinsic Value

Based on the simple intuition obtained in Section 1 that the bond price is formed from an intrinsic value by incorporating the market microstructure noises that arise from the trading activities, we construct an FM model with an observable economic factor - the buyer-seller initiation dummy.

In general, there are three steps in constructing the tick-by-tick bond price process Y from the intrinsic value process X. Since the intrinsic value can only be partially observed through prices, (X, Y) forms a partially-observed system. In order to prepare for parameter estimation, we would like to add in another process θ. To further allow other observable economic factor, we would like to include an observable vector process V. Therefore, we would like to consider an enlarged partially-observed model (θ, X, Y, V). Assume (θ, X, Y, V) is defined in a complete probability space $(\Omega, \mathcal{F}, P)$ with a filtration $\{\mathcal{F}_t\}_{0 \le t \le \infty}$ satisfying the usual conditions (see Protter[22]).

Step 1: We consider a specific model for the 5-year bond's intrinsic value process, which is assumed as below:

$$\frac{dX_t}{X_t} = \mu dt + (\sigma + \kappa V_t)dB_t \tag{1}$$

where B_t is a standard Brownian motion and V_t is the observable buyer-seller initiation dummy defined as below. We assume trading times are denoted by $t_1, t_2, \ldots, t_i, \ldots$.

$$V(t) = V(t_i) \quad \text{if } t_i \le t < t_{i+1},$$

where $V(t_i)$ takes value 1 if a trade is buyer-initiated (namely, the trade is a "Take", take the ask quote to sell); and takes value 0 if seller-initiated (namely, the trade is a "Hit", hit the bid quote to buy); and the value of the dummy stays the same until next trade. The instantaneous expected return is μ. The instantaneous volatility is $\sigma + \kappa V_t$. When κ is positive (negative), the volatility increases (decreases) from σ to $\sigma + \kappa$ for a buyer-initiated trade. Modelling the volatility structure in this fashion captures the possibility that price moves in the market exhibit asymmetric volatility as a function of transacting dealers' inventory position (Fleming and Rosenberg,[10]).

In the general case considered in Hu, Kuipers and Zeng,[16] a mild assumption on (θ, X), stated below, is invoked so that all relevant stochastic processes are included.

Assumption 1. (θ, X) *is the unique solution of a martingale problem for a generator* $\mathbf{A}_v$ *such that for a function f in the domain of* $\mathbf{A}_v$,

$$M_f(t) = f(\theta(t), X(t)) - \int_0^t \mathbf{A}_v f(\theta(s), X(s)) ds$$

is a $\mathcal{F}_t^{\theta, X, V}$-*martingale, where* $\mathcal{F}_t^{\theta, X, V}$ *is the σ-algebra generated by* $(\theta(s), X(s), V(s))_{0 \leq s \leq t}$.

The generator and martingale problem approach originated by Strook and Varadhan in a series of papers and their book[23] (see also Ethier and Kurtz[7] for an excellent book on this topic) furnishes a powerful tool for the characterization of Markov processes. Assumption 1 includes all relevant stochastic processes such as diffusion, jump, regime-switching, spike, and their combinations for modeling asset price.

Assumption 1 is more general than that in Zeng[25] by allowing other observable economic factors to affect (θ, X), namely, allowing other observable factors in the generator, $\mathbf{A}_v$.

Let $\theta = (\mu, \sigma, \kappa, \rho)$ where ρ is a parameter for non-clustering noise to be described. The generator of the 5-year bond's intrinsic value process described in Equation (1) is

$$\mathbf{A}_v f(\theta, x) = \mu x \frac{\partial}{\partial x} f(\theta, x) + \frac{1}{2}(\sigma + \kappa v)^2 x^2 \frac{\partial^2}{\partial x^2} f(\theta, x). \qquad (2)$$

However, in UHF data, the price can not be observed continuously in time, neither can it moves continuously. Therefore, we need two more steps. Step 2 takes care of the trading times and Step 3 takes care of the trading noise.

In Step 2: We assume trading times $t_1, t_2, \ldots, t_i, \ldots$, are driven by an *Exponential Autoregressive Conditional Duration* (EACD) model proposed by Engle and Russell.[6]

We use a couple ways to describe EACD model. The first way is from the viewpoint of point process. We view trading times $t_1, t_2, \ldots, t_i, \ldots$ as a conditional Poisson process specified by its stochastic intensity

$$\bar{\lambda}(t) = \sum_{i \geq 0} \mathbf{I}\{t_i < t \leq t_{i+1}\}(t)\psi_{i+1}^{-1} \tag{3}$$

where ψ_{i+1} is the conditional expectation of the $i + 1$ duration given the past information and ψ_{i+1} is given in Equation (7).

In the other way, EACD model is specified in terms of the conditional density of the durations. Let $\Delta t_{i+1} = t_{i+1} - t_i$ be the time interval between two trades, which is called duration. We will specify the density of Δt_{i+1} directly. Let η be a vector of parameters in EACD model. Recall

$$E(\Delta t_{i+1}|\Delta_i, ..., \Delta_1; \eta) = \psi(\Delta_i, ..., \Delta_1; \eta) = \psi_{i+1}. \tag{4}$$

Then, the EACD class of models consists of parameterizations of ψ_{i+1} and the assumption that

$$\Delta_{i+1} = \psi_{i+1}\varepsilon_{i+1} \tag{5}$$

where

$$\{\varepsilon_i\} \sim \text{i.i.d. with exponential distribution of mean one.} \tag{6}$$

A general specification of ψ_{i+1}, which is similar to ARMA model in time series literature to accommodate the unlimited past memory characteristic of Δt, is

$$\psi_{i+1} = \omega + \sum_{j=1}^{m} \alpha_j \Delta_{i+1-j} + \sum_{k=1}^{q} \beta_k \psi_{i+1-k}. \tag{7}$$

Equations (4) - (7) consist of an EACD(m, q) model where the m and q refer to the orders of the lags.

In Step 3: $Y(t_i)$, the price at time t_i, is corrupted from $X(t_i)$, the intrinsic value, with trading noise. We model noise by a random transformation, $y = F(x)$, with the transition probability $p(y|x)$. Namely,

$$Y(t_i) = F(X(t_i)).$$

The random transformation, $F(x)$, is flexible and is able to accommodate different kinds of noise. As demonstrated in Section 1, there are three important types of noise that have been identified in bond tick price : discrete, clustering, and nonclustering. First, intraday prices move discretely, resulting in "discrete noise". Second, because prices do not happen evenly on all 128ths ticks, but more concentrate on some coarser ticks such as odd 64ths and 32ths or coarser, "price clustering" is obtained. Third, the "non-clustering noise" contains all other unspecified noise. Below we follow the approach in Zeng[25] to construct $F(x)$ to accommodate these three noise. There are three steps. For simple notation, at a trading time t_i, let $x = X(t_i)$, $y = Y(t_i)$, and $y' = Y'(t_i) = R[X(t_i) + W_i, \frac{1}{M}]$, where W_i is to be defined as the non-clustering noise.

- **Step (a):** Add non-clustering noise W; $x' = x + W$, where W is the non-clustering noise at trade i. We assume $\{W_i\}$, are independent of the value process, and they are *i.i.d.* with a doubly geometric distribution:

$$P\{W = w\} = \begin{cases} (1 - \rho) & \text{if } w = 0 \\ \frac{1}{2}(1 - \rho)\rho^{128|w|} & \text{if } w = \pm\frac{1}{128}, \pm\frac{2}{128}, \cdots \end{cases}.$$

- **Step (b):** Incorporate discrete noise by rounding off x' to its closest tick, $y' = R[x', \frac{1}{128}]$.
- **Step (c):** Incorporate clustering noise by biasing y' through a random biasing function $b_i(\cdot)$ at trade i. $\{b_i(\cdot)\}$ is assumed independent of $\{y_i'\}$. To be consistent with the 5-year bond price data analyzed, we construct a simple random biasing function only for the tick of 1/128 percentage. For other tick size, it can be done similarly. The data to be fitted has this clustering phenomenon: odd thirty-seconds or coarser are most likely and have about the same frequencies; odd sixty-fourths are the second most likely and have about the same frequencies; and odd one hundred and twenty-eighths are least likely and have about the same frequencies. To generate such clustering, a random biasing function is constructed based on the following rules: if the fractional part of y' is an even one hundred twenty-eighths, then y stays on y' with probability one; if the fractional part of y' is an odd one hundred twenty-eighths, then y stays on y' with probability $1 - \alpha - \beta$, y moves to the closest odd sixty-fourth with probability α, and moves to the closest odd thirty-seconds or coarser with probability β.

In brief,

$$Y(t_i) = b_i(R[X(t_i) + V_i, \frac{1}{M}]) = F(X(t_i)).$$

The detail of $b_i(\cdot)$, and the explicit $p(y|x)$ for F can be found in Appendix A. A simulation example in Section 4 demonstrates that the constructed $F(x)$ are able to capture the tick-level sample characteristics of bond price data.

Before moving to the second representation of the FM model, we give a couple remarks for this representation about its connections to other models in finance and statistics literature.

Remark 1. Under this construction, information influences $X(t)$, the value of an asset, and has a permanent impact on the price while noise modeled by $F(x)$ (or $p(y|x)$) only has a transitory impact on price. The formulation is analogous to the time series VAR structural models used in many market microstructure papers (see a survey paper[13] by Hasbrouck and a recent paper Hasbrouck[14]).

Remark 2. Furthermore, the formulation is closely connected to the two-scale frameworks combining market microstructure noises in recent literature of realized volatility estimators. See Zhang, Mykland and Ait-Sahalia,[27] Ait-Sahalia, Mykland and Zhang,[1] Bandi and Russell,[2] and Fan and Wang.[8] Especially, Li and Mykland in[20] demonstrates that rounding noise in UHF data may seriously distort even the two-scale estimators of realized volatility, and the error could be infinite. Note that price discreteness exists in the bond price data also.

2.2. *Filtering with Counting Process Observations*

Because of bond price discreteness, we can formulate the bond prices as a collection of counting processes in the following form:

$$\vec{Y}(t) = \begin{pmatrix} Y_1(t) \\ Y_2(t) \\ \vdots \\ Y_n(t) \end{pmatrix}, \tag{8}$$

where $Y_j(t)$ is the counting process recording the cumulative number of trades that have occurred at the jth bond price level (denoted by y_j) up to time t. We make additional four more mild assumptions on the model so that both representations are equivalent. We assume that there exists a reference measure Q with certain properties.

Assumption 2. *Under the reference measure Q, $\{Y_j\}_{j=1}^n$ are independent unit Poisson processes.*

Assumption 3. *Under the reference measure Q, $(\theta, X), Y_1, Y_2, \ldots, Y_n$ are independent.*

Assumption 4. *The intensities are of the form:* $\lambda_j(t) = \bar{\lambda}(t)p(y_j|x)$, *where* $\bar{\lambda}(t) = \sum_i \mathbf{I}\{t_i < t \le t_{i+1}\}(t)\psi_{i+1}^{-1}$ *is the total trading intensity at time t and* $p(y_j|x)$ *is the transition probability from* x *to* y_j, *the jth price level.*

This assumption imposes a desirable structure for the intensities of the model. It means that the total trading intensity $\bar{\lambda}(t)$ determines when the next trade will occur and $p(y_j|x)$, which is the same as $p(y|x)$ of $y = F(x)$, determines at which price level the next trade will occur given the value is x. Note that $p(y_j|x)$ models how the trading noise enters the price process. The intensity structure plays an essential role in the equivalence of the two approaches of modeling.

Under probability measure P, $Y_j(t)$ is a conditional Poisson process with the stochastic intensity, $\lambda_j(t) = \bar{\lambda}(t)p(y_j|x)$. Given $\mathcal{F}_t^{\theta,X,Y}$, $Y_j(t)$ has a Poisson distribution with parameter $\int_0^t \lambda_j(s)ds = \int_0^t \bar{\lambda}(s)p(y_j|X(s))ds$. Moreover, $Y_j(t) - \int_0^t \lambda_j(s)ds$ is a $\mathcal{F}_t^{\theta,X,Y}$ - martingale and $\int_0^t \lambda_j(s)ds$ is called the compensator of $Y_j(t)$ for each j.

Assumption 5. $\int_0^T E^P[\bar{\lambda}(s)]ds < \infty$, *for* $t > 0$.

Under very mild conditions, the reference measure Q exists. Assumption 5 is a mild technical condition to ensure the existence of Q. The Girsanov type change of measure theorem for Poisson process can be found in (T2 and T3 Theorems in Chapter 4.2 of Bremaud[3]).

Remark 3. Under this representation, (θ, X) becomes the signal, which cannot be observed directly, but can be partially observed through the counting processes, $\vec{Y}$ with the observable factor, V. The counting process observations $\vec{Y}(t)$ is distorted observations of $X(t)$ by trading noise, modeled by $p(y_j|x)$. Hence, $(\theta, X, \vec{Y}, V)$ is framed as a *filtering model with counting process observations and an observable factor*.

Remark 4. To solve problems in mathematical finance such as the option pricing and hedging and the portfolio selection, the stochastic differential or integral equation form of the most recent price is needed. However, the previous two representations do not provide such a form of price. And a third representation is given in Lee and Zeng[19] and Xiong and Zeng.[24] In those two papers, the option pricing and hedging through local risk minimizing criterion and the portfolio selection problem are studied respectively.

2.3. *The Equivalence of the Two Representation*

We construct the price from the intrinsic value in Representation I and frame it as a filtering problem with counting process observations in Representation II. The following proposition states their equivalence in distribution. This guarantees the statistical inference based on the second representation is also valid to the first one.

Proposition 2.1. The two representations of the model in Sections 2.1 and 2.2, respectively, have the same probability law.

The proof is exactly the same as that of the proposition in,[26] where the main idea is to show both representations have the same stochastic intensity kernel.

Since the observed prices can be equivalently represented by Y or $\vec{Y}$ in Representations I or II, respectively, we use Y or $\vec{Y}$ in exchange of one another in the rest of the paper.

3. Bayes Estimation via Filtering

This section presents the filtering equations for the FM model with an observable factor, a robustness theorem, and a theorem for the consistency of the Bayes estimates without proofs. A more general version of these results with proofs can be found in Hu, Kuipers and Zeng.[16] For the FM model considered in this paper, we construct an efficient recursive algorithm to compute the joint posteriors and the Bayes estimates. The robustness theorem not only provides a blueprint through the Markov chain approximation method to construct recursive algorithms, but also ensures the robustness of such algorithms.

3.1. *The Statistical Foundations*

We first study the continuous-time joint likelihood, the likelihood function (from frequentists' viewpoint), the integrated likelihood (from Bayesians' viewpoint), and the posterior of the proposed model. They are characterized by the unnormalized and normalized filtering equations. For those who are interested in the continuous-time likelihood ratio and Bayes factors for hypotheses testing or model selections, and the system of evolution equations characterizing the evolution of Bayes factors, we refer to Kouritzin and Zeng[17] for details.

3.1.1. *The Continuous-time Joint Likelihood*

The probability measure P of $(\theta, X, \vec{Y}, V)$ can be written as $P = P_{\theta,x,v} \times P_{y|\theta,x,v}$, where $P_{\theta,x,v}$ is the probability measure for (θ, X, V) such that $M_f(t)$ in Assumption 1 is a $\mathcal{F}_t^{\theta,X,V}$-martingale, and $P_{y|\theta,x,v}$ is the conditional probability measure on $D_{R^n}[0,\infty)$ for $\vec{Y}$ given (θ, X, V) (where $D_{R^n}[0,\infty)$ is the space of right continuous with left limit functions). Under P, $\vec{Y}$ relies on (θ, X, V). Recall that there exists a reference measure Q such that under Q, $(\theta, X, V), \vec{Y}$ become independent, (θ, X, V) remains the same probability law and $Y_1, Y_2, \ldots, Y_n$ become unit Poisson processes. Therefore, Q can be decomposed as $Q = P_{\theta,x,v} \times Q_y$, where Q_y is the probability measure for n independent unit Poisson processes. One can obtain the Radon-Nikodym derivative of the model, that is the joint likelihood of $(\theta, X, \vec{Y}, V)$, $L(t)$, as (see[3] pg 166),

$$L(t) = \frac{dP}{dQ}(t) = \frac{dP_{\theta,x,v}}{dP_{\theta,x,v}}(t) \times \frac{dP_{y|\theta,x}}{dQ_y}(t) = \frac{dP_{y|\theta,x,v}}{dQ_y}(t)$$

$$= \prod_{j=1}^{n} \exp\left\{ \int_0^t \log \lambda_j(s-) dY_j(s) - \int_0^t \left[\lambda_j(s) - 1\right] ds \right\}, \tag{9}$$

or in stochastic differential equation(SDE) form:

$$L(t) = 1 + \sum_{j=1}^{n} \int_0^t \left[\lambda_j(s-) - 1\right] L(s-) d(Y_j(s) - s).$$

3.1.2. *The Continuous-time Likelihoods of $\vec{Y}$*

Note that X are not observable. To do parameter estimation, we would like to have the likelihood of $\vec{Y}$ alone. Namely, we would like to integrate out X. The probability way to achieve this is to use conditional expectation. Let $\mathcal{F}_t^{Y,V} = \sigma\{(\vec{Y}(s), V(s)) | 0 \leq s \leq t\}$ be all the available information up to time t. We use $E^Q[X]$ and $E^P[X]$ to indicate that the expectation is taken with respect to the measures Q and P, respectively.

Definition 1. Let

$$\rho(f, t) = E^Q[f(\theta(t), X(t))L(t) | \mathcal{F}_t^{Y,V}] = \int f(\theta, x) \rho_t(d\theta, dx).$$

In the frequentists' paradigm, $(\theta(0), X(0))$ is fixed and the likelihood of Y is $E^Q[L(t) | \mathcal{F}_t^{Y,V}] = \rho(1, t)$. In Bayesian paradigm, we assume a prior on $(\theta(0), X(0))$ and the integrated (or marginal) likelihood of Y is also $\rho(1, t)$.

3.1.3. The Continuous-time Posterior

In order to do Bayes estimation, we would like to consider the conditional distribution and the posterior.

Definition 2. Let π_t be the conditional distribution of $(\theta(t), X(t))$ given $\mathcal{F}_t^{Y,V}$ and let

$$\pi(f, t) = E^P[f(\theta(t), X(t))|\mathcal{F}_t^{Y,V}] = \int f(\theta, x)\pi_t(d\theta, dx).$$

In Bayesian paradigm, we assume a prior on $(\theta(0), X(0))$, so π_t becomes the continuous-time posterior, which is determined by $\pi(f, t)$ for all continuous and bounded f. The relationship between $\rho(f, t)$ and $\pi(f, t)$ is provided in Bayes Theorem (see,[3] page 171), which states: $\pi(f, t) = \rho(f, t)/\rho(1, t)$. For this reason, the equation governing the evolution of $\rho(f, t)$ is called the *unnormalized filtering equation*, and that of $\pi(f, t)$ is called the *normalized filtering equation*.

3.2. Filtering Equations

Stochastic partial differential equations (SPDEs) provide an powerful machinery to characterize the infinite dimensional continuous-time likelihoods and posteriors. The likelihoods are characterized by the unnormalized filtering equation and the posteriors are characterized by the normalized filtering equation. The following theorem put together these two filtering equations.

Theorem 1. *Suppose that* $(\theta, X, \vec{Y}, V)$ *satisfies Assumptions 1 - 5. Then,* ρ_t *is the unique measure-valued solution of the SPDE, the unnormalized filtering equation,*

$$\rho(f, t) = \rho(f, 0) + \int_0^t \rho(\mathbf{A}f, s)ds + \sum_{j=1}^n \int_0^t \rho((\bar{\lambda}p_j - 1)f, s-)d(Y_j(s) - s), \quad (10)$$

for $t > 0$ *and* $f \in D(\mathbf{A})$, *the domain of generator* $\mathbf{A}$, *where* $\bar{\lambda} = \bar{\lambda}(t)$ *is the trading intensity and is assumed to be* $\mathcal{F}_t^{\theta, X, Y, V}$-*predictable, and* $p_j = p(y_j|x)$ *is the transition probability from* x *to* y_j.

π_t *is the unique measure-valued solution of the SPDE, the normalized filtering equation,*

$$\pi(f, t) = \pi(f, 0) + \int_0^t \pi(\mathbf{A}f, s)ds$$
$$+ \sum_{j=1}^n \int_0^t \Big[\frac{\pi(f\bar{\lambda}p_j, s-)}{\pi(\bar{\lambda}p_j, s-)} - \pi(f, s-)\Big]d\big(Y_j(s) - \pi(\bar{\lambda}p_j, s)s\big). \quad (11)$$

Moreover, when the trading intensity is $\bar{\lambda}(t)$ is only $\mathcal{F}_t^{Y,V}$-predictable, namely, $\bar{\lambda}(t)$ depends on only the past observations and the observable factors, the normalized filtering equation is simplified as

$$\pi(f,t) = \pi(f,0) + \int_0^t \pi(\mathbf{A}f,s)ds + \sum_{j=1}^n \int_0^t \left[\frac{\pi(fp_j,s-)}{\pi(p_j,s-)} - \pi(f,s-) \right] dY_j(s).$$

(12)

Remark 5. Note that $\bar{\lambda}(t)$ disappears in Equation (12). This not only greatly reduces the computation in of Bayes estimates, but also implies that the joint posterior and the Bayes estimates are independent or "model free" of the assumptions of trading times as long as $\bar{\lambda}(t)$ is $\mathcal{F}_t^{Y,V}$-predictable.

Remark 6. For the FM model studied in this paper, the trading intensity is given by Equation (3), which is, indeed, $\mathcal{F}_t^{Y,V}$-predictable. Therefore, we only use the simplified filtering equation (12) to construct a recursive algorithm to compute Bayes estimates.

Another nice property of the filtering equations for π_t is that they satisfy a separation principle in the following sense. Let the trading times be $t_1, t_2, \ldots$, then Equation (12) can be written in two parts. The first describes the evolution without trades and is called the *propagation equation*. The second describes the update when a trade occurs and is called the *updating equation*. The propagation equation is written as

$$\pi(f, t_{i+1}-) = \pi(f, t_i) + \int_{t_i}^{t_{i+1}-} \pi(\mathbf{A}f, s)ds.$$

(13)

It has no random component and this implies that the posterior evolves deterministically between trades.

When the price at time t_{i+1} happens at the jth price level, the updating equation is written as

$$\pi(f, t_{i+1}) = \frac{\pi(fp_j, t_{i+1}-)}{\pi(p_j, t_{i+1}-)}.$$

(14)

Note that the price level j, the observation, is random. Therefore, the updating equation is random.

3.3. *A Convergence Theorem and Recursive Algorithms*

Theorem 1 describes the evolutions of the continuous-time likelihoods and posteriors. They are both infinite dimensional. This means, in order to compute them, one needs to approximate the infinite dimensional system by a finite dimensional system, based on which algorithms are constructed. The algorithms, based on the filtering equations, are naturally recursive, handling one datum at a time. Moreover, the algorithms are easily parallelizable and thus can make real-time updates and handle large data sets. One basic requirement for the recursive algorithms is robustness (or consistency): The approximate likelihoods or posteriors, computed by the recursive algorithms, converges to the true ones. The following theorem ensures the robustness and provides a blueprint for constructing consistent algorithms through Kushner's Markov chain approximation methods.

Let $(\theta_\epsilon, X_\epsilon)$ be an approximation of (θ, X). Then, we define

$$\vec{Y}_\epsilon(t) = \begin{pmatrix} Y_{\epsilon,1}(t) \\ Y_{\epsilon,2}(t) \\ \vdots \\ Y_{\epsilon,n}(t) \end{pmatrix}, \tag{15}$$

where $Y_{\epsilon,j}(t)$ has the stochastic intensity $\bar{\lambda}(t)p(y_j|X_\epsilon(t),\theta_\epsilon)$. Let $\mathcal{F}_t^{\vec{Y}_\epsilon} = \sigma(\vec{Y}_\epsilon(s), 0 \leq s \leq t)$, and $L_\epsilon(t) = L\left(\left(\theta_\epsilon(s), X_\epsilon(s), Y_\epsilon(s)\right)_{0 \leq s \leq t}\right)$ as in Equation (9). We use the notation, $X_\epsilon \Rightarrow X$, to mean X_ϵ converges weakly to X in the Skorohod topology as $\epsilon \to 0$. We assume that $(\theta_\epsilon, X_\epsilon, \vec{Y}_\epsilon, V)$ are on $(\Omega_\epsilon, \mathcal{F}_\epsilon, P_\epsilon)$, and Assumptions 1 - 5 are also satisfied for $(\theta_\epsilon, X_\epsilon, \vec{Y}_\epsilon, V)$. Then, a reference measure Q_ϵ exists with similar properties. Next, we define the approximations of $\rho(f,t)$, and $\pi(f,t)$ and state the theorem of robustness.

Definition 3. Let

$$\rho_\epsilon(f,t) = E^{Q_\epsilon}\left[f\left(\theta_\epsilon(t), X_{\epsilon_x}(t)\right)L_\epsilon(t)|\mathcal{F}_t^{\vec{Y}_\epsilon}\right],$$

and

$$\pi_\epsilon(f,t) = E^{P_\epsilon}\left[f\left(\theta_\epsilon(t), X_{\epsilon_x}(t)\right)|\mathcal{F}_t^{\vec{Y}_\epsilon}\right].$$

Theorem 2. *Suppose that Assumptions 1 - 5 hold for the models $(\theta, X, \vec{Y}, V)$ and that Assumptions 1 - 5 hold for the approximate models $(\theta_\epsilon, X_\epsilon, \vec{Y}_\epsilon, V)$. Suppose $(\theta_\epsilon, X_\epsilon) \Rightarrow (\theta, X)$ as $\epsilon \to 0$. Then, as $\epsilon \to 0$, (i) $\vec{Y}_\epsilon \Rightarrow \vec{Y}$ under P_ϵ and P; for all bounded continuous functions, f (ii) $\rho_{\epsilon,c}(f,t) \Rightarrow \rho(f,t)$; and (iii) $\pi_{\epsilon,c}(f,t) \Rightarrow \pi(f,t)$.*

Remark 7. Part (i) implies the convergence of the approximate observations to the true ones under the physical measures; Part (ii) implies the robustness of the (integrated) likelihood; Part (iii) implies the robustness of posterior.

This theorem offers a three-step blueprint for constructing a robust recursive algorithm based on Kushner's Markov chain approximation method to compute the continuous-time likelihoods or posteriors. Since we focus on Bayes estimation in this paper, here, we first give a big picture as to how to construct such an algorithm for computing the posterior and Bayes estimates for a model. There are three steps. In Step 1, we produce $(\theta_\epsilon, X_\epsilon)$, the Markov chain approximation to (θ, X). Note that $(\theta_\epsilon, X_\epsilon)$ is restricted to the discrete finite-dimensional state space of $(\theta_\epsilon, X_\epsilon)$. We also obtain $p_{\epsilon,j} = p(y_j|\theta_\epsilon, x_\epsilon)$ as an approximation to $p_j = p(y_j|\theta, x)$. In Step 2, we obtain the filtering equation for $\pi_\epsilon(f, t)$ for the approximate model $(\theta_\epsilon, X_\epsilon, Y_\epsilon, V)$ by applying Theorem 1. Since the FM model considered in this paper has a trading intensity depending only on the past of observations, we only consider the simplified normalized filtering equation. Recall the separation principle and we can write the filtering equation for the approximate model as the propagation equation:

$$\pi_\epsilon(f, t_{i+1}-) = \pi_\epsilon(f, t_i) + \int_{t_i}^{t_{i+1}-} \pi_\epsilon(\mathbf{A}_\epsilon f, s)ds, \qquad (16)$$

and the updating equation (assuming that a trade at jth price level occurs at time t_{i+1}):

$$\pi_\epsilon(f, t_{i+1}) = \frac{\pi_\epsilon(f p_{\epsilon,j}, t_{i+1}-)}{\pi_\epsilon(p_{\epsilon,j}, t_{i+1}-)}. \qquad (17)$$

In Step 3, we turn Equations (16) and (17) to the recursive algorithm in discrete state space and in discrete times by two substeps: (a) expresses $\pi_\epsilon(\cdot, t)$ as a finite array using the components $\pi_\epsilon(f, t)$ with lattice-point indicator f and (b) use Euler scheme to approximate the time integral in (16). Details of the construction of the algorithm are given in the sub-section below.

3.4. *The Recursive Algorithms for Posteriors and Bayes Estimates*

In the FM model considered in this paper, we can classify the parameters into three groups. The parameters of clustering noise, α and β is Group I and can be estimated through the method of relative frequency. The parameters of EACD model for the trading times are Group II and they can be estimated using the maximum likelihood estimation proposed by Engle and

Russell.[6] The other parameters, μ, σ, κ and ρ, are Group III and they are estimated by Bayes estimation via filtering through the recursive algorithm to be constructed.

We follow the three-step blueprint for the Markov chain approximation method summarized in the end of Section 3.3 to construct the recursive algorithms. Finally, we show the consistency of the algorithms. For notational simplicity, let $\vec{\theta}_{\vec{\epsilon}} = (\mu_{\epsilon_\mu}, \sigma_{\epsilon_\sigma}, \kappa_{\epsilon_\kappa}, \rho_{\epsilon_\rho})$ to denote an approximate discretized parameter signal, which is random in Bayesian framework.

3.4.1. Step 1: Construct $(\vec{\theta}_{\vec{\epsilon}}, X_{\epsilon_x})$

First, we latticize the parameter spaces of $\mu, \sigma, \kappa, \rho$ and the state space of X. Suppose there are $n_\mu + 1$, $n_\sigma + 1$, $n_\kappa + 1$, $n_\rho + 1$ and $n_x + 1$ lattices in the latticized spaces of $\mu, \sigma, \kappa, \rho$ and X respectively. For instance,

$$\mu : [\alpha_\mu, \beta_\mu] \to \{\alpha_\mu, \alpha_\mu + \epsilon_\mu, \ldots, \alpha_\mu + (n_\mu - 1)\epsilon_a, \beta_a\}$$

where the number of lattices is $n_\mu + 1$. Define $\mu_k = \alpha_\mu + k\epsilon_\mu$, the kth element in the latticized parameter space of μ, and define $\sigma_l, \kappa_m, \rho_n$ and X_w similarly. Let

$$\vec{\theta}_{\vec{v}} = (\mu_k, \sigma_l, \kappa_m, \rho_n)$$

as where $\vec{v} = (k, l, m, n)$.

We construct a birth and death generator $\mathbf{A}_{\varepsilon,v}$, such that $\mathbf{A}_{\varepsilon,v} \to \mathbf{A}_v$ pointwisely. Namely, we construct a birth and death process $(\vec{\theta}_{\vec{\epsilon}}, X_{\epsilon_x})$, a simple example of Markov chain, to approximate $(\theta, X(t))$ using the generator for the GBM with observable factor. Recall the generator in Equation (2), which consists of the first and second derivatives. We employ the central difference approximation to the differentials to construct the approximate generator as below:

$$\mathbf{A}_\varepsilon f(\vec{\theta}_{\vec{v}}, x_w)$$

$$= \mu_k x_w \left(\frac{f(\vec{\theta}_{\vec{v}}, x_w + \epsilon_x) - f(\vec{\theta}_{\vec{v}}, x_w - \epsilon_x)}{2\epsilon_x} \right)$$

$$+ \frac{1}{2}(\sigma_l + \kappa_m v)^2 x_w^2 \left(\frac{f(\vec{\theta}_{\vec{v}}, x_w + \epsilon_x) + f(\vec{\theta}_{\vec{v}}, x_w - \epsilon_x) - 2f(\vec{\theta}_{\vec{v}}, x_w)}{(\epsilon_x)^2} \right)$$

$$= \beta(\vec{\theta}_{\vec{v}}, x_w)(f(\vec{\theta}_{\vec{v}}, x_w + \epsilon_x) - f(\vec{\theta}_{\vec{v}}, x_w))$$

$$+ \delta(\vec{\theta}_{\vec{v}}, x_w)(f(\vec{\theta}_{\vec{v}}, x_w - \epsilon_x) - f(\vec{\theta}_{\vec{v}}, x_w)), \tag{18}$$

where

$$\beta(\vec{\theta}_{\vec{v}}, x_w) = \frac{1}{2}\left(\frac{(\sigma + \kappa v)^2 x_w^2}{\epsilon_x^2} + \frac{\mu_k x_w}{\epsilon_x}\right),$$

and

$$\delta(\vec{\theta}_{\vec{v}}, x_w) = \frac{1}{2}\left(\frac{(\sigma + \kappa v)^2 x_w^2}{\epsilon_x^2} - \frac{\mu_k x_w}{\epsilon_x}\right).$$

$\beta(\vec{\theta}_{\vec{v}}, x_w)$ and $\delta(\vec{\theta}_{\vec{v}}, x_w)$ are the birth and death rates, respectively, and should be nonnegative. If necessary ϵ_x can be made small to ensure the nonnegativity.

Clearly, $\mathbf{A}_{\varepsilon,v} \to \mathbf{A}_v$. So, $(\vec{\theta}_{\vec{\varepsilon}}, X_{\epsilon_x}) \Rightarrow (\vec{\theta}, X)$ as $\varepsilon \to 0$ where $\varepsilon = \max(\epsilon_\mu, \epsilon_\sigma, \epsilon_\kappa, \epsilon_\rho, \epsilon_x)$. With the approximate model $(\vec{\theta}_{\vec{\varepsilon}}, X_{\epsilon_x}(t))$, we have the approximate $\vec{Y}_\varepsilon$ which is defined by Equation (15).

The counting process observations can be treated as $\vec{Y}(t)$ defined by Equation (8) or $\vec{Y}_\varepsilon(t)$ defined by Equation (15) conditioning on whether the driving process is $(\vec{\theta}, X(t))$ or $(\vec{\theta}_{\vec{\varepsilon}}, X_{\epsilon_x}(t))$. In the true model, the parameters and the bond value are $(\vec{\theta}, X(t))$, and the counting process observations of bond price are regarded as $\vec{Y}(t)$. In the approximate model where we intend to compute the posterior and Bayes estimates, we use $(\vec{\theta}_{\vec{\varepsilon}}, X_{\epsilon_x}(t))$ to approach $(\vec{\theta}, X(t))$ and the counting process observations of bond price are regarded as $\vec{Y}_\varepsilon(t)$. The recursive algorithm is to calculate the joint posterior and Bayes estimates of the approximate model $(\vec{\theta}_{\vec{\varepsilon}}, X_{\epsilon_x}, \vec{Y}_\varepsilon, V)$, which is close to the joint posterior and Bayes estimates of the true model $(\vec{\theta}, X, \vec{Y}, V)$, by Theorem 2, when ε is small.

3.4.2. *Step 2: Obtain the SPDEs of the Approximate Model*

When $(\vec{\theta}, , X)$ is approximated by $(\vec{\theta}_{\vec{\varepsilon}}, X_{\epsilon_x})$, $\mathbf{A}_v$ by $\mathbf{A}_{\varepsilon.v}$, and Y by Y_ε, there accordingly exist probability measures P_ε and Q_ε, which approximate P and Q. It can be checked that Assumptions 1 - 5 hold for $(\vec{\theta}_{\vec{\varepsilon}} X_{\epsilon_x}, Y_\varepsilon)$ satisfying the conditions of Theorem 1. Let $(\vec{\theta}_{\vec{\varepsilon}}, X_{\epsilon_x})$ denote the discretized random signal and $(\vec{\theta}_{\vec{v}}, x_w)$ a lattice point there.

Definition 4. Let $\pi_{\varepsilon,t}$ be the conditional probability mass function of $(\vec{\theta}_{\vec{\varepsilon}}, X_{\epsilon_x}(t))$ on the discrete state space given $\mathcal{F}_t^{\vec{Y}_\varepsilon}$. Let

$$\pi_\varepsilon(f, t) = E^{P_\varepsilon}[f(\vec{\theta}_{\vec{\varepsilon}}, X_\epsilon(t)) \,|\, \mathcal{F}_t^{\vec{Y}_\varepsilon}] = \sum_{\vec{\theta}_{\vec{v}}, x_w} f(\vec{\theta}_{\vec{v}}, x_w)\pi_{\varepsilon,t}(\vec{\theta}_{\vec{v}}, x_w),$$

where the summation goes over all lattices in the discretized state spaces.

Then, the normalized filtering equation for the approximate model is given by Equations (16) and (17).

3.4.3. Step 3: Convert to the Recursive Algorithm

We convert Equations (16) and (17) to the recursive algorithm for computing the approximate joint posterior. First, we define the posterior that the recursive algorithm calculates.

Definition 5. The posterior of the approximate model at time t is denoted by

$$p_\varepsilon(\vec{\theta}_{\vec{v}}, x_w; t) = \pi_{\varepsilon,t}\{\vec{\theta}_{\vec{\epsilon}} = \vec{\theta}_{\vec{v}}, X_\epsilon(t) = x_w\}.$$

There are two substeps. The key of the first substep is to specify f as the following lattice-point indicator function:

$$\mathbf{I}_{\{\vec{\theta}_{\vec{\epsilon}} = \vec{\theta}_{\vec{v}}, X_\epsilon(t) = x_w\}}(\vec{\theta}_{\vec{\epsilon}}, X_\epsilon(t)). \tag{19}$$

We observe:

$$\pi_\varepsilon\left(\beta(\vec{\theta}_{\vec{\epsilon}}, X_\epsilon(t))\mathbf{I}_{\{\vec{\theta}_{\vec{\epsilon}} = \vec{\theta}_{\vec{v}}, X_\epsilon(t) + \epsilon_x = x_w\}}(\vec{\theta}_{\vec{\epsilon}}, X_\epsilon(t) + \epsilon_x), t\right) = \beta(\vec{\theta}_{\vec{v}}, x_{w-1})p_\varepsilon(\vec{\theta}_{\vec{v}}, x_{w-1}; t).$$

Along with suchlike results, Equation (16) turns into

$$p_\varepsilon(\vec{\theta}_{\vec{v}}, x_w; t_{i+1}-) = p_\varepsilon(\vec{\theta}_{\vec{v}}, x_w; t_i) + \int_{t_i}^{t_{i+1}-}\left(\beta(\vec{\theta}_{\vec{v}}, x_{w-1})p_\varepsilon(\vec{\theta}_{\vec{v}}, x_{w-1}; t)\right.$$

$$\left. - (\beta(\vec{\theta}_{\vec{v}}, x_w) + \delta(\vec{\theta}_{\vec{v}}, x_w))p_\varepsilon(\vec{\theta}_{\vec{v}}, x_w; t) + \delta(\vec{\theta}_{\vec{v}}, x_{w+1})p_\varepsilon(\vec{\theta}_{\vec{v}}, x_{w+1}; t)\right)dt. \tag{20}$$

Suppose that a trade at jth price level happens at time t_{i+1}, the updating Equation (17) becomes,

$$p_\varepsilon(\vec{\theta}_{\vec{v}}, x_w; t_{i+1}) = \frac{p_\varepsilon(\vec{\theta}_{\vec{v}}, x_w; t_{i+1}-)p(y_j|x_w, \rho_n)}{\sum_{\vec{v}', w'} p_\varepsilon(\vec{\theta}_{\vec{v}'}, x_{w'}; t_{i+1}-)p(y_j|x_{w'}, \rho_{n'})}, \tag{21}$$

where the summation goes over the total discretized space, and $p(y_j|x_w, \rho_n)$, which is the transition probability from x_w to y_j, is given by Equation (A.2) in Appendix A.

In the second substep, we use Euler scheme to approximate the time integral in Equation (20) in order to obtain a recursive algorithm discrete in time. Since the probability of the event that two or more jumps occur at the same time is zero, there are two possible cases for the length of duration.

Case 1, when $t_{i+1} - t_i \leq LL$, the length controller in the Euler scheme, we approach $p(\vec{\theta}_{\vec{v}}, x_w; t_{i+1}-)$ as

$$
\begin{aligned}
p(\vec{\theta}_{\vec{v}}, x_w; t_{i+1}-) \approx{} & p(\vec{\theta}_{\vec{v}}, x_w; t_i) + \Big[\beta(\vec{\theta}_{\vec{v}}, x_{w-1}) p(\vec{\theta}_{\vec{v}}, x_{w-1}; t_i) \\
& - \big(\beta(\vec{\theta}_{\vec{v}}, x_w) + \delta(\vec{\theta}_{\vec{v}}, x_w) \big) p(\vec{\theta}_{\vec{v}}, x_w; t_i) \\
& + \delta(\vec{\theta}_{\vec{v}}, x_{w+1}) p(\vec{\theta}_{\vec{v}}, x_{w+1}; t_i) \Big] (t_{i+1} - t_i).
\end{aligned}
\tag{22}
$$

Case 2, when $t_{i+1} - t_i > LL$, we select a finer partition $\{t_{i,0} = t_i, t_{i,1}, \ldots, t_{i,n} = t_{i+1}\}$ of $[t_i, t_{i+1}]$ such that $\max_j |t_{i,j+1} - t_{i,j}| < LL$ and then approximate $p(\vec{\theta}_{\vec{v}}, x_l; t_{i+1}-)$ by applying repeatedly Equation (22) from $t_{i,0}$ to $t_{i,1}$, then $t_{i,2}, \ldots$, until $t_{i,n} = t_{i+1}$.

The recursive algorithm consists of Equations (21) and (22). Using them, we calculate the approximate posterior at time t_{i+1} for $(\vec{\theta}, X(t_{i+1}))$ based on the posterior at time t_i. At time t_{i+1}, the Bayes estimates of $\vec{\theta}$ and $X(t_{i+1})$ are the expected values of the corresponding marginal posteriors.

To complete the algorithm for posterior, we choose a reasonable prior. Assume independence between $X(0)$ and $\vec{\theta}$. The prior for $X(0)$ can be set by $P\{X(0) = Y(t_1)\} = 1$ where $Y(t_1)$ is the first trade price of a data set because they are very close. For other parameters, we can simply assume a uniform prior over the discretized state space of $\vec{\theta}$. At $t = 0$, we select the prior as below:

$$
p(\vec{\theta}_{\vec{v}}, x_w; 0) = \begin{cases} \frac{1}{(1+n_\mu)(1+n_\sigma)(1+n_\kappa)(1+n_\rho)} & \text{if } x_w = Y(t_1) \\ 0 & \text{otherwise} \end{cases}.
$$

Finally, we mention two statistical and computational concerns for a prior on a parameter: suitable range and mesh size. Usually, the marginal posterior of a parameter obtained from a large data set is concentrated on a small area around the true value. Then, the uniform prior set on the small area is sufficient, because the posterior outside is of very small probability. After having a suitable range, we may choose a suitable mesh size, which ideally produces a posterior with a unique modal and bell-shaped distribution as shown in Table 5.1 of.[25]

3.4.4. *Robustness of the Recursive Algorithms*

There are two approximations in our recursive algorithms to compute the posteriors. One is to approximate the integral in the propagation equations by Euler scheme, whose convergence is well-known. The other one, which is more important, is the approximation of Equation (12) by Equations (16) and (17). Since, $(\vec{\theta}_{\vec{\epsilon}}, X_\epsilon) \Rightarrow (\vec{\theta}, X)$ by construction, Theorem 2 warrants these convergence in the sense of the weak convergence in the Skorohod topology, that is, the robustness of the posteriors and Bayes estimates.

3.4.5. *Consistency of the Bayes Estimates*

Theorem 3. *Under the setup of Section 2.1, suppose that the clustering parameters* α, β *are known, and* $(\mu, \sigma, \kappa, \rho)$ *has a prior. We assume that there are both infinite i such that* $V(t_i) = V(t_{i+1}) = 0$ *or* 1. *If* $\mu t - \frac{1}{2} \int_0^t (\sigma + \kappa V_s)^2 ds > 0$ *for any* $t > 0$, *then* $E[f(\mu, \sigma, \kappa, \rho)|\mathcal{F}_t^{\vec{Y}}] \to f(\mu, \sigma, \kappa, \rho)$ *a.s. as* $t \to \infty$ *for any bounded continuous function f.*

The condition "$\mu t - \frac{1}{2} \int_0^t (\sigma + \kappa V_s)^2 ds > 0$" keeps $X(t)$ strictly positive, and rules out bankruptcy.

Theorem 3 together with the robustness of the recursive algorithm implies that the computed Bayes estimates will converge to their true values. This is confirmed by the Monte Carlo results presented next.

4. A Monte Carlo Example

We provide a Monte Carlo example for two purposes: One is to show the tick sample characteristics are captured by the noise modeling and the other is to show the Bayes estimates of $(\mu, \sigma, \kappa, rho)$ are consistent. The latter also serves to test the correctness of the software developed.

4.1. *Tick Characteristics of Monte Carlo Data*

For simulation, we choose the parameter values close to the estimates from the 5-year Treasury note data. Let $\mu = 6.32 \times 10^{-8}$, $\sigma = 1.33 \times 10^{-5}$, $\kappa = -2.71 \times 10^{-6}$, (the above three parameters are in per second) $\rho = 0.023$, $\alpha = 0.250$, and $\beta = 0.385$. Since $\bar{\lambda}(t)$ has no impact in estimation and noise, we assume the trading intensity is constant: $a(t) = 0.06$ for all $t > 0$(i.e., one trade in about $1/0.06 = 450$ seconds). Using these parameters, we simulated 2,200 observations.

Figure 4 shows the pair of histograms of price changes and of the three ticks: odd 1/128, odd 1/64 and odd 1/32 or coarser. The similarity of Figures 4 and 3 shows the proposed modeling of noises is successful to capture the tick sample characteristics of the 5yr Treasury note price.

4.2. *Bayes Estimates for Monte Carlo Data*

Having constructed the recursive algorithm in Section 3.3 for calculating the Bayes estimates, we develop a Fortran program for the recursive algorithm

Fig. 4.

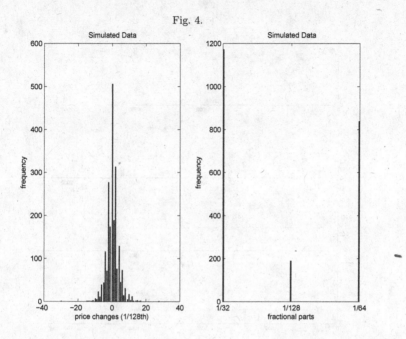

to calculate, at each trading time t_i, the joint posterior of $(\vec{\theta}, X(t))$, their marginal posteriors, their Bayes estimates and their standard errors (SE), respectively. The recursive algorithm and the software are fast enough to generate real-time Bayes estimates. We test the software extensively and verify it on Monte Carlo data, where we know the true parameters.

Figure 5 has four plots to display how the trade-by-trade Bayes estimates and the two standard errors (SE) bounds evolve in comparison with the true values for μ, σ, κ and ρ, respectively. The figure clearly shows that the estimates of μ, σ, κ and ρ converge to their true values, the two-SE bounds shrink, and the true values are within the two-SE bounds. Similarly, the Bayes estimates of $X(t)$ are close to their true values, which are always within two-SE bounds. Their final Bayes estimates, SEs, and true values are presented in Table 1. The true values are close to the Bayes estimates and all within two SE bounds.

5. An Empirical Study of 5-year Treasury Note

Our data are provided by GovPx, Inc, an organization owned by primary dealers and three inter-dealer brokers to collect and distribute real-time quotes and transaction data of U.S. Treasury Securities. GovPx, Inc. records

Fig. 5.

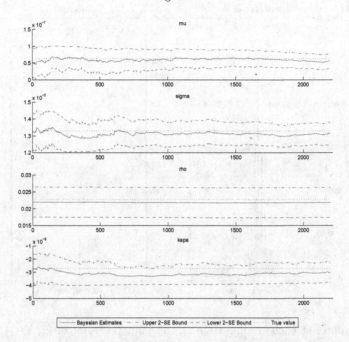

Table 1. Bayes Estimates for an Monte Carlo Data

Parameter	True Value	Bayes Estimate	St. Error
μ	6.32E-8	5.63E-8	1.09E-8
σ	1.33E-5	1.32E-5	3.25E-7
κ	-2.71E-6	-3.04E-6	3.94E-7
ρ	2.280E-2	2.19E-2	2.23E-3

The value of parameters are for per second.

all trades of Treasury Securities by the three major interdealer brokers except one major interdealer broker, Cantor Fitzgerald. Thus its data count
for a significant portion of the primary-dealer activity. We exam trades
of the active, "on-the-run", 5-year U.S. Treasury notes issued on November 15, 2000 and for the period 11/15-12/29, 2000. During this time period,
GovPx-brokered trades in Treasury notes count for roughly half of the trading volume in notes on a global basis. A big bulk of trading activity in the
treasury market is in the active instruments, so-called "on-the run" instru-
ment which is the most-recently auctioned security of that type. Secondary

trading in U.S. Treasury securities takes place in an over-the-counter market. Although the global market is open around-the-clock during the week, 95 percent of the trading occurs during New York trading hours, from around 7:30 A.M. to 5:00 P.M. Therefore, we restrict our series to the trades that are executed between 7:30 AM to 5:00 PM Eastern Time. The basic summary statistics can be found in Table 2.

Table 2. Summary Statistics for 5-year Treasury Note 11/15 - 12/29, 2000

	Size	Mean	Median	St. Dev.	Skewness	Kurtosis
5-year note	2,220	101.83%	101.84	1.03%	9.65E-3	1.745

The relative frequencies for odd 1/128, odd 1/64 and odd 1/32 or coarser are 18.24%, 37.52%, and 44.23%. Method of relative frequencies produces the estimates for the clustering parameters: $\alpha = .2505$ and $\beta = .3847$.

The daily pattern of treasury trading has been studied in details Fleming.[9] To model the daily duration pattern for the transactions of U.S. treasury, we follow the same approach as in Engle and Russell.[6] Figure 5 shows the nonparametric kernel estimation (run by supsmu function in Matlab) of the expected duration conditional on the time of the day. Consistent with previous studies, trading is most active between 8:30 a.m and 9:30 which is partly a result of the important macroeconomic news release time and the opening of U.S. Treasury futures market. The slight shorter durations around 2:30 p.m. and 3:00 p.m. coincides with the closing time of U.S. Treasury futures market. To capture this diurnal pattern, we use a cubic spline (run by spfit function in Matlab) to approximate the daily factor. Nodes are set on 7:30 a.m., 8:30 a.m., 9:00 a.m. and each hour after 9:00 a.m. until 5:00 p.m. when the trading activity falls.

After the diurnal adjustment for duration (namely, each Δt_i divided by a positive number produced by the cubic spline fit according to the time of day), we fit an $EACD(1,1)$ model using maximum likelihood approach. The parameter estimates are given in Table 3. All parameters are significant. The sum of α_1 and β_1 is 0.9377, indicating strong persistence in duration. We do diagnostic check on EACD(1,1) model. If the model is appropriate, the residues, or the standardized durations, are i.i.d. exponential distributed with mean one. Higher order moments should be independent also. The Ljung-Box Q-Statistics for residue and residue-squared suggest that the model does a very good job of accounting for the intertemporal dependence in durations. Next, we consider the exponential specification of

Fig. 6.

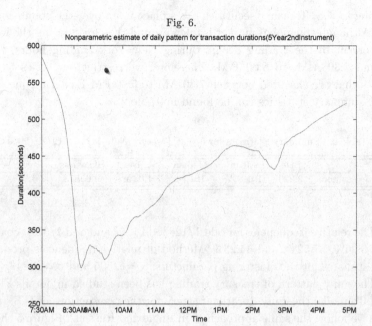

Nonparametric estimate of daily pattern for transaction durations(5Year2ndInstrument)

residue. For the null hypothesis that the residue is exponential, Engle and Russell[6] (page 1144) provided a simple test, whose limiting distribution is standard normal and the test statistic is reported in the last line of Table 3. Even though the exponential assumption is rejected, the test value is much smaller than those in Engle and Russell.[6] This also suggests other distributions such as Weibull or generalized Gamma distributions may be considered in the future research. Even for other ACD models, since they are $\mathcal{F}_t^{Y,V}$-predictable, the Bayes estimates of $(\mu, \sigma, \kappa, \rho)$ presented in Table 4 are not changed and these estimates are "model-free" of the assumptions of duration.

With the estimates of the non-clustering parameters, we apply the tested Fortran program to the transaction data of 5-year bond and obtain Bayes estimates for the extended FM model for two cases. One is GBM case and the other is GBM with initiation dummy in volatility. The Bayes estimates of both cases are presented in Table 4. All the estimates are statistically significant. Note that the estimates of μ and ρ changes quite significantly, suggesting the significant impact of including the initiation dummy. Moreover, the sign κ is negative, implying inventory theory is the main impact in the bond trading.

Table 3.   Parameter Estimates for EACD(1,1) Model

	Estimates	Std Error	T-Stat	Significance
α_0	0.06225	0.01177	5.290	1.2E-7
α_1	0.06542	7.56E-3	8.657	0
β_1	0.8723	0.01672	52.17	0
Ljung-Box	Q-Statistics	of Adjusted	Durations	
Q(10-0)	136.72			0
Q(20-0)	168.28			0
Ljung-Box	Q-Statistics	of Residue		
Q(10-0)	5.4943			0.8558
Q(20-0)	16.276			0.6993
Ljung-Box	Q-Statistics	of Residue-	Squared	
Q(10-0)	3.607			0.9633
Q(20-0)	13.607			0.8499
Test Statistic	4.2134			1.258E-5

Table 4.   Annualized Bayes Estimates for 5-year Bond, 11/15-12/29, 2000

Model	μ	σ	κ	ρ
GBM only	25.10%	3.41%	0	0.128
($\kappa \equiv 0$)	(10.01%)	(0.06 %)	(0)	(0.0147)
Extended GBM	54.50%	3.90%	-0.80%	0.0228
($\kappa \neq 0$)	(10.63%)	(0.11%)	(0.14%)	(0.0085)

Standard errors are in parentheses.

6. Conclusion

This paper reviews recent development of a rich class of filtering models with counting process observations for the micromovement of asset price with observable factors and the related Bayes estimation via filtering. A specific extended FM model built upon GBM and with a buyer-seller initiation dummy in volatility is constructed for the 5-year Treasury note transaction price data. Employing the developed filtering techniques, we perform Bayes estimation via filtering for the model. We find that the buyer-seller initiation dummy is statistically significantly negative. This supports the inventory theory for the bond trading.

This model can be further extended in different directions. For example, we can study whether other observable factors have impacts on volatility. Especially, in Hu, Kuipers and Zeng,[16] we add another macro-economic news dummy in the volatility and we find that macro-economic news has even higher impacts on the bond price volatility than the buyer-seller initiation dummy. With these two examples, we believe the method developed should have many financial applications in empirical market microstructure theory.

Appendix A.   More on Clustering Noise

To formulate the biasing rule, we first define a classifying function $r(\cdot)$,

$$r(y) = \begin{cases} 2 \text{ if the fractional part of } y \text{ is odd } 1/128 \\ 1 \text{ if the fractional part of } y \text{ is odd } 1/64 \\ 0 \text{ if the fractional part of } y \text{ is odd } 1/32 \text{ or coarser} \end{cases} \qquad \text{(A.1)}$$

The biasing rules specify the transition probabilities from y' to y, $p(y|y')$. Then, $p(y|x)$, the transition probability can be computed through $p(y|x) = \sum_{y'} p(y|y')p(y'|x)$ where $p(y'|x) = P\{V = y' - R[x, \frac{1}{8}]\}$. Suppose $D = 8|y - R[x, \frac{1}{8}]|$. Then, $p(y|x)$ can be calculated as, for example, when $r(y) = 2$,

$$p(y|x) = \begin{cases} (1-\rho)(1+\alpha\rho) & \text{if } r(y) = 2 \text{ and } D = 0 \\ \frac{1}{2}(1-\rho)[\rho + \alpha(2 + \rho^2)] & \text{if } r(y) = 2 \text{ and } D = 1 \\ \frac{1}{2}(1-\rho)\rho^{D-1}[\rho + \alpha(1 + \rho^2)] & \text{if } r(y) = 2 \text{ and } D \geq 2 \end{cases} \qquad \text{(A.2)}$$

Acknowledgments

Part of the work was done when Yong Zeng was on sabbatical, visited and taught a special course on *Statistical Analysis of Ultra-high Frequency Financial Data - An Overview and A New Filtering Approach* in Department of Operations Research and Financial Engineering (ORFE) in Princeton University in Spring semester of 2007. He would like to thank Savas Dayanik for the invitation and thank ORFE Department of Princeton for the hospitality. He also gratefully acknowledge the support of the National Science Foundation under grant DMS-0604722.

References

1. Ait-Sahalia, Y., P. A. Mykland and L. Zhang, How Often to Sample a Continuous-Time Process in the Presence of Market Microstructure Noise, *Review of Financial Studies*, **18** (2005), 351–416.
2. Bandi, F. M. and J. R. Russell, Separating microstructure noise from volatility, *Journal of Financial Economics*, **79** (2006), 655-692.
3. Bremaud, P., 1981. *Point Processes and Queues:Martingale Dynamics.* Springer-Verlag, New York.
4. Cvitanic, J., R. Liptser and B. Rozovskii, A filtering approach to tracking volatility from prices observed at random times. *Annals of Applied Probability*, **16** (2006), 1633-1652.

5. Engle, R. F., The econometrics of ultra-high-frequency data, *Econometrica*, **68** (2000), 1–22.
6. Engle, R. F. Engle and J. R. Russell, Autoregressive Conditional Duration: A New Model for Irregularly Spaced Transaction Data, *Econometrica*, **66** (2000), 1127–1162.
7. Ethier, S. N. and T. G. Kurtz, 1986. *Markov processes : Characterization and convergence.* Wiley, New York.
8. Fan, J. and Y. Wang, Multi-scale jump and volatility analysis for high-Frequency financial data, *Journal of American Statistical Association*, **102** (2007), 1349-1362.
9. Fleming, M. J., The Round-the-Clock Market for U.S. Treasury Securities, *Economic Policy Review - Federal Reserve Bank of New York*, **3** (1997), 9-32.
10. Fleming, M. J. and J. V. Rosenberg, How do Treasury dealers manage their positions? it SSRN Working Paper #972367, Federal Reserve Bank of New York, 32 pp.
11. Frey, R., Risk-minimization with incomplete information in a model for high-frequency data, *Mathematical Finance*, **10** (2000), 215-225.
12. Frey, R. and W. J. Runggaldier, A nonlinear filtering approach to volatility estimation with a view towards high frequency data, *International Journal of Theoretical and Applied Finance* **4** (2001), 199-210.
13. Hasbrouck, J., 1996. *Modeling market microstructure time series, in* G. Maddala and C. Rao, eds, 'Handbook of Statistics', Vol. 14, North-Holland, Amsterdam, pp. 647-692.
14. Hasbrouck, J., Stalking the "efficient price" in market microstructure specifications: an overview, *Journal of Financial Markets* **5** (2002), 329-339.
15. Harris, L., Stock price clustering and discreteness, *Review of Financial Studies*, **4** (1991), 389-415.
16. Hu, X, D. Kuipers and Y. Zeng, Econometric analysis via filtering for ultra-high frequency data, Working paper (2010). University of Missouri at Kansas City.
17. Kouritzin, M. and Y. Zeng, Bayesian model selection via filtering for a class of micro-movement models of asset price, *International Journal of Theoretical and Applied Finance*, **8** (1991), 97-121.
18. Lee, C. M. C. and M. J. Ready, Inferring Trade Direction from Intraday Data, *Journal of Finance*, **46** (1991), 733-746.
19. Lee K. and Y. Zeng, Risk minimization for a filtering micromovement model of asset price, *Applied Mathematical Finance*, **17** (2010), 177–199.
20. Li, Y. and P. A. Mykland, Are volatility estimators bobust with respect to modeling assumptions? *Bernoulli*, **13** (2007), 601-622.
21. O'Hara, M., 1997. *Market microstructure theory.* Blackwell Publishing Inc., Malden, MA.
22. Protter, P. 2003. *Stochastic Integration and Differential Equations*, Springer-Verlag, Berlin, 2nd Edition.
23. Stroock, D. W. and S. R. S Varadhan, 1979. *Multidimensional Diffusion Processes*, Springer-Verlag, Berlin.
24. Xiong, J. and Y. Zeng, Mean-variance portfolio selection for partially-

observed point process. Working paper (2010). University of Tennessee.

25. Zeng, Y., A partialy observed model for micromovement of asset prices with Bayes estimation via filtering, *Mathematical Finance*, **13** (2000), 411–444.

26. Zeng, Y., Bayesian inference via filtering for a class of counting processes: Application to the micromovement of asset price, *Statistical Inference for Stochastic Processes*, **8** (2005), 331 – 354.

27. Zhang, L., Mykland, P. A. and Aït-Sahalia, Y., A tale of two time scales: Determining integrated volatility with noisy high frequency data, *Journal of the American Statistical Association*, **100** (2005), 1394–1411.

Jump Bond Markets Some Steps towards General Models in Applications to Hedging and Utility Problems

Michael Kohlmann

Department of Mathematics and Statistics
University of Konstanz
D-78457, Konstanz, Germany
Email: michael.kohlmann@uni-konstanz.de

Dewen Xiong

Department of Mathematics
Shanghai Jiaotong University
Shanghai 200240, P.R.China
Email: xiongdewen@sjtu.edu.cn

In finance, a bond is a debt security in which the authorized issuer owes the holder a debt and, depending on the terms of the bond, is obliged to pay interest (the coupon) and/or to repay the principal a a later date, termed maturity. A bond is a formal contract to repay borrowed money with interest at fixed intervals.

Thus a bond is like a loan: The issuer is the borrower, the holder is the lender, and the coupon is the interest. Bonds provide the borrower with external funds to finance long-term investments, or, in the case of government bonds, to finance current expenditure. Bonds must be repaid a fixed intervals over a period of time.

Bonds and stocks are both securities, but the major difference between the two is that stockholders ave an equity stake in the company (i.e. they are owners), whereas bondholders have a creditor stake in the company (i.e. they are lenders). Another difference is that bonds have a defined term or maturity after which the bond is redeemed, whereas may be outstanding indefinitely. The simplest case of a bond is a bank account with fixed interest rate r_t:

$$dP(t, T) = r_t P(t, T)dt, \ P(T, T) = 1.$$

A short look at the chart of a banks interest rates over some years however shows that the interest rate is by no means fixed so that we should assume that it is a random variable. This leads to the well known classical bond models (in alphabetical order of the authors) by Black-Derman, Chen, Cox-Ingersoll-Ross (CIR), Heath-Jarrow-Morton (HJM), Ho-Lee, Hull-White, Vasicek (V), among many others, where those with an abbreviation after the names are the best known to my observation.

However, while models for stock markets are meanwhile quite satisfactory, the models for bonds/interest rates have diverse deficiencies, so that reality is not really well described. Also the models appear mathematically more difficult and technical than the classical stock market models: While researchers and practitioners are concerned with the fine-tuning in stock models, the basics on a general model for bond markets are still discussed without a commonly agreed result.

The first attempt to describe a general model is found in the seminal paper by Björk et al(1997)[7] and we are often referring to this article.

Here we will discuss two mainstreams of problems, namely mean variance hedging and utility optimization (exponential utility indifference) in a general jump bond market. The purpose of this paper then is to introduce some new techniques, especially techniques from the theory of backward stochastic differential equations (BSDEs) in mean variance hedging, and to contribute to a general model.

In the first part we will consider a model based on n maturities to apply recent results from MVH in jump stock markets. Carmona et al(2004)[10] impressively describe the shortcomings of the models based on a finite number of maturities. To make things short: In the corresponding continuous HJM model the market is infinite dimensional with only a finite number of random sources so that this market always is complete which is contrary to the observations. Further shortcomings caused by the infiniteness of the market are described in Cont (2004)[17].

There are several attempts to overcome these difficulties. Carmona et al(2004)[10] introduce an infinite dimensional Brownian motion and so use Malliavin calculus methods to treat hedging problems with hedging strategies derived from a Clark-Ocone formula. Baran et al.[2] consider generalized strategies in an infinite dimensional HJM-model. A similar approach is taken by DeDonno[18]. However these generalized strategies are not very useful to solve hedging problems more explicitly. In the second part of this paper we will here propose and extend an infinitely dimensional market where we consider measure valued strategies. Of course also these strategies have certain drawbacks when we come to the economical interpretation. And the main drawback is the fact that we always have to work in martingale markets to consider $(\mathbb{M}, Q^0)$-normalized martingales as approximate wealth processes. For many problems this is sufficient but e.g. to describe superhedging we needed the notion of semimartingale. This, however, still is a long-standing open problem already described by Schwartz(1994)[43]: *Il n'est pas facile de savoir exactement de qu'on doit appeler une semimartingale valeurs dans un espace vectoriel topologique.* So, the contents of this paper should be seen as one perhaps promising looking step towards a general model for which we then consider exponential hedging problems.

The paper is organized as follows: In the first part we construct a market of bonds with jumps driven by a general marked point process as well as by an $\mathbb{R}^n$-valued Wiener process based on Björk et al(1997)[7], in which there exists at least one equivalent martingale measure Q^0. Then we consider the mean-variance hedging of a contingent claim $H \in L^2(\mathcal{F}_{T_0})$ w.r.t. self-financing portfolios based on the given maturities $T_1, \cdots, T_n$ with $T_0 < T_1 < \cdots < T_n \le T^*$. We introduce the concept of variance-optimal martingale (VOM) and describe the VOM by a backward semimartingale equation (BSE). By making use of the concept of $\mathcal{E}^*$-martingales introduced by Choulli et al.(1998)[13], we obtain another BSE which has a unique solution. We derive an explicit solution of the optimal strategy and the optimal cost of the mean-variance hedging by the solutions of these two BSEs.

In the second section we consider the optimal exponential utility in a bond market with jumps basing on a model similar to Björk, Kabanov and Runggaldier(1997)[7] which is arbitrage-free. Similar to the normalized integral with respect to the cylindrical martingale first introduced in Mikulevicius and Rozovskii(1998)[39], we introduce the $(\mathbb{M}, Q^0)$-normalized martingale and local $(\mathbb{M}, Q^0)$-normalized martingale. For a given maturity $T_0 \in [0, T^*]$, we describe the minimal entropy martingale (MEM) based on $[T_0, T^*]$ by a backward semimartingale equation (BSE) w.r.t. the $(\mathbb{M}, Q^0)$-normalized martingale. Then we give an explicit form of the optimal approximate wealth to the optimal exputility problem by making use of the solution of the BSE. Finally, we describe the dynamics of the exp utility indifference valuation of a bounded contingent claim $H \in L^\infty(\mathcal{F}_{T_0})$ by another BSE under the minimal entropy martingale measure in the incomplete market.

The present paper strongly relies on unpublished works 45–48 of the authors and a seminar talk of the first author at the Nomura Institute of Oxford University in October 2009. Full proofs of results left out in this report are found in the cited preprints.

Keywords: Bond market with jumps; BSE; variance optimal martingale (VOM); $\mathcal{E}^*$-martingale; mean-variance hedging (MVH); minimal entropy martingale (MEM); $(\mathbb{M}, Q^0)$-normalized martingale; exp utility indifference valuation (exp-UIV).

Mathematics Subject Classification. 90A09, 60H30, 60G44.

1. Preliminaries: The Basic Model

We consider a financial market model on a stochastic basis $(\Omega, \mathcal{F}, \mathbb{F}, P)$, where $\mathbb{F} = \{\mathcal{F}_t; t \in [0, T^*]\}$ is a filtration satisfying the usual conditions with $\mathcal{F}_{T^*} = \mathcal{F}$. The basis is assumed to carry an n-dimensional Brownian motion $W = (W_1, \cdots, W_n)'$ as well as an integer-valued random measure $\mu(du, dy)$ on $\mathbb{R}_+ \times \mathbb{R}$ with compensator $\nu(du, dy) = \lambda(u, dy)du$.

Assumption 1.1. *The filtration* $\mathbb{F} = \{\mathcal{F}_t; t \in [0, T^*]\}$ *is the natural filtration generated by* W *and* μ, *i.e.,*

$$\mathcal{F}_t = \sigma\{W(s), \mu([0, s] \times A), B; \quad 0 \le s \le t, \ A \in \mathcal{B}, \ B \in \mathcal{N}\}$$

where $\mathcal{N}$ *is the collection of* P-*null sets from* $\mathcal{F}$.

For example from remark 3.2 of Björk et al (1997)[7] the stochastic basis has the following predictable representation property: any local martingale M is of the form

$$M_t = M_0 + \int_0^t f'_u dW(u) + \int_0^t \int_{\mathbb{R}} \psi(u, y)(\mu(du, dy) - \nu(du, dy))$$

where f is an $\mathbb{R}^n$-valued predictable process and ψ is a $\widetilde{\mathcal{P}}$-measurable function such that

$$\int_0^T |f_u|^2 du < \infty, \quad \int_0^T \int_{\mathbb{R}} |\psi(u,y)|\lambda(u,dy)du < \infty.$$

In this paper, we let the 4-tuple $(\sigma, \delta, \phi, \varphi)$ be given such that

(i) $\sigma(u,T)$ is a bounded $\mathbb{R}^n$-valued predictable process on $[0,T]$, and $\delta(u,y,T)$ is a bounded $\widetilde{\mathcal{P}}$-measurable function on $[0,T] \times E$ for every fixed $T \in [0,T^*]$; we assume that both $\sigma(u,T)$ and $\delta(u,y,T)$ are continuously differentiable in the T-variable, and we set

$$D(u,y,T) = -\int_u^T \delta(u,y,l)dl;$$

(ii) ϕ is a bounded $\mathbb{R}^n$-valued predictable process on $[0,T^*]$;
(iii) φ is a bounded strictly positive $\widetilde{\mathcal{P}}$-measurable function on $[0,T^*] \times E$.

We introduce 16

$$\Psi(u,T;\sigma,\delta,\phi,\varphi) := \sigma(u,T)' \int_u^T \sigma(u,s)ds - \sigma(u,T)'\phi(u)$$
$$- \int_E \delta(u,y,T)e^{D(u,y,T)}\varphi(u,y)\lambda(u,dy), \quad (1)$$

we let the forward rate dynamics be given by 15

$$llf(t,T) = f(0,T) + \int_0^t \Psi(u,T;\sigma,\delta,\phi,\varphi)du \quad (2)$$
$$+ \int_0^t \sigma(u,T)'dW(u) + \int_0^t \int_E \delta(u,y,T)\mu(du,dy)$$

From Proposition 2.2 of Björk et al(1997)[7] the short rate $r_t := f(t,t)$ is then given by

$$r_t = r_0 + \int_0^t \alpha_u du + \int_0^t \sigma(u,u)'dW(u) + \int_0^t \int_E \delta(u,x,u)\mu(du,dy),$$

where

$$\alpha_u = f_T(u,u) - \sigma(u,u)'\phi(u) - \int_E \delta(u,y,u)e^{D(u,y,u)}\varphi(u,y)\lambda(u,dy)$$

and

$$f_T(u,u) := \left.\frac{\partial f(u,T)}{\partial T}\right|_{T=u}.$$

Since

$$
\begin{aligned}
A(u,T) &:= -\int_u^T \Psi(u,l;\sigma,\delta,\phi,\varphi)dl \\
&= -\int_u^T \sigma(u,l)'\Big\{\int_u^l \sigma(u,s)ds\Big\}dl + \int_u^T \sigma(u,l)'\phi(u)dl \\
&\quad + \int_E \Big\{\int_u^T \delta(u,y,l)e^{D(u,y,l)}dl\Big\}\varphi(u,y)\lambda(u,dy) \\
&= -\frac{1}{2}\|S(u,T)\|^2 - S(u,T)'\phi(u) - \int_E \big\{e^{D(u,y,T)} - 1\big\}\varphi(u,y)\lambda(u,dy),
\end{aligned}
$$

where $S(u,T) = (S_1(u,T),\cdots,S_n(u,T))'$ and where $S_i(u,T) := -\int_u^T \sigma_i(u,l)dl$, one sees from Björk, Kabanov and Runggaldier(1997)[7] that the bond price process $p(t,T) := e^{-\int_t^T f(t,s)ds}$ can be rewritten as

$$
\begin{aligned}
p(t,T) &= p(0,T) + \int_0^t p(u-,T)\big\{r_u + A(u,T) + \frac{1}{2}\|S(u,T)\|^2\big\}du \\
&\quad + \int_0^t p(u-,T)S(u,T)'dW(u) + \int_0^t\int_E p(u-,T)\big\{e^{D(u,y,T)} - 1\big\}\mu(du,dy) \\
&= p(0,T) + \int_0^t p(u-,T)r_u du + \int_0^t p(u-,T)S(u,T)'d\tilde{W}(u) \\
&\quad + \int_0^t\int_E p(u-,T)\big\{e^{D(u,y,T)} - 1\big\}\big\{\mu(du,dy) - \varphi(u,y)\lambda(u,dy)du\big\},
\end{aligned}
$$

where $\tilde{W}(t) := W(t) - \int_0^t \phi(u)du$. As ϕ is a bounded $\mathbb{R}^n$-valued process and φ a bounded strictly positive $\tilde{\mathcal{P}}$-measurable function, $\varphi(u,y) - 1 > -1$, we have that

$$
\begin{aligned}
Z_t^0 &= \mathcal{E}\Big\{\int_0^{\cdot} \phi(u)'dW(u) + \int_0^{\cdot}\int_E (\varphi(u,y) - 1)(\mu(du,dy) - \nu(du,dy))\Big\}_t \\
&= 1 + \int_0^t Z_{u-}^0 \phi(u)'dW(u) + \int_0^t\int_E Z_{u-}^0 (\varphi(u,y) - 1)(\mu(du,dy) - \nu(du,dy)),
\end{aligned}
$$

is a strictly positive uniformly integrable martingale on $[0,T^*]$. So we can define a new probability $Q^0 \sim P$ by

$$
\left.\frac{dQ^0}{dP}\right|_{\mathcal{F}} := Z_{T^*}^0.
$$

It follows from Girsanov's theorem that $\tilde{W}$ is a Brownian motion under Q^0 and

$$
\tilde{\nu}(du,dy) := \varphi(u,y)\lambda(u,dy)du
$$

is the compensator of $\mu(du, dy)$ under Q^0, thus the dynamics of the bond price under Q^0 are given by

$$p(t,T) = p(0,T)\mathcal{E}\left\{ \int_0^{\cdot} r_u du + \int_0^{\cdot} S(u,T)'d\tilde{W}(u) \right.$$
$$\left. + \int_0^{\cdot} \int_E \left\{ e^{D(u,y,T)} - 1 \right\}(\mu(du,dy) - \tilde{\nu}(du,dy)) \right\}_t.$$

The discounted bond price $\bar{p}(t,T) = e^{-\int_0^t r_u du} p(t,T)$ is then given by

$$\bar{p}(t,T) = \bar{p}(0,T)\mathcal{E}\left\{ \int_0^{\cdot} S(u,T)'d\tilde{W}(u) \right.$$
$$\left. + \int_0^{\cdot} \int_E \left\{ e^{D(u,y,T)} - 1 \right\}(\mu(du,dy) - \tilde{\nu}(du,dy)) \right\}_t,$$
(3)

which is a uniformly integrable Q^0-martingale, by the boundedness of S and D.

Definition. (also see 7) We say that a probability Q on $(\Omega, \mathcal{F})$ is a *martingale measure* if $Q \sim P$ and if the discounted bond price $\{\bar{p}(t,T); t \in [0,T^*]\}$ is a local martingale under Q for each $T \in [0,T^*]$.

Remark. As Q^0 is a martingale measure the market is arbitrage free (also see Proposition 3.9 of Björk et al(1997)[7]). From 7, we adopt the following lemma

Lemma 1.1. *Let Q be a martingale measure and Z be the density process of Q with respect to Q^0, then Z can be represented in the following form*

$$Z_t = \mathcal{E}\left\{ \int_0^{\cdot} z_1(u)'d\tilde{W}(u) + \int_0^{\cdot} \int_E z_2(u,y)(\mu(du,dy) - \tilde{\nu}(du,dy)) \right\}_t \quad (4)$$

where z_1 is an $\mathbb{R}^n$-valued predictable process and z_2 is a $\tilde{\mathcal{P}}$-measurable function with $z_2(u,y) > -1$ such that for all $u \in [0,T]$ and $T \in [0,T^*]$ the following structure condition holds

$$z_1(u)'S(u,T) + \int_E \left\{ e^{D(u,y,T)} - 1 \right\} z_2(u,y)\varphi(u,y)\lambda(u,dy) = 0. \quad (5)$$

We have:

Corollary 1.1. *When $\sigma(u,T)$ is independent on the maturity T, i.e., $\sigma(u,T) \equiv \sigma(u)$, where $\{\sigma(u); u \in [0,T^*]\}$ is a bounded $\mathbb{R}^n$-valued predictable process, and $\delta(u,y,T)$ is of the following form*

$$\delta(u,y,T) = -\frac{\tilde{\delta}(u,y)}{1 + \tilde{\delta}(u,y)(T_1 - u)}$$

for a bounded $\widetilde{\mathcal{P}}$-measurable function $\tilde{\delta}(u, y)$ with $\tilde{\delta}(u, y) > -\dfrac{1}{T^*}$, then (z_1, z_2) in the density process must satisfy

$$-z_1(u)'\sigma(u) + \int_E \tilde{\delta}(u, y) z_2(u, y) \varphi(u, y) \lambda(u, dy) = 0. \qquad (6)$$

Obviously the market is incomplete.

2. MVH in Jump Bond Markets

The mean-variance hedging in stock jump markets has already been discussed in e.g. Arai(2005)[1], Černý-Kallsen(2007)[11] and Kohlmann-Xiong-Ye(2007)[31,35]. Arai(2005)[1] adapted the Gouriéroux-Laurent-Pham (1998)[25] (GLP-)approach to the discontinuous case mainly under the assumption that the VOMM exists as a probability measure. A decomposition of H on S is derived by the same sort of argument as in the alternative approach in Section 4 of Rheinländer and Schweizer (1997)[42], so that the decomposition of H on S, in general, is not an orthogonal one, that is, not a GKW one. Another work about mean-variance hedging in a general semimartingale setting is Černý-Kallsen(2007)[11]. They introduce a new measure (called *'opportunity-neutral measure'*) P^* such that the minimal martingale measure relative to P^* coincides with the variance-optimal martingale measure relative to the original probability measure P. Then a so-called *opportunity process* and an *adjustment process* are introduced which give the densities of the VOMM and P^*. Basing on these tools methods are developed to carry over projection methods -inspired by the GKW- or GPL-method in simpler settings- to derive a formula for the optimal hedge in feedback form (Lemma 4.9). However they avoid to use control theoretical tools -although the new tools have their well understood counterparts there- and do not show the connection to the meanwhile well developed BSDE approach to the MVH-problem. In a different approach, Kohlmann-Xiong-Ye(2007)[31,35] relate the variance optimal martingale measure (VOMM) to a backward semimartingale equation (BSE) and show that the (VOMM) is equivalent to the original measure P if and only if the BSE has a solution. For a general contingent claim, they derive an explicit solution of the optimal strategy and the optimal cost of the mean-variance hedging by means of another backward semimartingale equation and an appropriate predictable process δ.

In this part of the paper, we continue the work of Kohlmann-Xiong-Ye(2007)[31,35] and discuss the mean-variance hedging in a market of bonds with jumps, in which the variance optimal martingale measure may be negative, from the point of view of stochastic control. Based on some adapted facts about the adjustment process (see Černý-Kallsen(2007)[11] or Schweizer(1996)[44]), we introduce the variance-optimal martingale M^*, a Q^0-local martingale whose jumps may even be equal to -1. Then we derive a nonlinear backward semimartingale equation (see (6)), and give an explicit form of the variance-optimal martingale M^* by the bounded solution of the BSE (6). In section (2.2), we introduce the $\mathcal{E}^*$-martingale, which is similar to the concept of $\mathcal{E}$-martingales introduced in Choulli-Krawczyk-Stricker(1998)[13]. By applying the representation of an $\mathcal{E}^*$-martingale, we derive another linear BSE (see (7)) whose parameters depend on the BSE (6), and we derive an explicit solution of the optimal strategy and the optimal cost of the mean-variance hedging by means of this solution of the BSE (7). In section (2.3), we consider a special bond market where σ in (15) is independent of the maturity T and δ is of the following form

$$\delta(u, y, T) = -\frac{\tilde{\delta}(u, y)}{1 + \tilde{\delta}(u, y)(T_1 - u)}$$

for a bounded $\widetilde{\mathcal{P}}$-measurable function $\tilde{\delta}(u, y)$ with $\tilde{\delta}(u, y) > -\dfrac{1}{T^*}$. In this special case we derive the optimal strategy in a very explicit form.

2.1. Hedging on the Basis of n Maturity Times and the Variance Optimal Martingale Measure

We consider a market as described above in section 1.

Now fix a $T_0 \in (0, T^*)$ and let $T_1, \cdots, T_n$ be n maturities with $T_0 < T_i < T^*$ for each i and $T_1 < T_2 < \cdots < T_n$. We are mainly concerned with the question whether a contingent claim given by $H \in L^2(\mathcal{F}_{T_0})$ can be replicated or hedged by a self-financing portfolio based on

$$\bar{p}(t, T_1), \ldots, \bar{p}(t, T_n),$$

which is defined in the following way:

Definition. A *self-financing portfolio based on the maturities* $T_1, \cdots, T_n$ is an $\mathbb{R}^n$-valued predictable process $\pi = (\pi_1, \cdots, \pi_n)'$ such that the corresponding discounted value process

$$V_t^{x,\pi} = x + \int_0^t \sum_{j=1}^n \pi_j(u) S(u; T_j)' d\tilde{W}(u) \int_0^t \int_E \sum_{j=1}^n \pi_j(u) \{e^{D(u,y,T_j)} - 1\}$$
$$(\mu(du, dy) - \tilde{\nu}(du, dy)),$$

$$(1)$$

$t \in [0, T_0]$, is a local Q^0-martingale with $V_T^{x,\pi} \in L^2(\mathcal{F}_{T_0}, P)$. The family of all self-financing portfolios based on the maturities $T_1, \cdots, T_n$ is denoted by $\mathcal{A}(T_1, \cdots, T_n)$.

Of course in general, a contingent claim $H \in L^2(\mathcal{F}_{T_0}, P)$ can not be fully replicated by a self-financing portfolio based on the maturities $T_1, \cdots, T_n$, thus we discuss the following minimization question

$$\min_{\pi \in \mathcal{A}(T_1, \cdots, T_n)} E\left[\left(H - V_{T_0}^{x,\pi}\right)^2\right],$$

and give an explicit solution of the optimal strategy by BSDEs.

Let

$$\mathbb{Z}_s^2(T_1, \cdots, T_n) := \left\{ Z \left| \begin{array}{l} Z \text{ is a local } Q^0\text{-martingale on } [0, T_0] \text{ with } Z_0 = 1 \text{ and} \\ E_{Q^0}\left[Z_{T_0}^0(Z_{T_0})^2\right] < \infty \text{ such that } Z_t \bar{p}(t, T_i) \text{ is a local } Q^0\text{-} \\ \text{martingale on } [0, T_0] \text{ for each } i \end{array} \right. \right\},$$

$$\mathbb{Z}_{\exp}^2(T_1, \cdots, T_n) := \left\{ Z; \ Z \in \mathbb{Z}_s^2(T_1, \cdots, T_n), Z \text{ is a stochastic exponential} \right\},$$

$$\mathbb{Z}_e^2(T_1, \cdots, T_n) := \left\{ Z; \ Z \in \mathbb{Z}_s^2(T_1, \cdots, T_n), Z \text{ is strictly positive} \right\},$$

Obviously we have $\mathbb{Z}_e^2(T_1, \cdots, T_n) \subset \mathbb{Z}_{\exp}^2(T_1, \cdots, T_n) \subset \mathbb{Z}_s^2(T_1, \cdots, T_n)$. For any $Z \in \mathbb{Z}_e^2(T_1, \cdots, T_n)$, one can define a measure Q by $\frac{dQ}{dQ^0}\big|_{\mathcal{F}_{T_0}} := Z_{T_0}$, and then Q is an equivalent martingale measure for each $\bar{p}(t, T_i)$, $i = 1, \cdots, n$.

Remark. From Jacod(1979)[26] the following is well known: for any $Z \in \mathbb{Z}_{\exp}^2(T_1, \cdots, T_n)$, let $\tau := \inf\{t : Z_t = 0\}$ then $Z_{t-} \neq 0$ a.s. for $0 \leq t \leq \tau$, and $Z_t = 0$ a.s. for $\tau \leq t \leq T_0$. We note that Z **may be negative**.

We define for any $t, u \in [0, T_0]$ and $y \in E$

$$\mathbb{S}(t) := \begin{pmatrix} S_1(t, T_1) & \cdots & S_n(t, T_1) \\ \cdots & \cdots & \cdots \\ S_1(t, T_n) & \cdots & S_n(t, T_n) \end{pmatrix}_{n \times n} = (S_j(t, T_i))_{n \times n},$$

$$\Upsilon(u, y) := \left(e^{D(u,y,T_1)} - 1, \cdots, e^{D(u,y,T_n)} - 1\right)',$$

$$\bar{P}(t) := (\bar{p}(t, T_1), \cdots, \bar{p}(t, T_n))',$$

and, similar to Assumption (H2) in Biagini(2001)[4], we assume in this section that

(H) $\mathbb{S}(t)$ is invertible for all $t \in [0, T_0]$.

Lemma 2.1. $Z \in \mathcal{Z}_{\exp}^2$ if and only if Z can be represented as

$$Z_t = 1 + \int_0^t Z_{u-} z_1(u)' d\tilde{W}(u) + \int_0^t \int_E Z_{u-} z_2(u, y)(\mu(du, dy) - \tilde{\nu}(du, dy))$$

where z_1 is a $\mathbb{R}^n$-valued predictable process and z_2 is a $\tilde{\mathcal{P}}$-measurable function such that for all $u \in [0, T_0]$

$$\mathbb{S}(u)z_1(u) + \int_E \Upsilon(u, y)z_2(u, y)\varphi(u, y)\lambda(u, dy) = 0. \tag{2}$$

We denote by $\mathcal{T}_{T_0}$ the family of stopping times τ with $\tau \leq T_0$. For any given stopping time $\tau \in \mathcal{T}_{T_0}$ and (z_1, z_2) satisfying (2), let

$$\begin{aligned}
\mathcal{E}^\tau(z_1, z_2) := \mathcal{E} \Bigg\{ & \int_0^\cdot \mathbb{I}_{]\tau, T_0]}(u)z_1(u)' d\tilde{W}(u) \\
& + \int_0^\cdot \int_E \mathbb{I}_{]\tau, T_0]}(u)z_2(u, y)(\mu(du, dy) - \tilde{\nu}(du, dy)) \Bigg\},
\end{aligned}$$

and we write $\mathcal{Z}_{\exp}^{2,\tau}$ for the family of (z_1, z_2) satisfying (2) with $E\left\{ \left(Z_{\tau, T_0}^0 \mathcal{E}^\tau(z_1, z_2)_{T_0} \right)^2 \right\} < \infty$. We consider the following kind of minimization problem

$$\Psi(\tau) := \operatorname*{ess\,inf}_{(z_1, z_2) \in \mathcal{Z}_{\exp}^{2,\tau}} E\left[\left(Z_{\tau, T_0}^0 \mathcal{E}^\tau(z_1, z_2)_{T_0} \right)^2 \Big| \mathcal{F}_\tau \right]. \tag{3}$$

The following lemma is adapted from 11,34,35

Lemma 2.2. There exists an $\mathbb{R}^n$-valued predictable process $\tilde{a} = (\tilde{a}_1, \cdots, \tilde{a}_n)'$ (called **adjustment process**) on $[0, T_0]$ such that for any $\tau \in \mathcal{T}_{T_0}$, $\mathcal{E}\left((-\tilde{a}' \mathbb{I}_{]\tau, T_0]}) \cdot \bar{P} \right)_{T_0} \in L^2(P)$ and such that

(i) $\left\{ \mathcal{E}\left((-\tilde{a}' \mathbb{I}_{]\tau, T_0]}) \cdot \bar{P} \right)_s Z_{\tau, s}^0 \mathcal{E}^\tau(z_1, z_2)_s; \ s \in [\tau, T_0] \right\}$ is a uniformly integrable martingale for any $\tau \in \mathcal{T}_{T_0}$ and $(z_1, z_2) \in \mathcal{Z}_{\exp}^{2,\tau}$;

(ii) $E\left[\mathcal{E}\left((-\tilde{a}' \mathbb{I}_{]\tau, T_0]}) \cdot \bar{P} \right)_{T_0} \Big| \mathcal{F}_\sigma \right]$

$= E\left[\left(\mathcal{E}\left((-\tilde{a}' \mathbb{I}_{]\tau, T_0]}) \cdot \bar{P} \right)_{T_0} \right)^2 \Big| \mathcal{F}_\sigma \right] \in (0, 1]$, a.s., for any stopping times $\sigma \leq \tau$;

(iii) for any $\pi \in \mathcal{A}(T_1, \cdots, T_n)$

$$E\left((V_{T_0}^{x, \pi} - V_\tau^{x, \pi}) \mathcal{E}\left((-\tilde{a}' \mathbb{I}_{]\tau, T_0]}) \cdot \bar{P} \right)_{T_0} \right) = 0.$$

From Lemma 2.2, it appears to be natural to assume that

Assumption 2.1. *For any* $\tau \in \mathcal{T}_{T_0}$, $\left\{ M_t^{(\tau)}; t \in [\tau, T_0] \right\}$ *is a stochastic exponential, where*

$$M_t^{(\tau)} := \frac{E\left[\mathcal{E}\left((-\tilde{a}'\mathbb{I}_{]\tau, T_0]}\right) \cdot \bar{P}\right)_{T_0} \big| \mathcal{F}_t\right]}{E\left[\mathcal{E}\left((-\tilde{a}'\mathbb{I}_{]\tau, T_0]}\right) \cdot \bar{P}\right)_{T_0} \big| \mathcal{F}_\tau\right]}.$$

Under Assumption 2.1, $\{M_t^{(\tau)}(\bar{p}(t, T_i) - p(\tau, T_i)); t \in [\tau, T_0]\}$ is a local martingale, thus $\left\{\frac{1}{Z_t^0} M_t^{(\tau)}(\bar{p}(t, T_i) - p(\tau, T_i)); t \in [\tau, T_0]\right\}$ is a local Q^0-martingale, therefore there exists a pair $(z_1^{(\tau)}, z_2^{(\tau)})$ depending on τ satisfying (2) such that

$$M_t^{(\tau)} = Z_{\tau,t}^0 \mathcal{E}^\tau\left(z_1^{(\tau)}, z_2^{(\tau)}\right)_t, \text{ a.s., for } t \in [\tau, T_0], \tag{4}$$

and

$$\begin{aligned}
\Psi(\tau) &= E\left[\left(Z_{\tau,T_0}^0 \mathcal{E}^\tau\left(z_1^{(\tau)}, z_2^{(\tau)}\right)_{T_0}\right)^2 \Big| \mathcal{F}_\tau\right] \\
&= E\left[\left(M_{T_0}^{(\tau)}\right)^2 \big| \mathcal{F}_\tau\right] \\
&= \frac{1}{E\left[\mathcal{E}\left((-\tilde{a}'\mathbb{I}_{]\tau, T_0]}\right) \cdot \bar{P}\right)_{T_0} \big| \mathcal{F}_\tau\right]}.
\end{aligned}$$

Lemma 2.3. *Under Assumption 2.1, there exists a pair* (z_1^*, z_2^*) *satisfying (2) such that for all* $\tau \in \mathcal{T}_{T_0}$, $(z_1^*, z_2^*) \in \mathcal{Z}_{\exp}^{2,\tau}$ *and*

$$\Psi(\tau) = E_{Q^0}\left[Z_{\tau,T_0}^0\left(\mathcal{E}^\tau(z_1^*, z_2^*)_{T_0}\right)^2 \Big| \mathcal{F}_\tau\right], \quad Q^0\text{-a.s.}$$

Proof. We only need to show that $\left(z_1^{(\tau)}, z_2^{(\tau)}\right)$ in (4) is independent of τ, i.e., for any $\tau_1, \tau_2 \in \mathcal{T}_{T_0}$ with $\tau_1 < \tau_2$, we need to show that $z_1^{\tau_1}\mathbb{I}_{]\tau_2, T_0]} = z_1^{\tau_2}\mathbb{I}_{]\tau_2, T_0]}$ and $z_2^{\tau_1}\mathbb{I}_{]\tau_2, T_0]} = z_2^{\tau_2}\mathbb{I}_{]\tau_2, T_0]}$, which follows from the definition of $M^{(\tau)}$ and the fact that for any $t \geq \tau_2$

$$\begin{aligned}
M_t^{(\tau_1)} &= \frac{\mathcal{E}\left((-\tilde{a}'\mathbb{I}_{]\tau_1, \tau_2]}\right) \cdot \bar{P}\right)_{\tau_2} \times E\left[\mathcal{E}\left((-\tilde{a}'\mathbb{I}_{]\tau_2, T_0]}\right) \cdot \bar{P}\right)_{T_0} \big| \mathcal{F}_t\right]}{E\left[\mathcal{E}\left((-\tilde{a}'\mathbb{I}_{]\tau_1, T_0]}\right) \cdot \bar{P}\right)_{T_0} \big| \mathcal{F}_{\tau_1}\right]} \\
&= \frac{\mathcal{E}\left((-\tilde{a}'\mathbb{I}_{]\tau_1, \tau_2]}\right) \cdot \bar{P}\right)_{\tau_2} \times E\left[\mathcal{E}\left((-\tilde{a}'\mathbb{I}_{]\tau_2, T_0]}\right) \cdot \bar{P}\right)_{T_0} \big| \mathcal{F}_{\tau_2}\right]}{E\left[\mathcal{E}\left((-\tilde{a}'\mathbb{I}_{]\tau_1, T_0]}\right) \cdot \bar{P}\right)_{T_0} \big| \mathcal{F}_{\tau_1}\right]} M_t^{(\tau_2)} \cdot \\
&= M_{\tau_2}^{(\tau_1)} M_t^{(\tau_2)},
\end{aligned}$$

Note here that $M_{\tau_2}^{(\tau_2)} = 1$. $\qquad\qquad\qquad\qquad\qquad\qquad\qquad\square$

We have the following corollaries

Corollary 2.1. *Let Assumption 2.1 to be satisfied, then Ψ is a bounded RCLL semimartingale satisfying*

$$1 \leq \Psi(t) \leq C < \infty$$

for some positive constant C.

Corollary 2.2. *Let Assumption 2.1 be satisfied and let (z_1^*, z_2^*) be given as in Lemma 2.3, then for any $\tau \in \mathcal{T}_{T_0}$ and for any $t \geq \tau$*

$$Z_{\tau,t}^0 \mathcal{E}^\tau(z_1^*, z_2^*)_t \Psi(t) = \Psi(\tau)\mathcal{E}\left(\left(-\tilde{a}'\mathbb{I}_{]\tau, T_0]}\right) \cdot \bar{P}\right)_t, \quad Q^0\text{-}a.s.$$

Definition. Let Assumption 2.1 be satisfied and let (z_1^*, z_2^*) be as in Lemma 2.3. Let

$$M_t^* = \int_0^t z_1^*(u)' d\tilde{W}(u) + \int_0^t \int_E z_2^*(u,y)(\mu(du,dy) - \tilde{\nu}(du,dy)),$$

then M^* is a Q^0-local martingale, which is called the **variance-optimal martingale**. We introduce an increasing sequence of stopping times $(\tau_n^*)_{n\in\mathbb{N}}$ defined by

$$\tau_{n+1}^* = \inf\left\{t > \tau_n^* | \Delta M_t^* = -1\right\} \wedge T_0.$$

Remark. The following properties are straightforward

(1) $\mathcal{E}^\tau(z_1^*, z_2^*)_t = \mathcal{E}\left(M^* - (M^*)^\tau\right)_t$;

(2) $\mathcal{E}^{\tau_n^*}(z_1^*, z_2^*) = \left(\mathcal{E}^{\tau_n^*}(z_1^*, z_2^*)\right)^{\tau_{n+1}^*}$;

(3) $\mathcal{E}^{\tau_n^*}(z_1^*, z_2^*)_t \neq 0$ for $t \in [\![\tau_n^*, \tau_{n+1}^*[\![$ on $\{\tau_{n+1}^* < T\}$.

We easily derive that

$$Z_{\tau_n^*, t}^0 = 1 + \int_{\tau_n^*}^t Z_{\tau_n^*, u-}^0 \phi(u)' dW(u)$$

$$+ \int_{\tau_n^*}^t \int_E Z_{\tau_n^*, u-}^0 \left(\varphi(u,y) - 1\right)\left(\mu(du,dy) - \nu(du,dy)\right).$$

For simplicity, we introduce $Z^*_{\tau^*_n,t} := \mathcal{E}^{\tau_n}(z^*_1, z^*_2)_t$ and $\tilde{Z}_{\tau^*_n,t} := \Psi(\tau^*_n)\mathcal{E}\left(\left(-\tilde{a}'\mathbb{I}_{]\tau,T_0]}\right)\cdot\bar{P}\right)_t$, so that

$$Z^*_{\tau^*_n,t} = 1 + \int_{\tau^*_n}^t Z^*_{\tau^*_n,u-} z^*_1(u)' d\tilde{W}(u) + \int_{\tau^*_n}^t \int_E Z^*_{\tau^*_n,u-} z^*_2(u,y)(\mu(du,dy) - \tilde{\nu}(du,dy))$$

$$\tilde{Z}_{\tau^*_n,t} = \Psi(\tau^*_n) - \int_{\tau^*_n}^t \tilde{Z}_{\tau^*_n,u-} \hat{a}(u)'\mathbb{S}(u)d\tilde{W}(u)$$
$$- \int_{\tau^*_n}^t \int_E \tilde{Z}_{\tau^*_n,u-} \hat{a}(u)'\Upsilon(u,y)(\mu(du,dy) - \tilde{\nu}(du,dy)),$$

where $\hat{a} = (\hat{a}_1,\cdots,\hat{a}_n)'$ and $\hat{a}_i(u) = \bar{p}(u-,T_i)\tilde{a}_i(u)$ for each $i = 1,\cdots,n$.

A lengthy, though not very difficult computation shows that the following representations hold: Let

$$\varrho(u,y) := \frac{1 - \hat{a}(u)'\Upsilon(u,y)}{z^*_2(u,y) + 1}.$$

As Ψ is strictly positive, also $\varrho(u,y)$ is strictly positive, thus

$$z^*_2(u,y) = \frac{1 - \hat{a}(u)'\Upsilon(u,y)}{\varrho(u,y)} - 1$$

and

$$z^*_1(u) = -\int_E \mathbb{S}(u)^{-1}\Upsilon(u,y)z^*_2(u,y)\varphi(u,y)\lambda(u,dy)$$
$$= -\int_E \mathbb{S}(u)^{-1}\Upsilon(u,y)\left\{\frac{1 - \hat{a}(u)'\Upsilon(u,y)}{\varrho(u,y)} - 1\right\}\varphi(u,y)\lambda(u,dy)$$
$$:= -\mathcal{I}_u(\hat{a},\varrho),$$

where $\mathbb{S}(u)^{-1}$ is the inverse of $\mathbb{S}(u)$. Therefore,

$$\Psi(t) = \Psi(\tau^*_n) - \int_{\tau^*_n}^t \Psi(u-)\left\{\phi(u) - \mathcal{I}_u(\hat{a},\varrho) + \mathbb{S}(u)'\hat{a}(u)\right\}'dW(u)$$
$$+ \int_{\tau^*_n}^t \Psi(u-)\left\{\|\phi(u) - \mathcal{I}_u(\hat{a},\varrho)\|^2 + 2\hat{a}(u)'\mathbb{S}(u)\phi(u)\right\}du$$
$$+ \int_{\tau^*_n}^t \int_E \Psi(u-)\left\{\frac{(1 - \hat{a}(u)'\Upsilon(u,y))^2}{\varrho(u,y)}\varphi(u,y)\right.$$
$$\left. + 2\hat{a}(u)'\Upsilon(u,y)\varphi(u,y) - 1\right\}\lambda(u,dy)du \tag{5}$$
$$+ \int_{\tau^*_n}^t \int_E \Psi(u-)\left\{\frac{\varrho(u,y)}{\varphi(u,y)} - 1\right\}\mu(du,dy)$$

for each n. Thus we consider the following backward semimartingale equation

$$
\begin{cases}
Y_t = Y_0 - \int_0^t Y_{u-}\big\{\phi(u) - \mathcal{I}_u(a,\rho) + \mathbb{S}(u)'a(u)\big\}' dW(u) \\
\qquad + \int_0^t Y_{u-}\Big\{\|\phi(u) - \mathcal{I}_u(a,\rho)\|^2 + 2a(u)'\mathbb{S}(u)\phi(u)\Big\} du \\
\qquad + \int_0^t \int_E Y_{u-}\Big\{\dfrac{(1 - a(u)'\Upsilon(u,y))^2}{\rho(u,y)}\varphi(u,y) \\
\qquad\qquad\qquad\qquad + 2a(u)'\Upsilon(u,y)\varphi(u,y) - 1\Big\}\lambda(u,dy)du \\
\qquad + \int_0^t \int_E Y_{u-}\Big\{\dfrac{\rho(u,y)}{\varphi(u,y)} - 1\Big\}\mu(du,dy), \\
Y_{T_0} = 1.
\end{cases}
\tag{6}
$$

A **bounded solution** of the BSE (5) is a triple (Y, a, ρ) satisfying (5) such that

(1) a is an $\mathbb{R}^n$-valued predictable process such that for any stopping time $\tau \in \mathcal{T}_{T_0}$,

$$
E\Big[\mathcal{E}\Big(-\sum_{i=1}^n \int \frac{a_i(u)\mathbb{I}_{\llbracket\tau,T_0\rrbracket}(u)}{\bar{p}(u-,T_i)} d\bar{p}(u,T_i)\Big)^2_{T_0}\Big] < \infty;
$$

(2) $\rho(u,y)$ is a strictly positive $\widetilde{\mathcal{P}}$-measurable function such that for any $\tau \in \mathcal{T}_{T_0}$

$$
(z_1^*, z_2^*) \in \mathcal{Z}_{\exp}^{2,\tau}
$$

where $z_1^*(u) := -\mathcal{I}_u(a, \rho)$ and $z_2^*(u,y) := \dfrac{1 - a(u)'\Upsilon(u,y)}{\rho(u,y)} - 1$;

(3) Y is a bounded RCLL semimartingale with $0 < c_1 \leq Y \leq c_2 < \infty$ for some constants c_1 and c_2.

Remark. Although $\rho(u,y)$ is strictly positive, z_2^* may be less than -1, thus $\mathcal{E}^\tau(z_1^*, z_2^*)$ may be **negative**. From (5), we have the following lemma:

Lemma 2.4. *Under Assumptions 2.1 and (H), $(\Psi, \hat{a}, \varrho)$ is a bounded solution of the BSE (6).*

With these results we get the form of the variance-optimal martingale as a first step to the solution of the MVH-problem:

Theorem 2.1. *Assume Assumptions (H) and 2.1 be satisfied and (Y, a, ρ) be a bounded solution of the BSE (5), then for any $\tau \in \mathcal{T}_{T_0}$, (z_1^*, z_2^*) is the solution to the minimization problem (3), i.e.,*

$$M_t^* := \int_0^t z_1^*(u)' d\tilde{W}(u) + \int_0^t \int_E z_2^*(u, y)(\mu(du, dy) - \tilde{\nu}(du, dy))$$

is the variance-optimal martingale.

Proof. It is shown that for any $(z_1, z_2) \in \mathcal{Z}_{\exp}^{2,\tau}$, $\{Y_t Z_{\tau,t}^0 Z_{\tau,t}^* Z_{\tau,t}^0 \mathcal{E}^\tau(z_1, z_2)_t;$ $t \in [\tau, T_0]\}$ is a local martingale. As Y is bounded, one can easily derive that $\{Y_t Z_{\tau,t}^0 Z_{\tau,t}^* Z_{\tau,t}^0 \mathcal{E}^\tau(z_1, z_2)_t; t \in [\tau, T_0]\}$ is uniformly integrable, thus

$$E\left[Z_{\tau,T}^0 Z_{\tau,T}^* Z_{\tau,T}^0 \mathcal{E}^\tau(z_1, z_2)_T \big| \mathcal{F}_\tau\right] = E\left[Y_T Z_{\tau,T}^0 Z_{\tau,T}^* Z_{\tau,T}^0 \mathcal{E}^\tau(z_1, z_2)_T \big| \mathcal{F}_\tau\right] = Y_\tau.$$

Similar to the proof of Theorem 3.4 of Kohlmann-Xiong-Ye(2007)[31,35], one derives that (z_1^*, z_2^*) is the solution of the minimization problem (3) for any $\tau \in \mathcal{T}_{T_0}$ and Y_τ is the optimal cost, i.e.,

$$Y_\tau = E\left[\left(Z_{\tau,T_0}^0 \mathcal{E}^\tau(z_1^*, z_2^*)_{T_0}\right)^2 \bigg| \mathcal{F}_\tau\right] = \operatorname*{ess\,inf}_{(z_1, z_2) \in \mathcal{Z}_{\exp}^{2,\tau}} E\left[\left(Z_{\tau,T_0}^0 \mathcal{E}^\tau(z_1, z_2)_{T_0}\right)^2 \bigg| \mathcal{F}_\tau\right],$$

which completes the proof. $\square$

Corollary 2.3. *Under Assumptions (H) and 2.1, there exists a unique solution of the BSE (6).*

2.2. $\mathcal{E}^*$-Martingale and Mean-Variance Hedging

In this section, we assume that

Assumption (BS): the BSE (6) has a bounded solution denoted by (Y, a, ρ).

Recall that

$$z_1^*(u) \quad = -\mathcal{I}_u(a, \rho),$$

$$z_2^*(u, y) = \frac{1 - a(u)' \Upsilon(u, y)}{\rho(u, y)} - 1,$$

where

$$\mathcal{I}_u(a, \rho) = \int_E \mathbb{S}(u)^{-1} \Upsilon(u, y) \left\{ \frac{1 - a(u)' \Upsilon(u, y)}{\rho(u, y)} - 1 \right\} \varphi(u, y) \lambda(u, dy).$$

From Theorem 2.1

$$M_t^* := \int_0^t z_1^*(u)' d\tilde{W}(u) + \int_0^t \int_E z_2^*(u, y)(\mu(du, dy) - \tilde{\nu}(du, dy))$$

is the variance-optimal martingale. As above, we let

$$\tau_{n+1}^* = \inf\left\{t > \tau_n^* | \Delta M_t^* = -1\right\} \wedge T_0.$$

Also $\left\{Z_{\tau_n^*, t}^0 \mathcal{E}(M^* - (M^*)^{\tau_n^*})_t; t \in [\tau_n^*, T_0]\right\}$ is a square-integrable martingale, and one derives from Choulli-Krawczyk-Stricker(1998)[13] that $Z^0\mathcal{E}(M^*)$ is a regular stochastic exponential satisfying the reverse Hölder inequality (R_2), i.e., there exists a constant C such that

$$E\left[\left(Z_{\tau, T_0}^0 \mathcal{E}\left(M^* - (M^*)^\tau\right)_{T_0}\right)^2 \Big| \mathcal{F}_\tau\right] \leq C < \infty, \quad a.s.$$

Similar to Definition 3.11 of 13, we give the definition of $\mathcal{E}^*$-martingale

Definition.

(i) X is an $\mathcal{E}^*$-**martingale**, if for any n

$$E\left[Z_{\tau_n^*, T_0}^0 \left|\mathcal{E}\left(M^* - (M^*)^{\tau_n^*}\right)_{T_0} X_{\tau_n^*}\right|\right] < \infty$$

and $\left\{(X_t - X_{t \wedge \tau_n^*})Z_{\tau_n^*, t}^0 \mathcal{E}\left(M^* - (M^*)^{\tau_n^*}\right)_t; \ t \in [0, T_0]\right\}$ is a martingale. The class of $\mathcal{E}^*$-martingales is denoted by $\mathcal{M}(\mathcal{E}^*)$.

(ii) X is called as an $\mathcal{E}^*$-**local martingale** if $\left\{(X_t - X_{t \wedge \tau_n^*})Z_{\tau_n^*, t}^0 \mathcal{E}\left(M^* - (M^*)^{\tau_n^*}\right)_t; \ t \in [0, T_0]\right\}$ is a local martingale for each n. The class of $\mathcal{E}^*$-local martingales is denoted by $\mathcal{M}_{\text{loc}}(\mathcal{E}^*)$.

Lemma 2.5. *Under Assumption (BS), an RCLL process X is a $\mathcal{E}^*$-local martingale iff it is a semimartingale such that $X + [M^*, X]$ is a Q^0-local martingale.*

Proof. X is a local $\mathcal{E}^*$-martingale iff $\left\{(X_t - X_{t \wedge \tau_n^*})\mathcal{E}\left(M^* - (M^*)^{\tau_n^*}\right)_t; \ t \in [0, T_0]\right\}$ is a Q^0-local martingale for each n. The rest is completely the same as in the proof of Proposition 3.15 of 13. $\qquad\square$

Proposition 2.1. *Let Assumption (BS) be satisfied and let X be a special semimartingale, then X is an $\mathcal{E}^*$-local martingale iff there exist an $\mathbb{R}^d$-valued predictable process ξ_1 and a $\widetilde{\mathcal{P}}$-measurable function $\xi_2(u, y)$ such that*

$$X_t = X_0 + \int_0^t \xi_1(u)' dW(u) + \int_0^t \int_E \xi_2(u, y)\mu(du, dy)$$

$$- \int_0^t \xi_1(u)'\{z_1^*(u) + \phi(u)\}du$$

$$- \int_0^t \int_E \xi_2(u, y)(z_2^*(u, y) + 1)\varphi(u, y)\lambda(u, dy)du,$$

Q^0-a.s., for every $t \in [0, T_0]$.

Proof. Prove that $X_t + [X, M^*]_t$ is a Q^0-local martingale iff

$$A_t + \int_0^t \xi_1(u)'\{z_1^*(u) + \phi(u)\}du + \int_0^t \int_E \xi_2(u, y)\left\{\left(z_2^*(u, y) + 1\right)\varphi(u, y) - 1\right\}\lambda(u, dy)du = 0,$$

Q^0-a.s., for each $t \in [0, T_0]$, and hence

$$X_t = X_0 + \int_0^t \xi_1(u)'dW(u) + \int_0^t \int_E \xi_2(u, y)\mu(du, dy)$$
$$- \int_0^t \xi_1(u)'\{z_1^*(u) + \phi(u)\}du - \int_0^t \int_E \xi_2(u, y)\left(z_2^*(u, y) + 1\right)\varphi(u, y)\lambda(u, dy)du,$$

which completes the proof. □

Corollary 2.4. *Let Assumption (BS) be satisfied and let H be any contingent claim belonging to* $L^2(\mathcal{F}_{T_0}, P)$*, then the BSE*

$$\begin{cases} h_t = h_0 + \int_0^t \xi_1(u)'dW(u) + \int_0^t \int_E \xi_2(u, y)\mu(du, dy) \\ \qquad - \int_0^t \xi_1(u)'\{z_1^*(u) + \phi(u)\}du - \int_0^t \int_E \xi_2(u, y)\left(z_2^*(u, y) + 1\right)\varphi(u, y)\lambda(u, dy)du, \\ h_T = H \end{cases}$$

$$(7)$$

has a solution (h, ξ_1, ξ_2).

Proof. Let h be an RCLL process on $[0, T_0]$ satisfying

$$h_t = \frac{E_{Q^0}\left[H\mathcal{E}(M^* - (M^*)^{\tau_n^*})_{T_0} \big| \mathcal{F}_t\right]}{\mathcal{E}(M^* - (M^*)^{\tau_n^*})_t}$$

for $\tau_n^* \leq t < \tau_{n+1}^*$ and for each n. As $Z^0\mathcal{E}(M^*)$ satisfies the reverse Hölder inequality (R_2) and $H \in L^2(\mathcal{F}_{T_0}, P)$, one directly derives from the proof of Proposition 3.12(iii) of Choulli-Krawczyk-Stricker(1998)[13] that h is an $\mathcal{E}^*$-martingale. From Theorem 4.1 of 13, we see that

$$\left\| \max_{t \in [0, T_0]} |h_t| \right\|_2 \leq C\|H\|_2 < \infty$$

for some constant C, thus h is a special semimartingale. It follows from Proposition 2.1 that there exist an $\mathbb{R}^d$-valued predictable process ξ_1 and a $\widetilde{\mathcal{P}}$-measurable function $\xi_2(u, y)$ such that (h, ξ_1, ξ_2) is a solution of the BSE (7). □

With these results mainly adopted from recent results on mean variance hedging we now turn to the already above stated mean-variance hedging problem in the bond market and derive an analytical solution. Given a contingent claim $H \in L^2(\mathcal{F}_{T_0}, P)$, which might not be fully replicated by a self-financing portfolio based on the maturities $T_1, \cdots, T_n$, we will try to solve the following minimization question

Problem (MVH-n): $\qquad J = \min_{\pi \in \mathcal{A}(T_1, \cdots, T_n)} E\left[\left(H - V_{T_0}^{x,\pi}\right)^2\right].$

To give an explicit form of the optimal strategy, we need to introduce

$$\mathfrak{R}_u := \mathbb{S}(u)\mathbb{S}(u)' + \int_E \Upsilon(u,y)\Upsilon(u,y)'\frac{\varphi(u,y)}{\rho(u,y)}\lambda(u,dy) \text{ and}$$

$$\zeta_u := \mathbb{S}(u)\xi_1(u) + \int_E \Upsilon(u,y)\xi_2(u,y)\frac{\varphi(u,y)}{\rho(u,y)}\lambda(u,dy).$$

Since $\mathbb{S}(u)$ is a $n \times n$ invertible matrix for each u and both $\varphi(u,y)$ and $\rho(u,y)$ are strictly positive, $\mathfrak{R}_u$ is a symmetric matrix which is strictly positive. We have the following theorem

Theorem 2.2. *Let Assumptions (H) and 2.1 be satisfied and let (Y, a, ρ) be the bounded solution of the BSE (6). For a given contingent claim $H \in L^2(\mathcal{F}_{T_0}, P)$, let (h, ξ_1, ξ_2) be a solution of the BSE (7), let*

$$\pi^*(u) := \mathfrak{R}_u^{-1}\zeta_u + (h_{u-} - V_{u-}^{x,\pi^*})a(u), \tag{8}$$

then π^* is the optimal strategy of Problem (MHV-n). The optimal cost is given by

$$J = Y_0^{-1}(h_0 - x)^2$$
$$+ E\int_0^{T_0} Y_{u-}^{-1}\left\{\|\xi_1(u)\|^2 + \int_E \xi_2(u,y)^2\frac{\varphi(u,y)}{\rho(u,y)}\lambda(u,dy) - \zeta_u'\mathfrak{R}_u^{-1}\zeta_u\right\}du.$$

Remark. We let $V_t^* := V_t^{x,\pi^*}$ be the wealth process corresponding to (x, π^*), one can see that $h - V^*$ is the solution of the following stochastic differential equation (SDE)

$$h_t - V_t^* = h_0 - x + \mathcal{R}_t - \int_0^t (h_{u-} - V_{u-}^*)d\tilde{M}_u, \tag{9}$$

where

$$\mathcal{R}_t := \int_0^t \left\{\xi_1(u)' - \zeta_u'\mathfrak{R}_u^{-1}\mathbb{S}(u)\right\}dW^*(u)$$
$$+ \int_0^t \int_E \left\{\xi_2(u,y) - \zeta_u'\mathfrak{R}_u^{-1}\Upsilon(u,y)\right\}(\mu(du,dy) - \nu^*(du,dy)),$$
$$\tilde{M}_t := -\int_0^t a(u)'\mathbb{S}(u)dW^*(u) - \int_0^t a(u)'\Upsilon(u,y)(\mu(du,dy) - \nu^*(du,dy))$$

and where

$$W^*(t) = W(t) - \int_0^t \{z_1^*(u) + \phi(u)\} du$$
$$\nu^*(du, dy) = \big(z_2^*(u, y) + 1\big)\varphi(u, y)\lambda(u, dy)du.$$

To solve the SDE (9), we let $\tilde{Z}_t := \mathcal{E}(\tilde{M})_t$, one can see that $\tilde{Z}_{\tau_n^*} = 0$ and $\tilde{Z}_- I_{[\![\tau_n^*, \tau_{n+1}^*[\![} \neq 0$ for each n. It follows from Theorem 6.8 of Jacod(1979)[26] that the solution of the SDE (9) can be written as

$$h - V^* = \sum_{n \geq 0} (h - V^*)^{(n)} I_{[\![\tau_n^*, \tau_{n+1}^*[\![},}$$

where

$$(h - V^*)^{(n)} = {}^n\tilde{Z} \bigg\{ \Delta \mathcal{R}_{\tau_n^*} + \frac{1}{{}^n\tilde{Z}_-} \cdot (\mathcal{R}^{\tau_{n+1}^*} - \mathcal{R}^{\tau_n^*})$$
$$+ \big(\frac{1}{{}^n\tilde{Z}_-} I_{[\![0, \tau_{n+1}^*[\![}\big) \cdot \Big[\mathcal{R}, \tilde{M}^{\tau_{n+1}^*} - \tilde{M}^{\tau_n^*}\Big] \bigg\}$$

and where ${}^n\tilde{Z} = \mathcal{E}(\tilde{M} - \tilde{M}^{\tau_n^*}) = \mathcal{E}(\tilde{M}^{\tau_{n+1}^*} - \tilde{M}^{\tau_n^*})$.

Proof of Theorem 2.2. First prove that

$$\begin{aligned}
Y_t^{-1}(h_t - V_t^{x,\pi})^2 &= Y_0^{-1}(h_0 - x)^2 + m_t \\
&+ \int_0^t Y_{u-}^{-1} \Big\| \xi_1(u) - \mathbb{S}(u)'\pi(u) + (h_{u-} - V_{u-}^{x,\pi})\mathbb{S}(u)'a(u) \Big\|^2 du \\
&+ \int_0^t \int_E Y_{u-}^{-1} \Big\{ \xi_2(u, y) - \pi(u)'\Upsilon(u, y) + (h_{u-} - V_{u-}^{x,\pi}) \Big\}^2 \frac{\varphi(u, y)}{\rho(u, y)}\lambda(u, dy)du \\
&+ \int_0^t \int_E 2Y_{u-}^{-1}(h_{u-} - V_{u-}^{x,\pi})\{\pi(u)'\Upsilon(u, y) - \xi_2(u, y)\}\{1 - a(u)'\Upsilon(u, y)\} \frac{\varphi(u, y)}{\rho(u, y)}\lambda(u, dy)du \\
&- \int_0^t \int_E Y_{u-}^{-1}(h_{u-} - V_{u-}^{x,\pi})^2 \Big\{ 1 - (a(u)'\Upsilon(u, y))^2 \Big\} \frac{\varphi(u, y)}{\rho(u, y)}\lambda(u, dy)du,
\end{aligned}$$

where

$$\begin{aligned}
m_t^\pi &= \int_0^t Y_{u-}^{-1} \Big\{ 2\big(h_{u-} - V_{u-}^{x,\pi}\big)\{\xi_1(u) - \mathbb{S}(u)'\pi(u)\}' \\
&\quad + \big(h_{u-} - V_{u-}^{x,\pi}\big)^2 \{\phi(u) - \mathcal{I}_u(a, \rho) + \mathbb{S}(u)'a(u)\}' \Big\} dW(u) \\
&+ \int_0^t \int_E Y_{u-}^{-1} \Big\{ \{\xi_2(u, y) - \pi(u)'\Upsilon(u, y) \\
&\quad + (h_{u-} - V_{u-}^{x,\pi})\}^2 \frac{\varphi(u, y)}{\rho(u, y)} - (h_{u-} - V_{u-}^{x,\pi})^2 \Big\} \{\mu(du, dy) - \nu(du, dy)\}
\end{aligned}$$

is a local martingale under P. Therefore,

$$Y_t^{-1}(h_t - V_t^{x,\pi})^2 = Y_0^{-1}(h_0 - x)^2 + m_t^\pi$$
$$+ \int_0^t Y_{u-}^{-1} \left\| \xi_1(u) - \mathcal{S}(u)'\pi(u) + (h_{u-} - V_{u-}^{x,\pi})\mathcal{S}(u)'a(u) \right\|^2 du$$
$$+ \int_0^t \int_E Y_{u-}^{-1} \left\{ \xi_2(u,y) - \Upsilon(u,y)'\pi(u) + (h_{u-} - V_{u-}^{x,\pi})\Upsilon(u,y)'a(u) \right\}^2 \frac{\varphi(u,y)}{\rho(u,y)} \lambda(u,dy)du$$

$$= Y_0^{-1}(h_0 - x)^2 + m_t^\pi$$
$$+ \int_0^t Y_{u-}^{-1} \left\{ \left(\pi(u) - \Re_u^{-1}\zeta_u - (h_{u-} - V_{u-}^{x,\pi})a(u) \right)' \Re_u \left(\pi(u) - \Re_u^{-1}\zeta_u - (h_{u-} - V_{u-}^{x,\pi})a(u) \right) \right.$$
$$\left. + \|\xi_1(u)\|^2 + \int_E \xi_2(u,y)^2 \frac{\varphi(u,y)}{\rho(u,y)} \lambda(u,dy) - \zeta_u' \Re_u^{-1}\zeta_u \right\} du.$$

As $\pi \in \mathcal{A}(T_1, \cdots, T_n)$, we derive that $V^{x,\pi}$ is an $\mathcal{E}^*$-local martingale with $V_{T_0}^{x,\pi} \in L^2(P)$, thus $V^{x,\pi}$ is in fact an $\mathcal{E}^*$-martingale. It follows from Theorem 4.1 (ii) of Choulli-Krawczyk-Stricker(1998)[13] that

$$\left\| \max_{t \in [0,T_0]} |V_t^{x,\pi}| \right\|_2 \leq \|V_{T_0}^{x,\pi}\|_2 < \infty.$$

Similarly, one can see that

$$\left\| \max_{t \in [0,T_0]} |h_t| \right\|_2 \leq \|H\|_2 < \infty.$$

Since Y is a bounded process with $0 < c_1 \leq Y \leq c_2 < \infty$ for some constants c_1 and c_2, one can see that $Y^{-1}(h - V^{x,\pi})^2$ is a submartingale which is belongs to the class (D), thus m is a uniformly integrable martingale and

$$\int_0^t Y_{u-}^{-1} \left\{ \left(\pi(u) - \Re_u^{-1}\zeta_u - (h_{u-} - V_{u-}^{x,\pi})a(u) \right)' \Re_u \left(\pi(u) - \Re_u^{-1}\zeta_u - (h_{u-} - V_{u-}^{x,\pi})a(u) \right) \right.$$
$$\left. + \|\xi_1(u)\|^2 + \int_E \xi_2(u,y)^2 \frac{\varphi(u,y)}{\rho(u,y)} \lambda(u,dy) - \zeta_u' \Re_u^{-1}\zeta_u \right\} du$$

is an integrable increasing process. Therefore,

$$E(h_{T_0} - V_{T_0}^{x,\pi})^2 = E Y_{T_0}^{-1}(h_{T_0} - V_{T_0}^{x,\pi})^2$$
$$= Y_0^{-1}(h_0 - x)^2$$
$$+ E \int_0^{T_0} Y_{u-}^{-1} \left\{ \left(\pi(u) - \Re_u^{-1}\zeta_u - (h_{u-} - V_{u-}^{x,\pi})a(u) \right)' \Re_u \left(\pi(u) - \Re_u^{-1}\zeta_u - (h_{u-} - V_{u-}^{x,\pi})a(u) \right) \right.$$
$$\left. + \|\xi_1(u)\|^2 + \int_E \xi_2(u,y)^2 \frac{\varphi(u,y)}{\rho(u,y)} \lambda(u,dy) - \zeta_u' \Re_u^{-1}\zeta_u \right\} du.$$

From the definition of π^* in (8), one can see that

$$Y_t^{-1}(h_t - V_t^{x,\pi^*})^2 = Y_0^{-1}(h_0 - x)^2 + m_t^{\pi^*}$$
$$+ \int_0^t Y_{u-}^{-1} \left\{ \|\xi_1(u)\|^2 + \int_E \xi_2(u,y)^2 \frac{\varphi(u,y)}{\rho(u,y)} \lambda(u,dy) - \zeta_u' \Re_u^{-1}\zeta_u \right\} du.$$

Since m^{π^*} is only a local martingale, there exists a sequence of stopping times (τ_n) with $\uparrow \lim_{n \to \infty} \tau_n = T_0$ such that $\{m_{t \wedge \tau_n}^{\pi^*}; t \in [0, T_0]\}$ is a uniformly integrable martingale for each n. Since $Y\left(h - V^{x, \pi^*}\right)^2$ is left continuous at T_0, one can see from Fatou's lemma that

$$
\begin{aligned}
E\left[\left(h_{T_0} - V_{T_0}^{x, \pi^*}\right)^2\right] &= E\left[Y_{T_0}\left(h_{T_0} - V_{T_0}^{x, \pi^*}\right)^2\right] = E\left[\liminf_{n \to \infty} Y_{\tau_n}\left(h_{\tau_n} - V_{\tau_n}^{x, \pi^*}\right)^2\right] \\
&\leq \liminf_{n \to \infty} E\left[Y_{\tau_n}\left(h_{\tau_n} - V_{\tau_n}^{x, \pi^*}\right)^2\right] \\
&= Y_0^{-1}(h_0 - x)^2 + \liminf_{n \to \infty} E\left[\int_0^{\tau_n} Y_{u-}^{-1}\left\{\|\xi_1(u)\|^2 + \int_E \xi_2(u, y)^2 \frac{\varphi(u, y)}{\rho(u, y)} \lambda(u, dy) - \zeta_u' \Re_u^{-1} \zeta_u\right\} du\right] \\
&\leq Y_0^{-1}(h_0 - x)^2 + E\left[\int_0^{T_0} Y_{u-}^{-1}\left\{\|\xi_1(u)\|^2 + \int_E \xi_2(u, y)^2 \frac{\varphi(u, y)}{\rho(u, y)} \lambda(u, dy) - \zeta_u' \Re_u^{-1} \zeta_u\right\} du\right] \\
&\leq E\left[\left(h_{T_0} - V_{T_0}^{x, \pi}\right)^2\right],
\end{aligned}
$$

from which we can see that $\pi^* \in \mathcal{A}(T_1, \cdots, T_n)$ and π^* is the optimal strategy. $\qquad\square$

2.3. *The Case of not Invertible* $\mathbb{S}$

In this section, we consider the mean-variance hedging when $\sigma(u, T)$ is independent of the maturity time T, i.e., $\sigma(u, T) \equiv \sigma(u)$, where $\{\sigma(u); u \in [0, T^*]\}$ is a bounded $\mathbb{R}^n$-valued predictable process, and $\delta(u, y, T)$ is of the following form

$$
\delta(u, y, T) = -\frac{\tilde{\delta}(u, y)}{1 + \tilde{\delta}(u, y)(T - u)}
$$

for a bounded $\widetilde{\mathcal{P}}$-measurable function $\tilde{\delta}(u, y)$ with $\tilde{\delta}(u, y) > -\frac{1}{T^*}$. One can see that $S(u, T) = -\sigma(u)(T - u)$ and $e^{D(u, y, T)} - 1 = \tilde{\delta}(u, y)(T - u)$, thus the discounted bond price maturing at T is given by

$$
\bar{p}(t, T) = \bar{p}(0, T) \mathcal{E}\left\{-\int_0^\cdot \sigma(u)'(T - u) d\tilde{W}(u) \right.
$$
$$
\left. + \int_0^\cdot \int_E \tilde{\delta}(u, y)(T - u)(\mu(du, dy) - \tilde{\nu}(du, dy))\right\}_t.
$$
$$
\tag{10}
$$

Fix a $T_0 \in (0, T^*)$ and given any integer n_0, let $T_1, \cdots, T_{n_0}$ be n_0 maturities with $T_0 < T_i < T^*$ for each i and $T_1 < T_2 < \cdots < T_{n_0}$. Given a contingent claim $H \in L^2(\mathcal{F}_{T_0})$, we consider the mean-variance hedging problem of H by a self-financing portfolio based on $\bar{p}(t, T_1), \ldots, \bar{p}(t, T_{n_0})$. From the following Lemma, one can see that to hedge H, we only need to trade the bond maturing at T_1:

Lemma 2.6. *Assume that the discounted bond price $\bar{p}(t,T)$ is given by (10), then for any*

$$\pi = (\pi_1, \cdots, \pi_{n_0})' \in \mathcal{A}(T_1, \cdots, T_{n_0})$$

there exists $\hat{\pi} \in \mathcal{A}(T_1)$ such that for any $t \in [0, T_0]$

$$V_t^{x,\pi} = V_t^{x,\hat{\pi}}, \quad a.s.,$$

thus for a given contingent claim $H \in L^2(\mathcal{F}_{T_0}, P)$,

$$\min_{\pi \in \mathcal{A}(T_1, \cdots, T_{n_0})} E\left[\left(H - V_{T_0}^{x,\pi} \right)^2 \right] = \min_{\pi \in \mathcal{A}(T_1)} E\left[\left(H - V_{T_0}^{x,\pi} \right)^2 \right].$$

Proof. One can see from (1) that

$$
\begin{aligned}
V_t^{x,\pi} &= x - \int_0^t \sum_{j=1}^{n_0} \pi_j(u)(T_j - u)\sigma(u)' d\tilde{W}(u) \\
&\quad + \int_0^t \int_E \sum_{j=1}^{n_0} \pi_j(u)(T_j - u)\tilde{\delta}(u,y)\big(\mu(du,dy) - \tilde{\nu}(du,dy)\big) \\
&= x - \int_0^t \hat{\pi}(u)(T_1 - u)\sigma(u)' d\tilde{W}(u) \\
&\quad + \int_0^t \int_E \hat{\pi}(u)(T_1 - u)\tilde{\delta}(u,y)\big(\mu(du,dy) - \tilde{\nu}(du,dy)\big) \\
&= V_t^{x,\hat{\pi}}
\end{aligned}
$$

where $\hat{\pi}(u) = \sum_{j=1}^{n_0} \pi_j(u) \dfrac{T_j - u}{T_1 - u}$ belongs to $\mathcal{A}(T_1)$. $\qquad\square$

Thus it is natural to consider the following problem

Problem (MHV-1): $\qquad J = \min_{\pi \in \mathcal{A}(T_1)} E\left[\left(H - V_{T_0}^{x,\pi} \right)^2 \right].$

We now denote $\widehat{\mathcal{Z}_{\exp}^{2,\tau}}$ for the family of (z_1, z_2) satisfying (6) with $E\left\{ \left(Z_{\tau,T_0}^0 \mathcal{E}^\tau(z_1, z_2)_{T_0} \right)^2 \right\} < \infty$, and consider the following kind of minimization problem

$$\Psi(\tau) := \operatorname*{ess\,inf}_{(z_1, z_2) \in \mathcal{Z}_{\exp}^{2,\tau}} E\left[\left(Z_{\tau,T_0}^0 \mathcal{E}^\tau(z_1, z_2)_{T_0} \right)^2 \Big| \mathcal{F}_\tau \right]. \tag{11}$$

Note that the adjustment process $\tilde{a}$ in this case is only a 1-dimensional process. Instead of Assumption 2.1, we here directly assume that

Assumption 2.2. *There exists a pair* (z_1^*, z_2^*) *satisfying* (2) *such that for all* $\tau \in \mathcal{T}_{T_0}$, $(z_1^*, z_2^*) \in \widehat{\mathcal{Z}_{\exp}^{2,\tau}}$ *and*

$$\Psi(\tau) = E_{Q^0}\left[\left. Z_{\tau,T_0}^0\left(\mathcal{E}^\tau(z_1^*, z_2^*)_{T_0}\right)^2 \right| \mathcal{F}_\tau\right], \quad Q^0\text{-}a.s.,$$

and

$$M_t^* = \int_0^t z_1^*(u)' d\tilde{W}(u) + \int_0^t \int_E z_2^*(u, y)(\mu(du, dy) - \tilde{\nu}(du, dy))$$

is again called the **variance-optimal martingale**.

To describe the variance-optimal martingale, we consider the following BSE which is similar to (6)

$$
\begin{cases}
Y_t = Y_0 - \displaystyle\int_0^t Y_{u-}\{\phi(u) + \hat{z}_1(u) - a(u)\sigma(u)\}' dW(u) \\
\quad + \displaystyle\int_0^t \int_E Y_{u-}\left\{\frac{\varphi(u, y)}{\rho(u, y)}\left(1 - a(u)\tilde{\delta}(u, y)\right)^2 + 2a(u)\tilde{\delta}(u, y)\varphi(u, y) - 1\right\}\lambda(u, dy)du \\
\quad + \displaystyle\int_0^t Y_{u-}\left\{\|\phi(u) + \hat{z}_1(u)\|^2 - 2a(u)\sigma(u)'\phi(u)\right\}du \\
\quad + \displaystyle\int_0^t \int_E Y_{u-}\left\{\frac{\rho(u, y)}{\varphi(u, y)} - 1\right\}\mu(du, dy), \\
Y_{T_0} = 1.
\end{cases}
$$

$$(12)$$

A **bounded solution** of the BSE (12) is a tuple $(Y, a, \hat{z}_1, \rho)$ satisfying (12) such that

(1) a is an $\mathbb{R}^n$-valued predictable process such that for all stopping times $\tau \in \mathcal{T}_{T_0}$,

$$E\left[\mathcal{E}\left(-\int \frac{a(u)\mathbb{I}_{]\tau,T_0]}(u)}{\bar{p}(u-, T_1)(T_1 - u)} d\bar{p}(u, T_1)\right)_{T_0}^2\right] < \infty;$$

(2) $\hat{z}_1$ is an $\mathbb{R}^n$-valued W-integrable predictable process and $\rho(u, y)$ is a strictly positive $\widetilde{\mathcal{P}}$-measurable function such that for any $\tau \in \mathcal{T}_{T_0}$

$$(\hat{z}_1, \hat{z}_2) \in \widehat{\mathcal{Z}_{\exp}^{2,\tau}}$$

where $\hat{z}_2(u, y) := \dfrac{1 - a(u)\tilde{\delta}(u, y)}{\rho(u, y)} - 1$;

(3) Y is a bounded RCLL semimartingale with $0 < c_1 \leq Y \leq c_2 < \infty$ for some constants c_1 and c_2.

Similar to Lemma 2.4 and Theorem 2.1, we can show the following theorem

Theorem 2.3. *1) Under Assumption 2.2, the BSE (12) has a bounded solution;*

2) If the BSE (12) has a bounded solution denoted by $(Y, a, \hat{z}_1, \rho)$, then $Y_t = \Psi(t)$, a.s., for all $t \in [0, T_0]$ and

$$M_t^* := \int_0^t \hat{z}_1(u)' d\tilde{W}(u) + \int_0^t \int_E \hat{z}_2(u, y)(\mu(du, dy) - \tilde{\nu}(du, dy))$$

is the variance-optimal martingale.

We assume that the BSE (12) has a bounded solution denoted by $(Y, a, \hat{z}_1, \rho)$, let

$$M_t^* := \int_0^t \hat{z}_1(u)' d\tilde{W}(u) + \int_0^t \int_E \hat{z}_2(u, y)(\mu(du, dy) - \tilde{\nu}(du, dy))$$

be the variance-optimal martingale. In a similar way as above we can define $\mathcal{E}^*$-martingales and $\mathcal{E}^*$-local martingales in this case and show that the following BSE always has a solution for any given $H \in L^2(\mathcal{F}_{T_0}, P)$:

$$\begin{cases} h_t = h_0 + \int_0^t \xi_1(u)' dW(u) + \int_0^t \int_E \xi_2(u, y)\mu(du, dy) \\ \qquad - \int_0^t \xi_1(u)' \{\hat{z}_1(u) + \phi(u)\} du - \int_0^t \int_E \xi_2(u, y)(\hat{z}_2(u, y) + 1)\varphi(u, y)\lambda(u, dy) du, \\ h_T = H \end{cases}$$

(13)

has a solution (h, ξ_1, ξ_2).

Let

$$\hat{\Re}_u = \|\sigma(u)\|^2 + \int_E \frac{\varphi(u, y)}{\rho(u, y)} \tilde{\delta}(u, y)^2 \lambda(u, dy) \quad \text{and}$$

$$\hat{\zeta}_u = -\sigma(u)' \xi_1(u) + \int_E \frac{\varphi(u, y)}{\rho(u, y)} \tilde{\delta}(u, y) \xi_2(u, y)\lambda(u, dy).$$

In a similar way as in Theorem 2.2, we can derive the following theorem

Theorem 2.4. *Given a bounded $\mathbb{R}^n$-valued predictable process $\{\sigma(u); u \in [0, T^*]\}$ and a bounded $\tilde{\mathcal{P}}$-measurable function $\tilde{\delta}(u, y)$ with $\tilde{\delta}(u, y) > -\dfrac{1}{T^*}$ satisfying $\hat{\Re}_u \geq c > 0$ for some positive constant c, assume that the discounted bond price maturing at T is given by (10). Let Assumptions 12 be satisfied and let (Y, a, ρ) be the bounded solution of the BSE (12). For a given contingent claim $H \in L^2(\mathcal{F}_{T_0}, P)$, let (h, ξ_1, ξ_2) be a solution of the BSE (13), let*

$$\hat{\pi}^*(u) := \frac{1}{T_1 - u} \left\{ \frac{\hat{\zeta}_u}{\hat{\Re}_u} + (h_{u-} - V_{u-}^{x, \pi^*})a(u) \right\},$$

(14)

then π^* is the optimal strategy for Problem (MHV-1). The optimal cost is then given by

$$J = Y_0^{-1}(h_0 - x)^2 + E \int_0^{T_0} Y_{u-}^{-1} \Big\{ \|\xi_1(u)\|^2$$
$$+ \int_E \xi_2(u,y)^2 \frac{\varphi(u,y)}{\rho(u,y)} \lambda(u,dy) - \frac{\hat{\zeta}_u^2}{\hat{\mathfrak{R}}_u} \Big\} du \,.$$

Remark. The reader should note that in Lemma (2.6) T_1 can be replaced by any T_i, $i \in \{1, \cdots, n_0\}$. So the agent is free to choose the bond with any admissible time of maturity. As a consequence, it is not surprising that the optimal cost is independent of the chosen maturity time what is directly seen from Theorem 2.4. A similar observation in a simpler market setting is already made in Proposition 4.5 in Björk (1997)[5]. We should however note that this is somewhat contradictory to observations in reality: There seems to be an extra risk in hedging a claim with exercise time one month with a bond expiring in ten years (see (17)).

3. Optimal Exponential Utility in Jump Bond Markets

Still we consider the continuous-time bond market with jumps driven by an n-dimensional Brownian motion $W = (W_1, \cdots, W_n)'$ as well as an integer-valued random measure $\mu(du, dy)$ on $\mathbb{R}_+ \times \mathbb{R}$ with compensator $\nu(du, dy) = \lambda_u(dy)du$ as in section 1. However we need some few more assumptions. The main difficulty arises from the fact stated in 7 that "in the continuous-time bond market model there is naturally a continuum of basic traded securities (zero-coupon bonds parameterized by their maturities) while in the standard model of stock market there is normally only a finite number of securities." So, again, we first construct an arbitrage-free market with given parameters $(\sigma, \delta, \phi, \varphi)$ such that

(i) $\sigma(u,T)$ is a bounded $\mathbb{R}^n$-valued predictable process on $[0,T]$, and $\delta(u,y,T)$ is a bounded $\widetilde{\mathcal{P}}$-measurable function on $[0,T] \times E$ for every fixed $T \in [0, T^*]$; We assume that both of $\sigma(u,T)$ and $\delta(u,y,T)$ are continuously differential in the T-variable, let $D(u,y,T) := -\int_u^T \delta(u,y,l)dl$;

(ii) ϕ is a bounded $\mathbb{R}^n$-valued predictable process on $[0, T^*]$;

(iii) φ is a strictly positive $\widetilde{\mathcal{P}}$-measurable function on $[0, T^*] \times E$ with $0 < k_1 \le \varphi(u,y) \le k_2 < \infty$ for two constants k_1 and k_2.

Again, let

$$f(t,T) = f(0,T) + \int_0^t \Psi(u,T;\sigma,\delta,\phi,\varphi)du$$
$$+ \int_0^t \sigma(u,T)'dW(u) + \int_0^t \int_E \delta(u,y,T)\mu(du,dy), \tag{15}$$

where

$$\Psi(u,T;\sigma,\delta,\phi,\varphi) := \sigma(u,T)' \int_u^T \sigma(u,s)ds - \sigma(u,T)'\phi(u)$$
$$- \int_E \delta(u,y,T)e^{D(u,y,T)}\varphi(u,y)\lambda_u(dy). \tag{16}$$

The bond market is arbitrage-free, since Q^0 defined by $\frac{dQ^0}{dP}\big|_{\mathcal{F}} := Z_{T^*}^0$ is a equivalent martingale measure, where

$$Z_t^0 = 1 + \int_0^t Z_{u-}^0 \phi(u)'dW(u) + \int_0^t \int_E Z_{u-}^0 \left(\varphi(u,y) - 1\right)\left(\mu(du,dy) - \nu(du,dy)\right),$$

is a strictly positive uniformly integrable martingale on $[0,T^*]$, i.e., for each maturity T, the discounted bond price process $\bar{p}(t,T) = e^{-\int_0^t r_u du}p(t,T)$ $(:= e^{-\int_0^t r_u du}e^{-\int_t^T f(t,s)ds})$ is a local martingale on $[0,T]$ under Q^0. Of course, there may be many other equivalent martingale measure. In this paper, we consider the so called 'minimal entropy martingale' (the density process of the minimal entropy martingale measure) based on the bonds maturing in $[T_0, T^*]$ and then consider the following exp-utility $U(x) = -\exp\{-k_0 x\}$ optimization problem:

$$\text{Problem (exp-utility opt):} \qquad \sup_{X \in \mathcal{W}(x)} EU(X_{T_0})$$

where $\mathcal{W}(x)$ is the set of all admissible self-financing approximate wealth processes based on the bonds maturing in $[T_0, T^*]$ with initial wealth x. Then we consider the dynamic exp utility indifference value process for a given $H \in L^\infty(\mathcal{F}_{T_0})$, which is defined as a bounded RCLL semimartingale $C(H) = \{C_t(H); t \in [0, T_0]\}$ satisfying

$$\operatorname*{ess\,sup}_{X \in \mathcal{W}} E\left(U(X_{T_0} - X_t)\big|\mathcal{F}_t\right) = \operatorname*{ess\,sup}_{X \in \mathcal{W}} E\left(U\left(C_t(H) + X_{T_0} - X_t - H\right)\big|\mathcal{F}_t\right)$$

(see Mania and Schweizer(2005)[38]). Similar problems have been considered in Becher(2006)[3], Mania and Schweizer(2005)[38], or Morlais(2008)[40]. The main difference between this paper and those papers is that we have a continuum of traded securities and the wealth process is described by an $(\mathbb{M}, Q^0)$-normalized martingale.

The minimal entropy martingale measure (MEMM) has been investigated in various settings by several authors among them: Delbaen et al. (2002)[22], Frittelli (2000)[23], Fujiwara and Miyahara (2003)[24], Kohlmann and Xiong (2008)[33], Mania, Santacroce and Tevzadze (2003)[37], Mania and Schweizer (2005)[38], etc. But in the bond market, as the system of bonds maturing in $[T_0, T^*]$ is a continuum of securities, it is tempting to consider portfolios given by measures on maturities as in [6] or [7]. A crucial difficulty then arises from the fact that this set of portfolios is not complete. So any limiting procedure will make us leave a given set of admissible strategies and corresponding wealths, that is, by taking limits we always arrive at only approximate values. So, in a natural setting the notions of approximate completeness, approximately attainable claim, approximate wealth etc are found.

De Donno and Pratelli(2004)[19] consider the theory of cylindrical integration and the normalized integral, which was first introduced in Mikulevicius and Rozovskii(1998)[39], to generalize the concept of measure-valued strategies as measure valued portfolios are not sufficient to describe all financial portfolios (see the discussion in De Donno and Pratelli(2004)[19] and De Donno and Pratelli(2005)[20]). In this paper, similar to De Donno and Pratelli(2004)[19], we introduce the concept of '$(\mathbb{M}, Q^0)$-normalized martingale' and view the local $(\mathbb{M}, Q^0)$-normalized martingale as an approximate wealth process. As a main contribution to the mathematical theory we add to the technical results in De Donno and Pratelli(2004)[19] a representation for a given $(\mathbb{M}, Q^0)$-normalized martingale and describe the dynamics of the minimal entropy martingale (MEM) based on the bonds maturing in $[T_0, T^*]$ by a backward semimartingale equation (BSE) based on an $(\mathbb{M}, Q^0)$-normalized martingale, see Section 3. Then we give an example in Section 4, in which we can give the explicit solution to the BSE, and in Section 5, we give the explicit form of the optimal approximate wealth for the Problem (exp-utility opt) by using the solution of the BSE. Similar problems are considered in De Donno and Pratelli(2005)[20] in a different setting based on generalized strategies. Here we make direct use of the normalized martingales instead which turns out to be advantageous in hedging problems.

Finally, in Section 6 we discuss the exp utility indifference valuation (exp UIV) of a bounded contingent claim $H \in L^\infty(\mathcal{F}_{T_0})$, we introduce the concept of 'exp-orthogonality' and then describe the dynamics of the exp UIV of H by another BSE under the minimal entropy martingale measure Q^* in the incomplete market.

We are aware of the fact that from a practical point of view our results are not very satisfying. However, in our minds, the results might bring forward the general theory of bond markets which still needs a lot of research.

For a given maturity T_0, let $\mathbb{Z}$ be the set of strictly positive martingales on $[0, T_0]$ with $Z_0 = 1$ such that for any maturity $T \in [T_0, T^*]$, $Z_t \bar{p}(t, T)$ is a local martingale on $[0, T_0]$. Then we know that for any $Z \in \mathbb{Z}$, there exists a pair (z_1, z_2) satisfying (5) for all $T \in [T_0, T^*]$ and

$$
\begin{aligned}
Z_t = Z_t^{z_1, z_2} &:= Z_t^0 \widetilde{Z}_t^{z_1, z_2} \\
&= \mathcal{E} \bigg\{ \int_0^{\cdot} (\phi(u) + z_1(u))' dW(u) \\
&\quad + \int_0^{\cdot} \int_E \big\{ \varphi(u, y)(1 + z_2(u, y)) - 1 \big\} (\mu(du, dy) - \nu(du, dy)) \bigg\}_t .
\end{aligned}
$$

Definition. A martingale $\hat{Z} \in \mathbb{Z}$ is called the *minimal entropy martingale (MEM) based on* $[T_0, T^*]$ if it is the solution of the following optimization problem:

$$
\min_{Z \in \mathbb{Z}} E \left\{ Z_T \ln (Z_T) \right\}.
$$

Remark. One can easily see that $Z^0 \in \mathbb{Z}$ satisfies $E \left\{ Z_T^0 \ln (Z_T^0) \right\} < \infty$. Furthermore, since $D(u, y, T)$ is bounded for each fixed $T \in (0, T^*]$, one gets from (3) that $\bar{p}(\cdot, T)$ is a locally bounded process for each fixed T. From Theorem 2.1 and Theorem 2.2 of Frittelli(2000)[23] that **a MEM based on** $[T_0, T^*]$ **always exists!** In this paper, we will describe $\hat{Z}$ by a backward semimartingale equation based on the normalized martingale.

3.1. *The Minimal Entropy Martingale Based on* $[T_0, T^*]$

In this section, we will describe the dynamics of the MEM by a backward semimartingale equation. As usual, we denote by $\hat{Z}$ the minimal entropy martingale based on $[T_0, T^*]$, and as we have seen, there exists a pair (z_1^*, z_2^*) satisfying (5) for all $T \in [T_0, T^*]$ and

$$
\hat{Z}_t = Z_t^{z_1^*, z_2^*} := Z_t^0 \widetilde{Z}_t^{z_1^*, z_2^*} .
$$

As we do not want (z_1^*, z^*) be too 'strange' to be able to write $\widetilde{Z}^{z_1^*, z_2^*}$ as a stochastic exponential, we need the following assumption

Assumption 3.1. *Assume that* (z_1^*, z^*) *satisfy*

$$
\int_0^{T_0} \|z_1^*(u)\|^2 du + \int_0^{T_0} \int_E \varphi(u, y) z_2^*(u, y)^2 \lambda_u(dy) du < \infty, \quad Q^0\text{-a.s..}
$$

In other words,

$$\widetilde{Z}^{z_1^*, z_2^*} = \mathcal{E}\left(M^{z_1^*, z_2^*}\right),$$

where $M_t^{z_1^*, z_2^*} := \int_0^t z_1^*(u)' d\widetilde{W}(u) + \int_0^t \int_E z_2^*(u, y)(\mu(du, dy) - \tilde{\nu}(du, dy))$ *is a locally square integrable martingale under* Q^0.

It is natural to introduce

$$\mathbb{Z}_{\mathcal{E}nt} := \left\{ Z^{z_1, z_2} \, \middle| \, \begin{array}{l} Z^{z_1, z_2} \in \mathbb{Z} \text{ with } Z_0 = 1 \text{ and } E\left[Z_{T_0} \ln\left(Z_{T_0}\right)\right] < \infty \text{ and} \\ \int_0^{T_0} \|z_1(u)\|^2 du + \int_0^{T_0} \int_E \varphi(u, y) z_2(u, y)^2 \lambda_u(dy) du < \infty, \quad Q^0\text{-a.s.} \end{array} \right\}.$$

We have the following lemma

Lemma 3.1. Z^0 *belongs to* $\mathbb{Z}_{\mathcal{E}nt}$ *satisfies the reverse Hölder inequality* $R_{\mathcal{E}nt}(P)$:

$$E\left(Z_{\tau, T_0}^0 \ln\left(Z_{\tau, T_0}^0\right) \middle| \mathcal{F}_\tau\right) \le k < \infty, \quad a.s.$$

for all stopping times τ *with* $\tau \le T_0$, *where* $Z_{\tau, T_0}^0 = \dfrac{Z_{T_0}^0}{Z_\tau^0}$ *and* k *is a positive constant.*

Remark. Obviously, under Assumption 3.1 the MEM $\hat{Z} = Z^{z_1^*, z_2^*}$ is also the solution of the following optimization problem:

$$\text{Problem (MEM):} \quad \min_{Z^{z_1, z_2} \in \mathbb{Z}_{\mathcal{E}nt}} E\left\{Z_T^{z_1, z_2} \ln\left(Z_T^{z_1, z_2}\right)\right\}.$$

Note that now **we can trade all bonds with maturities in** $[T_0, T^*]$! Since there exists a continuum of basic traded securities, we introduce the cylindrical martingale and the normalized martingale which is related to the normalized integral in De Donno and Pratelli(2004)[19], which was first introduced in Mikulevicius-Rozovskii(1998)[39] as a mathematical tool without relation to financial problems.

3.2. *The Locally Square Integrable (LSI) Cylindrical Martingales and the Normalized Martingale*

We consider the family $\left\{(M^T)_{t \in [0, T_0]}; \, T \in [T_0, T^*]\right\}$ of square integrable martingale under Q^0, where M^T is given by

$$M_t^T := \int_0^t S(u, T)' d\widetilde{W}(u) + \int_0^t \int_E \left\{e^{D(u, y, T)} - 1\right\}(\mu(du, dy) - \tilde{\nu}(du, dy)),$$

one can see that for any $T_1, T_2 \in [T_0, T^*]$,

$$\left\langle M^{T_1}, M^{T_2} \right\rangle_t = \int_0^t \left\{ S(u, T_1)' S(u, T_2) + \int_E (e^{D(u,y,T_1)} - 1)(e^{D(u,y,T_2)} - 1)\varphi(u, y)\lambda_u(dy) \right\} du$$
$$= \int_0^t Q_u(T_1, T_2) du,$$

where

$$Q_u(T_1, T_2) = S(u, T_1)' S(u, T_2) + \int_E (e^{D(u,y,T_1)} - 1)(e^{D(u,y,T_2)} - 1)\varphi(u, y)\lambda_u(dy).$$

Let $\mathcal{C} = C([T_0, T^*])$, the space of all continuous function on $[T_0, T^*]$, then its topological dual $\mathcal{R}$ is the space of all Radon measures on $[T_0, T^*]$. Note that $(\mathcal{C}, \mathcal{R})$ is a Schwartz pair. We have the following lemma:

Lemma 3.2. *For any Radon measure $\ell \in \mathcal{R}$, let*

$$M_t^\ell := \int_0^t \ell(S(u, \cdot))' d\widetilde{W}(u) + \int_0^t \int_E \ell(e^{D(u,y,\cdot)} - 1)(\mu(du, dy) - \tilde{\nu}(du, dy))$$

where

$$\ell(S(u, \cdot)) := \int_{T_0}^{T^*} S(u, T)\ell(dT) = \left(\int_{T_0}^{T^*} S_1(u, T)\ell(dT), \cdots, \int_{T_0}^{T^*} S_n(u, T)\ell(dT) \right)',$$
$$\ell(e^{D(u,y,\cdot)} - 1) := \int_{T_0}^{T^*} \left\{ e^{D(u,y,T)} - 1 \right\} \ell(dT),$$

then the family of random processes $\mathbb{M} = \{M^\ell; \ell \in \mathcal{R}\}$ is a locally square integrable (LSI) cylindrical martingale with covariance operator function Q (see Mikulevicius and Rozovskii(1998)[39]), i.e., for any $\ell_1, \ell_2 \in \mathcal{R}$,

$$\left\langle M^{\ell_1}, M^{\ell_2} \right\rangle_t = \int_0^t \langle \ell_1, Q_u \ell_2 \rangle du$$

where $\langle \ell_1, Q_u \ell_2 \rangle = \ell_1(S(u, \cdot))' \ell_2(S(u, \cdot)) + \int_E \ell_1(e^{D(u,y,\cdot)} - 1)\ell_2(e^{D(u,y,\cdot)} - 1)\varphi(u, y)\lambda_u(dy).$

Similar to De Donno and Pratelli(2004)[19], we consider $\mathcal{R}$-valued predictable processes: An $\mathcal{R}$-valued predictable process $\ell = \{\ell_t(dT); t \in [0, T_0]\}$ is a process such that for each $f \in \mathcal{C}$, $\{ \int_{T_0}^{T^*} f(T)\ell_t(dT); t \in [0, T_0]\}$ is a predictable process. We denote by $L^2(\mathbb{M}, \mathcal{R})$ the set of $\mathcal{R}$-valued predictable processes ℓ satisfying

$$E_{Q^0}\left[\int_0^{T_0} \langle \ell_u, Q_u \ell_u \rangle du \right]$$
$$:= E_{Q^0}\left[\int_0^{T_0} \left\{ \|\ell_u(S(u, \cdot))\|^2 + \int_E \{\ell_u(e^{D(u,y,\cdot)} - 1)\}^2 \varphi(u, y)\lambda_u(dy) \right\} du \right] < \infty.$$

Note that different from De Donno and Pratelli(2004)[19], for any $\ell \in L^2(\mathbb{M}, \mathcal{R})$, we can define the stochastic integral directly by

$$\mathcal{I}_t(\ell) = \int_0^t \ell_u d\mathbb{M}(u) := \int_0^t \ell_u(S(u,\cdot))' d\widetilde{W}(u)$$
$$+ \int_0^t \int_E \ell_u(e^{D(u,y,\cdot)} - 1)(\mu(du, dy) - \tilde{\nu}(du, dy)),$$

and by stopping we can extend the stochastic integral $\int_0^t \ell_u d\mathbb{M}(u)$ to any $\ell \in L^2_{\text{loc}}(\mathbb{M}, \mathcal{R})$. As Mikulevicius and Rozovskii(1998)[39] pointed out, $L^2_{\text{loc}}(\mathbb{M}, \mathcal{R})$ is not complete. To overcome this we need to introduce the normalized martingale.

Definition. $M \in \mathcal{H}^2(Q^0)$ is called an $(\mathbb{M}, Q^0)$-**normalized martingale** if there exists a sequence $(\ell^n) \subset L^2(\mathbb{M}, \mathcal{R})$ such that

$$\lim_{n \to \infty} \|M - \mathcal{I}(\ell^n)\|_{\mathcal{H}^2(Q^0)} = 0,$$

and $M \in \mathcal{H}^2_{\text{loc}}(Q^0)$ is called a **local** $(\mathbb{M}, Q^0)$-**normalized martingale** if there exists a sequence of stopping times (τ_n) such that $M^{\tau_n} = \{M_{t \wedge \tau_n}; t \in [0, T_0]\}$ is a normalized Q^0-martingale for each n.

Remark. Mikulevicius and Rozovskii(1998)[39] show that every local $(\mathbb{M}, Q^0)$-normalized martingale can be viewed as the normalized integral of a predictable process with respect to the LSI cylindrical martingale $\mathbb{M} = \{M^\ell; \ell \in \mathcal{R}\}$ with values in the covariance spaces.

From the representation property of Q^0-martingales, we have the following proposition

Proposition 3.1. *Let M be an $(\mathbb{M}, Q^0)$-normalized martingale, then there exists an $\mathbb{R}^d$-valued predictable process $\{l_1(u); u \in [0, T_0]\}$ and a $\widetilde{\mathcal{P}}$-measurable function $l_2(u, y)$ and a sequence $(\ell^n) \subset L^2(\mathbb{M}, \mathcal{R})$ such that*

$$M_t = M_0 + M_t^{l_1, l_2} := M_0 + \int_0^t l_1(u) d\widetilde{W}(u) + \int_0^t \int_E l_2(u, y)\{\mu(du, dy) - \nu(du, dy)\}$$

and

$$\lim_{n \to \infty} E_{Q^0}\left[\int_0^{T_0} \left\{ (\ell_u^n(S(u,\cdot)) - l_1(u))^2 + \int_E \left(\ell_u^n(e^{D(u,y,\cdot)} - 1) - l_2(u, y)\right)^2 \varphi(u, y)\lambda_u(dy) \right\} du \right] = 0.$$

Proposition 3.2. *Let M be an $(\mathbb{M}, Q^0)$-normalized martingale, then M is a martingale under any EMM Q of the discounted bond price maturing at $[T_0, T^*]$ whose density process Z with respect to Q^0 belongs to $\mathcal{H}^2(Q^0)$.*

Proof. One proves that ZM is a uniformly integrable martingale under Q^0 and hence M is a uniformly integrable martingale under Q. $\square$

Corollary 3.1. *Let M^{l_1, l_2} be a local $(\mathbb{M}, Q^0)$-normalized martingale, then for any $Z^{z_1, z_2} \in \mathbb{Z}_{\mathcal{E}nt}$*

$$z_1(u)'l_1(u) + \int_E \varphi(u, y) z_2(u, y) l_2(u, y) \lambda_u(dy) = 0.$$

Proof. For any $Z^{z_1, z_2} \in \mathbb{Z}_{\mathcal{E}nt}$, one can see that Z^{z_1, z_2} can be rewritten as

$$Z_t^{z_1, z_2} = Z_t^0 \mathcal{E}\big(M^{z_1, z_2}\big)_t,$$

where

$$M_t^{z_1, z_2} := \int_0^t z_1(u)' d\widetilde{W}(u) + \int_0^t \int_E z_2(u, y)(\mu(du, dy) - \tilde{\nu}(du, dy)).$$

Since (z_1, z_2) satisfies

$$\int_0^{T_0} \|z_1(u)\|^2 du + \int_0^{T_0} \int_E \varphi(u, y) z_2(u, y)^2 \lambda_u(dy) du < \infty, \quad Q^0\text{-a.s.},$$

one can see that M^{z_1, z_2} is a locally square integrable martingale under Q^0, thus there exists a sequence of stopping times $(\tau_n)_{n \in \mathbb{N}}$ with $\tau_n \nearrow T_0$ as $n \to \infty$ such that for each n, $\left\{ M_{t \wedge \tau_n}^{l_1, l_2}; \ t \in [0, T_0] \right\}$ is a $(\mathbb{M}, Q^0)$-normalized martingale and $\left\{ M_{t \wedge \tau_n}^{z_1, z_2}; \ t \in [0, T_0] \right\}$ is a square integrable martingale under Q^0. Similar to the proof of Proposition 3.2, one can show that $\left\{ M_{t \wedge \tau_n}^{l_1, l_2} M_{t \wedge \tau_n}^{z_1, z_2}; \ t \in [0, T_0] \right\}$ is a uniformly integrable martingale under Q^0 for each n, thus $\left\{ M_t^{l_1, l_2} M_t^{z_1, z_2}; \ t \in [0, T_0] \right\}$ is a local martingale under Q^0 and

$$\left\langle M^{l_1, l_2}, M^{z_1, z_2} \right\rangle_t^{Q^0} = z_1(u)'l_1(u) + \int_E \varphi(u, y) z_2(u, y) l_2(u, y) \lambda_u(dy) = 0. \ \square$$

From Theorem 4.2 of De Donno and Pratelli[19], we may adapt the following corollary

Corollary 3.2. *If the martingale measure is unique, i.e. there is no nonzero pair (z_1, z_2) satisfying (5) for each $T \in [T_0, T^*]$, then the market is **approximately complete**, i.e., for every $\xi \in L^2(Q^0, \mathcal{F}_{T_0})$, there is an $(\mathbb{M}, Q^0)$-normalized martingale M such that $M_{T_0} = \xi$.*

Remark. Similar to De Donno and Pratelli[19], a **self-financing portfolio** is an $\mathcal{R}$-valued predictable process $\pi \in L^2_{\mathrm{loc}}(\mathbb{M}, \mathcal{R})$ and the corresponding **wealth process** is defined as

$$
\begin{aligned}
X^{x,\pi}_t &= x + \int_0^t \pi_u d\mathbb{M}_u \\
&= x + \int_0^t \pi_u (S(u, \cdot))' d\widetilde{W}(u) \\
&\quad + \int_0^t \int_E \pi_u (e^{D(u,y,\cdot)} - 1)(\mu(du, dy) - \tilde{\nu}(du, dy)),
\end{aligned}
$$

then every $(\mathbb{M}, Q^0)$-normalized martingale can be viewed as an **approximate wealth process**.

3.3. *The Dynamics of the MEM Based on $[T_0, T^*]$*

We now consider Problem (MEM) and we will describe the dynamics of the MEM based on $[T_0, T^*]$ by a backward semimartingale equation (BSE) based on a $(\mathbb{M}, Q^0)$-normalized martingale. We need to adapt the optimality principle (see e.g. Mania, Santacroce and Tevzadze(2003)[37]) as the following

Lemma 3.3. *1) There exists an RCLL semimartingale denoted by $V(t)$ such that for each $t \in [0, T_0]$*

$$
V(t) = \operatorname*{ess\,inf}_{Z \in \mathbb{Z}_{\mathcal{E}nt}} E\left(Z_{t,T_0} \ln(Z_{t,T_0}) \big| \mathcal{F}_t\right), \quad a.s.,
$$

which is the largest RCLL process with $V(T_0) = 0$ such that $\{Z_t\{V(t) + \ln(Z_t)\}; t \in [0, T_0]\}$ is a submartingale for each $Z \in \mathbb{Z}_{\mathcal{E}nt}$.

2) The following three properties are equivalent:

(i) $\hat{Z}$ is the minimal entropy martingale based on $[T_0, T^*]$;

(ii) $\hat{Z}$ is optimal for all conditional criteria, i.e., for each $t \in [0, T_0]$

$$
V(t) = E\left(\hat{Z}_{t,T_0} \ln(\hat{Z}_{t,T_0}) \big| \mathcal{F}_t\right), \quad a.s.;
$$

(iii) $\left\{\hat{Z}_t\left(V(t) + \ln(\hat{Z}_t)\right); t \in [0, T_0]\right\}$ is a uniformly integrable martingale.

We consider the following BSE

$$
\begin{cases}
a(t) = a(0) + \displaystyle\int_0^t \left\{ l_1^*(u)' - (\phi(u) + \theta_1(u))' \right\} dW(u) \\
\qquad + \displaystyle\int_0^t \left\{ \frac{1}{2} \|\phi(u) + \theta_1(u)\|^2 - l_1^*(u)'\phi(u) \right\} du \\
\qquad + \displaystyle\int_0^t \int_E \left\{ l_2^*(u,y) - \ln\left(\varphi(u,y)(1 + \theta_2(u,y))\right) \right\} \mu(du, dy) \\
\qquad + \displaystyle\int_0^t \int_E \left\{ \varphi(u,y)\left(1 + \theta_2(u,y) - l_2^*(u,y)\right) - 1 \right\} \nu(du, dy). \\
a(T_0) = 0 \, .
\end{cases}
$$

(17)

A **solution** of the BSE (17) is a 5-tuple $(a; l_1^*, l_2^*; \theta_1, \theta_2)$ satisfying (17) such that

(1) l_1^* is an $\mathbb{R}^d$-valued predictable process and l_2^* is a $\widetilde{\mathcal{P}}$-measurable function such that

$$
M_t^{l_1^*, l_2^*} := \int_0^t l_1^*(u)' d\widetilde{W}(u) + \int_0^t \int_E l_2^*(u,y)\{\mu(du, dy) - \tilde{\nu}(du, dy)\}
$$

is a **local** $(\mathbb{M}, Q^0)$**-normalized martingale**;

(2) θ_1 is an $\mathbb{R}^d$-valued predictable process, while θ_2 is a $\widetilde{\mathcal{P}}$-measurable function with $\theta_2(u, y) > -1$ such that $Z^{\theta_1, \theta_2} \in \mathbb{Z}_{\mathcal{E}nt}$;

(3) for any $Z \in \mathbb{Z}_{\mathcal{E}nt}$, $\{Z_t\{a(t) + \ln(Z_t)\}; \ t \in [0, T_0]\}$ belongs to the class (D).

Theorem 3.1. *Let Assumption 3.1 be satisfied, and assume that the BSE (17) has a solution denoted by $(a; l_1^*, l_2^*; \theta_1, \theta_2)$, then $V(t) = a(t)$, a.s. for every $t \in [0, T_0]$ and the MEM based on $[T_0, T^*]$ is given by Z^{θ_1, θ_2}.*

Proof. The idea is the same as the proof of Theorem 3.3 of Kohlmann and Xiong(2008)[33], but the computation is different. For any pair (z_1, z_2) such that

$Z^{z_1,z_2} \in \mathbb{Z}_{\varepsilon nt}$, by applying Itô's formula, one can see that

$$
\begin{aligned}
a(t) + \ln\left(Z_t^{z_1,z_2}\right) = a(0) &+ \int_0^t \left\{l_1^*(u) + z_1(u) - \theta_1(u)\right\}' dW(u) \\
&+ \int_0^t \left\{\frac{1}{2}\|\phi(u) + \theta_1(u)\|^2 - l_1^*(u)'\phi(u) - \frac{1}{2}\|\phi(u) + z_1(u)\|^2\right\} du \\
&+ \int_0^t \int_E \left\{l_2^*(u,y) - \ln\left(1 + \theta_2(u,y)\right) + \ln\left(1 + z_2(u,y)\right)\right\} \mu(du,dy) \\
&+ \int_0^t \int_E \varphi(u,y)\left(\theta_2(u,y) - l_2^*(u,y) - z_2(u,y)\right)\nu(du,dy),
\end{aligned}
$$

thus

$$
\begin{aligned}
Z_t^{z_1,z_2}&\left\{a(t) + \ln\left(Z_t^{z_1,z_2}\right)\right\} \\
= a(0) &+ \int_0^t Z_{u-}^{z_1,z_2}\left\{a(u-) + \ln\left(Z_{u-}^{z_1,z_2}\right)\right\}(\phi(u) + z_1(u))' dW(u) \\
&+ \int_0^t \int_E Z_{u-}^{z_1,z_2}\left\{a(u-) + \ln\left(Z_{u-}^{z_1,z_2}\right)\right\} \\
&\qquad\qquad \times\left\{\varphi(u,y)\left(1 + z_2(u,y)\right) - 1\right\}(\mu(du,dy) - \nu(du,dy)) \\[6pt]
&+ \int_0^t Z_{u-}^{z_1,z_2}\left\{l_1^*(u) + z_1(u) - \theta_1(u)\right\}' dW(u) \\
&+ \int_0^t Z_{u-}^{z_1,z_2}\left\{\frac{1}{2}\|\phi(u) + \theta_1(u)\|^2 - l_1^*(u)'\phi(u) - \frac{1}{2}\|\phi(u) + z_1(u)\|^2\right\} du \\
&+ \int_0^t \int_E Z_{u-}^{z_1,z_2}\left\{l_2^*(u,y) - \ln\left(1 + \theta_2(u,y)\right) + \ln\left(1 + z_2(u,y)\right)\right\} \mu(du,dy) \\
&+ \int_0^t \int_E Z_{u-}^{z_1,z_2}\varphi(u,y)\left(\theta_2(u,y) - l_2^*(u,y) - z_2(u,y)\right)\nu(du,dy) \\[6pt]
&+ \int_0^t Z_{u-}^{z_1,z_2}(\phi(u) + z_1(u))'\left\{l_1^*(u) + z_1(u) - \theta_1(u)\right\} du \\
&+ \int_0^t \int_E Z_{u-}^{z_1,z_2}\left\{\varphi(u,y)\left(1 + z_2(u,y)\right) - 1\right\} \\
&\qquad\qquad \times\left\{l_2^*(u,y) - \ln\left(1 + \theta_2(u,y)\right) + \ln\left(1 + z_2(u,y)\right)\right\} \mu(du,dy) \\[10pt]
= a(0) + m_t &+ \int_0^t Z_{u-}^{z_1,z_2}\left\{\frac{1}{2}\|\theta_1(u) - z_1(u)\|^2 + z_1(u)'l_1^*(u)\right\} du \\
&+ \int_0^t \int_E Z_{u-}^{z_1,z_2}\varphi(u,y)\Big\{z_2(u,y)l_2^*(u,y) + \theta_2(u,y) - z_2(u,y) \\
&\qquad\qquad +\left(1 + z_2(u,y)\right)\left\{\ln\left(1 + z_2(u,y)\right) - \ln\left(1 + \theta_2(u,y)\right)\right\}\Big\}\nu(du,dy)
\end{aligned}
$$

where

$$m_t := \int_0^t Z_{u-}^{z_1,z_2} \Big\{ \big\{a(u-) + \ln\big(Z_{u-}^{z_1,z_2}\big)\big\} (\phi(u) + z_1(u))'$$

$$+ \{l_1^*(u) + z_1(u) - \theta_1(u)\}' \Big\} dW(u)$$

$$+ \int_0^t \int_E Z_{u-}^{z_1,z_2} \Big\{ \big\{a(u-) + \ln\big(Z_{u-}^{z_1,z_2}\big)\big\} \{\varphi(u,y)\big(1 + z_2(u,y)\big) - 1\}$$

$$+ \varphi(u,y)\big(1 + z_2(u,y)\big)\{l_2^*(u,y) - \ln\big(1 + \theta_2(u,y)\big)$$

$$+ \ln\big(1 + z_2(u,y)\big)\} \Big\}(\mu(du,dy) - \nu(du,dy))$$

is a martingale under P. Since $M^{l_1^*,l_2^*}$ is a local $(\mathbb{M}, Q^0)$-normalized martingale, according to Corollary 3.1, one can see that

$$z_1(u)'l_1^*(u) + \int_E \varphi(u,y)z_2(u,y)l_2^*(u,y)\lambda_u(dy) = 0,$$

thus

$$Z_t^{z_1,z_2}\big\{a(t) + \ln\big(Z_t^{z_1,z_2}\big)\big\} = a(0) + m_t + \frac{1}{2}\int_0^t Z_{u-}^{z_1,z_2}\|\theta_1(u) - z_1(u)\|^2 du$$

$$+ \int_0^t \int_E Z_{u-}^{z_1,z_2}\varphi(u,y)\Big\{\theta_2(u,y) - z_2(u,y)$$

$$+ \big(1 + z_2(u,y)\big)\Big\{\ln\big(1 + z_2(u,y)\big) - \ln\big(1 + \theta_2(u,y)\big)\Big\}\Big\}\nu(du,dy)$$

is a local submartingale under P, since for all $z_2(u,y) > -1$,

$$\{\theta_2(u,y) - z_2(u,y)\} + (1 + z_2(u,y))\big\{\ln\big(1 + z_2(u,y)\big) - \ln\big(1 + \theta_2(u,y)\big)\big\} \geq 0$$

and equality holds when $z_2(u,y) = \theta_2(u,y)$. Also $\big\{Z_t^{\theta_1,\theta_2}\{a(t) + \ln\big(Z_t^{\theta_1,\theta_2}\big)\}$; $t \in [0,T_0]\big\}$ is a local martingale. By the definition of the solution, for any $Z^{z_1,z_2} \in \mathbb{Z}_{\mathcal{E}nt}$, $\{Z_t^{z_1,z_2}\{a(t) + \ln(Z_t^{z_1,z_2})\}$; $t \in [0,T_0]\}$ belongs to class (D), thus $\big\{Z_t^{l_1,l_2}\{a(t) + \ln\big(Z_t^{l_1,l_2}\big)\}$; $t \in [0,T_0]\big\}$ is a submartingale. From Lemma 3.3 we thus have

$$V_t \geq a(t), \quad a.s.$$

Furthermore, since $Z^{\theta_1,\theta_2} \in \mathbb{Z}_{\mathcal{E}nt}$ and $\big\{Z_t^{\theta_1,\theta_2}\{a(t) + \ln\big(Z_t^{\theta_1,\theta_2}\big)\}$; $t \in [0,T_0]\big\}$ is a uniformly integrable martingale,

$$Z_t^{\theta_1,\theta_2}\big\{a(t) + \ln\big(Z_t^{\theta_1,\theta_2}\big)\big\} = E\Big[Z_T^{\theta_1,\theta_2}\big\{a(T) + \ln\big(Z_T^{\theta_1,\theta_2}\big)\big\}\big|\mathcal{F}_t\Big]$$

$$= E\big[Z_T^{\theta_1,\theta_2}\ln\big(Z_T^{\theta_1,\theta_2}\big)\big|\mathcal{F}_t\big],$$

thus

$$a(t) = E\Big[Z_{t,T}^{\theta_1,\theta_2}\ln\big(Z_{t,T}^{\theta_1,\theta_2}\big)\big|\mathcal{F}_t\Big]$$

$$\geq \text{ess}\inf_{Z\in\mathbb{Z}_{\mathcal{E}nt}} E\big(Z_{t,T_0}\ln(Z_{t,T_0})\big|\mathcal{F}_t\big) = V(t),$$

thus $a(t) = V(t)$, a.s., and Z^{θ_1,θ_2} is the MEM based on $[T_0, T^*]$. $\square$

The following theorem improves Theorem 3.5 of Kohlmann and Xiong(2008)[33].

Theorem 3.2. *Let Assumption 3.1 be satisfied, then the BSE (17) has a solution if and only if there exists a $Z^{z_1,z_2} \in \mathbb{Z}_{\mathcal{E}nt}$ satisfying the reverse Hölder inequality $R_{\mathcal{E}nt}(P)$, which is of the following form*

$$Z_{T_0}^{z_1,z_2} = \exp\left\{c + M_{T_0}^{l_1,l_2}\right\}, \quad a.s.,$$

where M^{l_1,l_2} is a local $(\mathbb{M}, Q^0)$-normalized martingale such that

$$\left\{Z_t^{z_1,z_2} M_t^{l_1,l_2}; \ t \in [0, T_0]\right\}$$

is a uniformly integrable martingale.

Similar to the proof of Lemma 3.3 of Delbaen et al (2002)[22], we can show that the following corollary holds:

Corollary 3.3. *Let Assumption 3.1 be satisfied, and assume that the BSE (17) has a solution denoted by $(a; l_1^*, l_2^*; \theta_1, \theta_2)$, then for any $Z^{z_1,z_2} \in \mathbb{Z}_{\mathcal{E}nt}$,*

$$Z^{z_1,z_2} M^{l_1^*,l_2^*}$$

is a uniformly integrable martingale.

When the martingale measure is unique, i.e., there is no nonzero pair (z_1, z_2) satisfying (5) for each $T \in [T_0, T^*]$, by the boundedness of ϕ and φ, we see that $\ln(Z_{T_0}^0) \in L^2(Q^0, \mathcal{F}_{T_0})$, and from Corollary 3.2 that there exists an $(\mathbb{M}, Q^0)$-normalized martingale $M^{l_1^0,l_2^0}$ such that

$$M_{T_0}^{l_1^0,l_2^0} = \ln(Z_{T_0}^0).$$

Thus we have the following corollary:

Corollary 3.4. *If the martingale measure is unique, set*

$$a(t) = E_{Q^0}[\ln(Z_{t,T_0}^0)|\mathcal{F}_t],$$

then $(a; l_1^0, l_2^0; 0, 0)$ is the solution of the BSE (17).

Given n maturities $T_1, \cdots, T_n$ with $T_0 \le T_1 < T_2 < \cdots < T_n \le T^*$. Let $\mathbb{Z}^{T_1, \cdots, T_n}$ be the set of strictly positive martingale Z^{z_1,z_2} on $[0, T_0]$ with

$Z_0^{z_1,z_2} = 1$ such that for any maturity $T_i \in [T_0, T^*]$, $Z_t^{z_1,z_2} \bar{p}(t, T_i)$ is a local martingale on $[0, T_0]$. Similarly, we let

$$\mathbb{Z}_{\mathcal{E}nt}^{T_1,\cdots,T_n} := \left\{ Z^{z_1,z_2} \, \middle| \, \begin{array}{l} Z^{z_1,z_2} \in \mathbb{Z}^{T_1,\cdots,T_n} \text{ satisfies } E\left[Z_{T_0}^{z_1,z_2} \ln\left(Z_{T_0}^{z_1,z_2} \right) \right] < \infty \text{ and} \\ \int_0^{T_0} \|z_1(u)\|^2 du + \int_0^{T_0} \int_E \varphi(u,y) z_2(u,y)^2 \lambda_u(dy) du < \infty, \ Q^0\text{-a.s.} \end{array} \right\},$$

and consider the **minimal entropy martingale (MEM) based on** $T_1, \cdots, T_n$ which is defined as the solution of the following optimization problem:

$$\min_{Z \in \mathbb{Z}_{\mathcal{E}nt}^{T_1,\cdots,T_n}} E\left\{ Z_T \ln\left(Z_T \right) \right\}.$$

In general, the MEM based on $[T_0, T^*]$ may **not** be the MEM based on $T_1, \cdots, T_n$, however, we have the following corollary

Corollary 3.5. *Let Assumption 3.1 be satisfied. Assume that the BSE (17) has a solution denoted by $(a; l_1^*, l_2^*; \theta_1, \theta_2)$. If the local $(\mathbb{M}, Q^0)$-normalized martingale $M^{l_1^*, l_2^*}$ satisfies*

$$M_t^{l_1^*, l_2^*} = \int_0^t \ell_u^* d\mathbb{M}(u),$$

where $\ell_u^*(dT) = \sum_{j=1}^n l_u^j \delta_{T_j}(dT)$, and where l^j is a predictable process and $\delta_{T_j}(dT)$ is the Dirac measure at the point T_j for each j, then Z^{θ_1, θ_2} is the minimal entropy martingale based on $T_1, \cdots, T_n$.

This corollary allows to compare the results on bond markets with a finite number of maturities and the more general situation here.

3.4. An Example

In a situation which should be compared to (2.3), it is possible to give an example in which we can solve the BSE (17) explicitly. Assume that $S(u, T)$ is given by

$$S(u, T) = \mathfrak{s}_1(T - u) + \mathfrak{s}_2(T - u)^2$$

where $\mathfrak{s}_1$ and $\mathfrak{s}_2$ are two deterministic $\mathbb{R}^d$-valued vector with $\mathfrak{s}_1' \mathfrak{s}_2 = 0$, and $e^{D(u,y,T)} - 1$ is given by

$$e^{D(u,y,T)} - 1 = \mathfrak{d}(y)(T - u)$$

for some deterministic continuous function (y). To simplify the presentation, we again assume that ϕ is a deterministic $\mathbb{R}^d$-valued vector and $\varphi(y)$ is a deterministic bounded strictly positive continuous function on $\mathbb{R}$, while

$\lambda_u(dy) = \lambda(du)$ is a deterministic measure on $\mathbb{R}$ which is independent of u, thus $\nu(du, dy) = \lambda(dy)du$. Then the density process of Q^0 with respect to P is given by

$$Z_t^0 = \mathcal{E}\left\{ \phi' W + \int \int_{\mathbb{R}} (\varphi(y) - 1)\{\mu(du, dy) - \lambda(dy)du\} \right\}_t,$$

and under Q^0, the discounted price process of the bond maturing at T is given by

$$\bar{p}(t, T) = \bar{p}(0, T) + \int_0^t \bar{p}(u-, T) \left\{ \mathsf{s}_1'(T-u) + \mathsf{s}_2'(T-u)^2 \right\} d\widetilde{W}(u)$$
$$+ \int_0^t \int_{\mathbb{R}} \bar{p}(u-, T)\mathrm{d}(y)(T-u)\{\mu(du, dy) - \tilde{\nu}(du, dy)\}$$

where

$$\widetilde{W}(t) \quad = W(t) - \phi t$$
$$\tilde{\nu}(du, dy) = \varphi(y)\lambda(dy)du.$$

The following theorem can then be proved:

Theorem 3.3. *If the following deterministic equations*

$$\begin{cases} \|\mathsf{s}_1\|^2\gamma_1 + \mathsf{s}_1'\mathsf{s}_2\gamma_2 + \int_{\mathbb{R}} \mathrm{d}(y)e^{\mathrm{d}(y)\gamma_1}\lambda(dy) = \mathsf{s}_1'\phi + \int_{\mathbb{R}} \varphi(y)\mathrm{d}(y)\lambda(dy), \\ \mathsf{s}_1'\mathsf{s}_2\gamma_1 + \|\mathsf{s}_2\|^2\gamma_2 \qquad\qquad\qquad = \mathsf{s}_2'\phi, \end{cases}$$

(18)

has a solution denoted by (γ_1, γ_2), let

$$\begin{cases} l_1^*(u) \quad := \gamma_1\mathsf{s}_1 + \gamma_2\mathsf{s}_2, \\ l_2^*(u, y) := \gamma_1\mathrm{d}(y), \\ \theta_1(u) \quad := \gamma_1\mathsf{s}_1 + \gamma_2\mathsf{s}_2 - \phi, \\ \theta_2(u, y) := \dfrac{e^{\gamma_1\mathrm{d}(y)}}{\varphi(y)} - 1 \end{cases}$$

*and $a(0) := T_0 \times \left\{ \dfrac{1}{2}\|\gamma_1\mathsf{s}_1 + \gamma_2\mathsf{s}_2\|^2 + \int_{\mathbb{R}} \left\{ e^{\gamma_1\mathrm{d}(y)}(\gamma_1\mathrm{d}(y) - 1) + 1 \right\}\lambda(dy) \right\}$
and assume that a is given by*

$$a(t) = a(0) + \int_0^t \left\{ l_1^*(u)' - (\phi(u) + \theta_1(u))' \right\}dW(u)$$
$$+ \int_0^t \left\{ \frac{1}{2}\|\phi(u) + \theta_1(u)\|^2 - l_1^*(u)'\phi(u) \right\}du$$
$$+ \int_0^t \int_E \left\{ l_2^*(u, y) - \ln\left(\varphi(u, y)(1 + \theta_2(u, y))\right) \right\}\mu(du, dy)$$
$$+ \int_0^t \int_E \left\{ \varphi(u, y)\left(1 + \theta_2(u, y) - l_2^*(u, y)\right) - 1 \right\}\nu(du, dy),$$

then $(a; l_1^*, l_2^*; \theta_1, \theta_2)$ *is the solution of the BSE (17) and the minimal entropy martingale is given by* Z^{θ_1, θ_2}.

Proof. From Itô's formula applied to $\ln Z_t^{\theta_1, \theta_2}$ and the equations (18), we have

$$\ln Z_t^{\theta_1, \theta_2} = (\gamma_1 \mathbb{s}_1 + \gamma_2 \mathbb{s}_2)' \widetilde{W}(t) + \int_0^t \int_{\mathbb{R}} \gamma_1 \mathrm{d}(y) \{\mu(du, dy) - \tilde{\nu}(du, dy)\}$$
$$+ \left\{\frac{1}{2} \|\mathbb{s}_1 \gamma_1 + \gamma_2 \mathbb{s}_2\|^2 + \int_{\mathbb{R}} \left\{(\gamma_1 \mathrm{d}(y) - 1)e^{\gamma_1 \mathrm{d}(y)} + 1\right\} \lambda(dy)\right\} t$$

and hence

$$\ln Z_{T_0}^{\theta_1, \theta_2} = c + (\gamma_1 \mathbb{s}_1 + \gamma_2 \mathbb{s}_2)' \widetilde{W}(T_0) + \int_0^{T_0} \int_{\mathbb{R}} \gamma_1 \mathrm{d}(y) \{\mu(du, dy) - \tilde{\nu}(du, dy)\},$$

where

$$c = \left\{\frac{1}{2} \|\mathbb{s}_1 \gamma_1 + \gamma_2 \mathbb{s}_2\|^2 + \int_{\mathbb{R}} \left\{(\gamma_1 \mathrm{d}(y) - 1)e^{\gamma_1 \mathrm{d}(y)} + 1\right\} \lambda(dy)\right\} \times T_0.$$

Furthermore, trivially there exists an $\mathcal{R}$-valued function $\ell_u^*(dT)$ such that

$$\begin{cases} \gamma_1 = \int_{T_0}^{T^*} (T - u)\ell_u^*(dT), \\ \gamma_2 = \int_{T_0}^{T^*} (T - u)^2 \ell_u^*(dT), \end{cases} \tag{19}$$

thus $(\gamma_1 \mathbb{s}_1 + \gamma_2 \mathbb{s}_2)' \widetilde{W} + \int \int_{\mathbb{R}} \gamma_1 \mathrm{d}(y) \{\mu(du, dy) - \tilde{\nu}(du, dy)\}$ is a local $(\mathbb{M}, Q^0)$-normalized martingale, and from Theorem 3.2 we get that Theorem 3.3 holds. $\qquad \square$

3.5. *Optimal Exponential Utility*

In this section, we assume that the BSE (17) has a solution denoted by $(a; l_1, l_2; \theta_1, \theta_2)$. As an application, we now consider the optimal utility of an investor with exponential utility

$$U(x) = -\exp\{-k_0 x\},$$

where k_0 is a positive constant.

Definition. An **admissible self-financing approximate wealth process** X with the initial wealth $x > 0$ is a local $(\mathbb{M}, Q^0)$-normalized martingale with $X_0 = x$ such that XZ^{z_1, z_2} is a uniformly integrable martingale for any $Z^{z_1, z_2} \in \mathbb{Z}_{\mathcal{E}nt}$.

The set of all admissible self-financing approximate wealth processes with the initial wealth x is denoted by $\mathcal{W}(x)$.

We consider the following optimization problem:

$$\text{Problem (exp-utility opt):} \qquad \sup_{X \in \mathcal{W}(x)} EU(X_{T_0}).$$

As $U'(x) = k_0 \exp(-k_0 x)$ and $I(y) := (U')^{-1}(y)$ is given by

$$I(y) = -\frac{\ln y - \ln k_0}{k_0},$$

similar to Theorem 5.2 of Kohlmann and Xiong(2008)[33], we have the following theorem

Theorem 3.4. *Assume that the BSE (17) has a solution denoted by $(a; l_1^*, l_2^*; \theta_1, \theta_2)$, then the optimal exp-approximate wealth process (i.e., the solution of Problem* (exp-utility opt)) *is given by*

$$X_t^* = x - \frac{1}{k_0} M_t^{l_1^*, l_2^*}.$$

Proof. From Theorem 3.2 we get $X^* \in \mathcal{M}(x)$ and from the proof of Theorem 3.2, we derive that

$$\ln Z_{T_0}^{\theta_1, \theta_2} = a(0) + M_{T_0}^{l_1^*, l_2^*}.$$

Let $\lambda^* := k_0 e^{-k_0 x - a(0)}$, then one can see that

$$
\begin{aligned}
I(\lambda^* Z_{T_0}^{\theta_1, \theta_2}) &= -\frac{\ln(\lambda^* Z_{T_0}^{\theta_1, \theta_2}) - \ln k_0}{k_0} \\
&= -\frac{\ln(\lambda^*) + a(0) - \ln k_0}{k_0} - \frac{1}{k_0} M_{T_0}^{l_1, l_2} \\
&= x - \frac{1}{k_0} M_{T_0}^{l_1, l_2} = X_{T_0}^*.
\end{aligned}
$$

Thus for any $X \in \mathcal{M}(x)$,

$$
\begin{aligned}
E\{U(X_{T_0}) - U(X_{T_0}^*)\} &\le E\left\{ U'(X_{T_0}^*) \left(X_{T_0} - X_{T_0}^* \right) \right\} \\
&= E\left\{ U'(I(\lambda^* Z_{T_0}^{\theta_1, \theta_2})) \left(X_{T_0} - X_{T_0}^* \right) \right\} \\
&= E\left\{ \lambda^* Z_{T_0}^{\theta_1, \theta_2} \left(X_{T_0} - X_{T_0}^* \right) \right\} \\
&= \lambda^* (x - x) = 0,
\end{aligned}
$$

thus X^* is the solution of Problem (exp-utility opt). $\qquad\square$

From the discussion in the papers on this subject and from the short summary of this in the introduction it makes no sense to look for an admissible self-financing portfolio replicating X^*. Also the direct valuation problem of a contingent claim in this setting leads to problems with the intuitive economic interpretation. Thus we turn to the indifference valuation which really does make sense.

3.6. The Dynamic Exp Utility Indifference Valuation

We now consider the dynamic exp utility indifference valuation (UIV) for a given bounded contingent claim $H \in L^\infty(\mathcal{F}_{T_0})$. Let $\mathcal{W}$ be the set of all local $(\mathbb{M}, Q^0)$-normalized martingales X such that XZ^{z_1, z_2} is a uniformly integrable martingale for any $Z^{z_1, z_2} \in \mathbb{Z}_{\mathcal{E}nt}$, and let $\mathcal{W}^b := \{X \in \mathcal{W}; X$ is bounded $\}$. Similar to Mania and Schweizer(2005)[38], we can define dynamic exp UIV as the following

Definition. The dynamic exp utility indifference value process for a given $H \in L^\infty(\mathcal{F}_{T_0})$ is a bounded RCLL semimartingale, denoted by $C(H) = \{C_t(H); t \in [0, T_0]\}$, satisfying

$$\operatorname*{ess\,sup}_{X \in \mathcal{W}} E\left(U(X_{T_0} - X_t)\big|\mathcal{F}_t\right) = \operatorname*{ess\,sup}_{X \in \mathcal{W}} E\left(U\big(C_t(H) + X_{T_0} - X_t - H\big)\big|\mathcal{F}_t\right),$$
(20)

where $U(x) = -\exp\{-k_0 x\}$.

Remark. If the bond market is approximately complete or H is approximately hedgeable, (i.e., $H = X_T$, for some $X \in \mathcal{W}$), then $C(H) = 0$.

Let Assumption 3.1 be satisfied and assume that the BSE (17) has a solution denoted by $(a; l_1, l_2; \theta_1, \theta_2)$, one can define the minimal entropy martingale measure Q^* by

$$\frac{dQ^*}{dP}\big|_{\mathcal{F}_{T_0}} = Z_{T_0}^{\theta_1, \theta_2}.$$

Similar to 38, we have the following lemma from (3.2) and (20)

Lemma 3.4. *Let Assumption 3.1 be satisfied and assume that the BSE (17) has a solution denoted by $(a; l_1^*, l_2^*; \theta_1, \theta_2)$, then the exp utility indifference value process $C(H)$ can be determined by*

$$e^{k_0 C_t(H)} = \operatorname*{ess\,inf}_{X \in \mathcal{W}} E_{Q^*}\left[e^{-k_0(X_{T_0} - X_t - H)}\big|\mathcal{F}_t\right]$$
$$= \operatorname*{ess\,inf}_{X \in \mathcal{W}^b} E_{Q^*}\left[e^{-k_0(X_{T_0} - X_t - H)}\big|\mathcal{F}_t\right].$$

The following lemma is a version of Girsanov's theorem

Lemma 3.5. *Let Assumption 3.1 be satisfied and assume that the BSE (17) has a solution denoted by $(a; l_1^*, l_2^*; \theta_1, \theta_2)$. Then*

$$W^*(t) := \widetilde{W}(t) - \int_0^t \theta_1(u)du = W(t) - \int_0^t \{\phi(u) + \theta_1(u)\}du$$

is a $\mathbb{R}^d$-valued Brownian motion under Q^* and

$$\nu^*(du, dy) := (1 + \theta_2(u, y))\tilde{\nu}(du, dy) = (1 + \theta_2(u, y))\varphi(u, y)\lambda_u(dy)du$$

is the compensator of μ under Q^*.

Remark. For any $M^{l_1, l_2} \in \mathcal{W}$, since M^{l_1, l_2} is a local $(\mathbb{M}, Q^0)$-normalized martingale and thus M^{l_1, l_2} a local martingale under Q^*, M^{l_1, l_2} can also be written as

$$M_t^{l_1, l_2} = \int_0^t l_1(u)'dW^*(u) + \int_0^t \int_E l_2(u, y)(\mu(du, dy) - \nu^*(du, dy)).$$

For any $H \in L^\infty(\mathcal{F}_{T_0})$, we consider the following BSE

$$\begin{cases} h_t = h_0 - \int_0^t \frac{1}{2}k_0\|\xi(u)\|^2 du - \int_0^t \int_E \left\{ \frac{e^{k_0\zeta(u,y)} - 1}{k_0} - \zeta(u, y) \right\}\nu^*(du, dy) \\ \qquad + \int_0^t \{\hat{l}_1(u) + \xi(u)\}'dW^*(u) \\ \qquad + \int_0^t \int_E \{\hat{l}_2(u, y) + \zeta(u, y)\}(\mu(du, dy) - \nu^*(du, dy)), \\ h_T = B. \end{cases}$$

$$(21)$$

The **bounded solution** of the BSE (21) is a 5-tuple $(h; \hat{l}_1, \hat{l}_2; \xi, \zeta)$ which satisfies (21) and

i) $\hat{l}_1$ is a predictable process and $\hat{l}_2$ is a $\widetilde{\mathcal{P}}$-measurable function such that $M^{\hat{l}_1, \hat{l}_2} \in \mathcal{W}^b$;

ii) ξ is a predictable process and ζ is a $\widetilde{\mathcal{P}}$-measurable function such that **for any $M^{l_1, l_2} \in \mathcal{W}^b$, (ξ, ζ) is 'exp-orthogonal' to (l_1, l_2) in the following sense**

$$\xi(u)'l_1(u) + \int_E \frac{e^{k_0\zeta(u,y)} - 1}{k_0} \frac{e^{k_0 l_2(u,y)} - 1}{k_0}(1 + \theta_2(u, y))\varphi(u, y)\lambda_u(dy) = 0;$$

(note that this is inspired by a construction in Morlais(2008)[40]

iii) h is a bounded RCLL semimartingale.

Then we get from the standard martingale optimality principle (see e.g. Proposition 12 of Mania and Schweizer(2005)[38]):

Theorem 3.5. *Let Assumption 3.1 be satisfied, assume that the BSE (17) has a solution denoted by* $(a; l_1^*, l_2^*; \theta_1, \theta_2)$ *and the BSE (21) has a bounded solution* $(h; \hat{l}_1, \hat{l}_2; \xi, \zeta)$ *for given* $H \in L^\infty(\mathcal{F}_{T_0})$*, then for any* $t \in [0, T_0]$*,* $C_t(H) = h_t$*, a.s.*

4. Outlook

It is seen that by leaving the basic somewhat unrealistic model of the first sections towards more realistic ones many technical and interpretational difficulties arise. One way out of these deficiencies might be to consider more structured models like affine term structures. To this end it will be necessary to study Wishart processes (see Bru(1991)[9]). There is a lot of research going on in this field, and some results look rather promising. A different approach might be to consider markets consisting of a stock and many tradable bonds. However, for such markets arbitrage and possibly completeness has to be reconsidered in the sense that the new objects have to be compatible with the standard market objects to be able to redefine risk-neutral markets. Some attempts to develop such concepts of "compatitibility" in different situations are found e.g. in Jacod and Protter(2006)[28] and in a manuscript on the compatible bond-stock market with jumps which we are working on right now.

References

1. Arai, T. (2005). An extension of mean-variance hedging to the discontinuous case. *Finance Stochast.* 9, 129-139 (2005)
2. Baran, M., Jakubowski J., Zabczyk J. (2008): On incompleteness of bond markets with infinite number of random factors, submitted ; Preprint IMPAN No. 690; arXiv:0809.2270.
3. Becher, D.(2006). Bounded solution to backward SDE's with jumps for utility optimization and indifference hedging, *The Annals of Applied Probability*, Vol. 16, No. 4, 2027-2054. DOI: 10.1214/105051606000000475.
4. Biagini, F.(2001): *Quadratic Hedging Approaches for Interest Rate Models with Stochastic Volatility*, PhD Thesis, Scuola Normale Superiore.
5. Björk,T.(1997): *Interest rate theory* in Biais, B. (ed.) et al., Financial mathematics. Lectures given at the 3rd session of the Centro Internazionale Matematico Estivo (CIME), held in Bressanone, Italy, July 8-13, 1996. Berlin: Springer. Lect. Notes Math. 1656, 53-122 (1997).
6. Björk, T., Di Masi, G., Kabanov, Y., Runggaldier, W.(1997): Towards a general theory of bond markets. *Finance Stoch.* 1, 141-174.

7. Björk, T., Kabanov, Y., Runggaldier, W.(1997): Bond market structure in the presence of marked point processes. *Math. Finance* 7, 211-223.

8. Bobrovnytska, O. and Schweizer, M. (2004). Mean-variance headging and stochatic control: Beyond the Brownian setting. *IEEE Transactions on Automatic Control* 49, 396-408(2004).

9. Bru, M.F. (1991). Wishart Processes, *J.Theoretical Probability* 4:4, 725-751.

10. Carmona,R. and Tehranchi, M. (2004): A Characterization of Hedging Portfolios for Interest Rate Contingent Claims. *Ann.Appl.Proba.*, Vol 14,3, 1267-1294.

11. Černý, A. and Kallsen, J.(2007). On the structure of general mean-variance hedging strategies. *The Annals of Probability* 35,No. 4, 1479-1531(2007).

12. Černý, A. and Kallsen, J.(2008). A counterexample concerning the variance-optimal martingale measure. Mathematical Finance, Vol. 18, Issue 2, pp. 305-316, April 2008.

13. Choulli, T., Krawczyk,L., and Stricker,C. (1998). $\mathcal{E}$-martingales and their applications in mathematical finance, *The Annals of Probability*, 1998, Vol.26, No.2,853-876(1998).

14. Choulli, T. and Stricker, C.(2005). Minimal entropy-Hellinger martingale measure in incomplete market. *Math. Finance* 15, 465-490(2005).

15. Choulli, T. and Stricker, C.(2006). More on minimal entropy-Hellinger martingale measure. Math. Finance 16, 1-19(2006).

16. Choulli, T, Stricker, C. and Li, J.(2007). Minimal Hellinger martingale measures of order q. Finance Stoch., Vol. 11, no.3, 399-422(2007).

17. Cont, R.(2004): Modeling term structure dynamics: An infinite dimensional approach. *Int. J. Theor. Appl. Finance* 8, No. 3, 357-380 (2005).

18. De Donno, M.(2004): The term structure of interest rates as a random field: A stochastic integration approach. In Stochastic Processes and Applications to Mathematical Finance, Proceedings of the Ritsumeikan International Symposium, 2751.

19. De Donno, M. and Pratelli, M. (2004). On the use of measure-valued strategies in bond markets. *Finance Stochast.* 8, 87-109. DOI: 10.1007/s00780-003-0102-7.

20. De Donno, M. and Pratelli, M. (2005). A Theory of Stochastic Integration for Bond Markets. *The Annals of Applied Probability*, Vol. 15, No. 4, 2773-2791.

21. Delbaen, F. and Schachermayer, W. (1996). The Variance-Optimal Martingale Measure for Continuous Processes. *BERNOULLI* 2, 81-105(1996).

22. Delbaen, F., Grandits, P., Rheinläder, T., Samperi, D., Schweizer, M. and Stricker, C. (2002). Exponential hedging and entropic penalties. *Math. Finance*, Vol. 12, 99-123.

23. Frittelli, M.(2000). The minimal entropy martingale measure and the valuation problem in incomplete markets. Math. Finance, 10, 39-52.

24. Fujiwara, T. and Miyahara, Y. (2003): The minimal entropy martingale measures for geometric Levy processes. Finance Stochast. 7, 509-531.

25. Gourieroux, C., Laurent, J.P., Pham, H. (1998): Mean-Variance Hedging and Numéraire. *Mathematical Finance*, 8 (3), 179-200.

26. Jacod, J. (1979). *Calcul Stochastique et Problèmes de Martingales*. Lecture Notes in Maths. 714. Berlin: Springer.

27. Jacod, J., Shiryaev, A.N. (2003). *Limit theorems for stochastic processes.* Berlin Heidelberg NewYork: Springer 2003.
28. Jacod, J., Protter, Ph. (2010). Risk-neutral compatibility with option prices. to appear in *Finance and Stochastics* 2010.
29. El Karoui, N. and Quenez, M. C. (1995). Dynamic programming and pricing of contingent claims in an incomplete market. *SIAM J. Control Optim.* 33, 29-66(1995).
30. Kohlmann, M. and Tang, S. (2003,a). Multidimensional Backward Stochastic Riccati Equations and Applications. *SIAM Journal on Control and Optimization* 41, 1696-1721(2003).
31. Kohlmann, M., Xiong, D., and Ye, Z. (2007). Change of filtrations and mean-variance hedging. *Stochastics: An International Journal of Probability and Stochastic Processes*, Vol.79, No. 6, 539-562(2007).
32. Kohlmann, M. and Xiong, D.(2007). The mean-variance hedging of a default-able option with partial information. *Stochastic Analysis and Applications*, 25: 869-893(2007).
33. Kohlmann, M. and Xiong, D. (2008). The minimal entropy and the convergence of the p-optimal martingale measures in a general jump model. *Stochastic Analysis and Applications*, Volume 26, Issue 5, 941-977(2008).
34. Kohlmann,M., Xiong,D., and Ye,Z.(2010). The mean variance hedging in a general jump model. *Applied Mathematical Finance*, 17: 1, 29 57.
35. Xiong, D., Kohlmann, M. (2010). Mean Variance Hedging in a General Jump Market to appear in *Int. Journ. Theory Appl. Finance* 2010.
36. Laurent, J.P. and Pham, H. (1999). Dynamic Programming and Mean-Variance Headging. *Finance and Stochastics* 3, 83-110.
37. Mania, M., Santacroce M. and Tevzadze, R. (2003). A semimartingale BSDE related to the minimal entropy martingale measure. *Finance Stochast.* 7, 385-402.
38. Mania, M. and Schweizer, M. (2005). Dynamic exponential utility indifference valuation. *The Annals of Applied Probability* 2005, Vol. 15, No. 3, 2113-2143.
39. Mikulevicius, R. and Rozovskii, B. L.(1998): Normalized stochastic integrals in topological vector spaces. In: *Séminaire de probabilités XXXII* (Lecture notes in Mathematics). Berlin Heidelberg New York: Springer 1998, 137-165.
40. Morlais, M.A.(2008). Utility maximization in a jump market model. *Preprint.*
41. Niethammer, C.R. (2007). Portfolio Optimization and Optimal Martingale Measures in Markets with Jumps. *ph.d.thesis University of Konstanz, Germany.*
42. Rheinländer, M. and Schweizer, M.(1997). On L^2-projections on a Space of Stochastic Integrals. *Annals of Probability* 25, 1810-1831(1997).
43. Schwartz, L.(1994): Semi-martingales banachiques: le theor'eme des trois operateurs. Seminaire de Probabilites XXVIII (Lect. Notes Math., vol. 1583) BerlinHeidelbergNew York: Springer 1994.
44. Schweizer, M. (1996). Approximation Pricing and the Variance-Optimal Martingale Measure.Annals of Probability 24, 206-236(1996).
45. Xiong D. and Kohlmann M. (2010). The mean-variance hedging in a bond market with jumps. *Preprint* University of Konstanz, Germany, to appear in *Stochastic Analysis and Applications.*

46. Xiong D. and Kohlmann M. (2010). The optimal exponential utility in a jump bond market. *Preprint* University of Konstanz, Germany, to appear in *Stochastic Analysis and Applications*.

47. Xiong D. and Kohlmann M. (2010). The affine term structure with jumps. *Preprint* University of Konstanz, Germany.

48. Xiong D. and Kohlmann M. (2010). The term structure compatible to the stock price system with jumps. *Preprint* University of Konstanz, Germany.

Recombining Tree for Regime-Switching Model: Algorithm and Weak Convergence

R. H. Liu

Department of Mathematics
University of Dayton
300 College Park, Dayton, OH 45469-2316, USA
Email: ruihua.liu@notes.udayton.edu

This paper is concerned with a multinomial tree method for option pricing when the underlying asset price follows a regime-switching model. A direct extension of the well-known CRR tree to the regime-switching model is examined first. Then an efficient recombining tree is presented that grows linearly as the number of time steps increases. The weak convergence of the discrete multinomial approximations to the continuous regime-switching process is proved via the method of martingale problem formulation.

Keywords: Regime-switching model; multinomial tree; option pricing; weak convergence.

1. Introduction

Lattice methods have been broadly adapted in computational finance ever since the innovative work by Cox, Ross and Rubinstein (Ref. 7) for the Black-Scholes-Merton option pricing model (the CRR binomial tree). See Refs. 2,6,8,10,14–16,21–23 and the references therein. From the computational perspective, a crucial feature for a successful tree design is that the number of nodes can not grow too fast as the number of time steps increases. According to Nelson and Ramaswamy (Ref. 21), a binomial approximation to a diffusion is "computationally simple" if the number of nodes grows at most linearly in the number of time intervals. One of the main reasons for the CRR method being practically popular is because it grows linearly and is therefore computationally simple.

Pricing derivative securities in regime-switching model has drawn considerable attention in recent decade. See Refs. 1,3–5,9,11–13,17,20,24, among others. In this setting, asset prices are dictated by a number of stochastic differential equations coupled by a finite-state Markov chain, which represents random switch among different regimes. Model parameters (drift and volatility coefficients) are assumed to depend on the Markov chain. In certain cases, closed-form solutions can be obtained for option prices. For instance, Guo (Ref. 11) provided an analytical formula for the European call option prices assuming there are two regimes; Buffington and Elliott (Ref. 5) provided a general formula for $m \geq 2$ regimes; Guo and Zhang (Ref. 12) studied a perpetual American put option with two regimes and derived an analytical solution. However, it is extremely difficult to obtain a closed-form solution for many other cases, in particular, the American options with finite expiration time and with $m \geq 2$ regimes. Hence seeking efficient numerical approximation schemes comes up as an important alternative.

In Ref. 19, the author has developed a multinomial tree approach for regime-switching models which grows linearly as the number of time steps increases. This recombining tree allows us to use large number of time steps to obtain accurate approximations for both European and American option prices. We have also developed an approximation regime-switching model for the well-known Heston's stochastic volatility model and then employed the tree approach for pricing options. It is noted (see, for example Ref. 28) that using a regime-switching model makes it possible to describe stochastic volatility in a relatively simple manner (simpler than the so-called stochastic volatility models). Various numerical examples are studied in Ref. 19. See Ref. 19 for those numerical results and further discussions of the method.

The present paper is concerned with the weak convergence of the multinomial tree approximations to the continuous regime-switching diffusion process. For models without regime-switching, results on convergence of the binomial tree approximations have been obtained in a number of articles. See Refs. 6,14,15,21,23 and the references therein. However, due to the added Markov chain in the underlying model, the methods used in these papers can not be directly applied to our regime-switching case. Instead, we employ the martingale problem formulation procedure (Ref. 25) to establish our convergence result. This method is developed by Yin, Song and Zhang in Ref. 25 for a class of discrete numerical approximations for diffusions with regime-switching. Also see Ref. 28 for detailed accounts and recent generalizations of the method.

The rest of the paper is organized as follows. Section 2 is concerned with the tree construction. The regime-switching model is introduced in Subsection 2.1. A direct extension of the CRR tree to the regime-switching model is examined in subsection 2.2 which is seen not computationally simple. Our new recombining tree is presented in subsection 2.3. Section 3 is devoted to the weak convergence of the tree approximations. We establish the convergence result under a general framework that includes both the completely recombining tree and the partially recombining CRR tree as special cases. Section 4 provides further remarks and concludes the paper.

2. Regime-Switching Trees and Option Pricing

2.1. *Regime-Switching Model*

Let α_t be a continuous-time Markov chain taking values among m_0 different states, where m_0 is the total number of states considered for the economy. Each state represents a particular regime and is labeled by an integer i between 1 and m_0. Hence the state space of α_t is given by $\mathcal{M} := \{1, \ldots, m_0\}$. For example, if $m_0 = 2$ (two regimes), then $\alpha_t = 1$ can indicate a bullish market and $\alpha_t = 2$ a bearish market. Let matrix $Q = (q_{ij})_{m_0 \times m_0}$ denote the generator of α_t. From Markov chain theory (see for example, Yin and Zhang (Ref. 26)), the entries q_{ij} in matrix Q satisfy: (I) $q_{ij} \geq 0$ if $i \neq j$; (II) $q_{ii} \leq 0$ and $q_{ii} = -\sum_{j \neq i} q_{ij}$ for each $i = 1, \ldots, m_0$.

We assume that the risk-neutral probability space $(\Omega, \mathcal{F}, \widetilde{\mathcal{P}})$ is given. Let B_t be a standard Brownian motion defined on $(\Omega, \mathcal{F}, \widetilde{\mathcal{P}})$ and assume it is independent of the Markov chain α_t. We consider the following regime-switching geometric Brownian motion for the risk-neutral process of the underlying asset price S_t:

$$\frac{dS_t}{S_t} = (r_{\alpha_t} - d_{\alpha_t})dt + \sigma_{\alpha_t} dB_t, \quad t \geq 0, \tag{1}$$

where r_{α_t} is the instantaneous risk-free interest rate, d_{α_t} and σ_{α_t} are dividend rate and volatility rate of the asset, respectively. Note that those parameters depend on the Markov chain α_t, indicating that they can take different values in different regimes. We assume that $\sigma_i > 0$, for each $i \in \mathcal{M}$.

2.2. *A Direct Extension of CRR Tree to Regime-Switching*

A direct extension of the CRR tree (Ref. 7) to the regime-switching model (1) proceeds as follows. See Refs. 1,17.

Let $T > 0$ denote the maturity of option under consideration. Divide the interval $[0, T]$ into N steps. Thus the time step size is given by $h = \frac{T}{N}$. Consider the joint Markov process (S_t, α_t), $0 \leq t \leq T$ and let $(S_k, \alpha_k) := (S_t, \alpha_t)_{t=kh}$ be the state at the kth step of the tree,

$k = 0, 1, \ldots, N$. Assuming $(S_k, \alpha_k) = (S, i)$. Then, at the $(k + 1)$th step, the asset price S_{k+1} can either move up to Su_i with probability p_i or move down to Su_i^{-1} with probability $1 - p_i$, where

$$u_i = e^{\sigma_i \sqrt{h}}, \quad p_i = \frac{e^{r_i h} - e^{-\sigma_i \sqrt{h}}}{e^{\sigma_i \sqrt{h}} - e^{-\sigma_i \sqrt{h}}}.$$

On the other hand, the Markov chain α_{k+1} at the $(k+1)$th step may stay at state i with probability p_{ii}^{α} or jump to any other state $j \neq i$ with probability p_{ij}^{α}, where the one-step transition probabilities p_{ij}^{α} of the Markov chain α_k are defined by

$$p_{ij}^{\alpha} = P\{\alpha_{k+1} = j | \alpha_k = i\}, \ 1 \leq i, j \leq m_0. \tag{2}$$

It then follows that at the $(k + 1)$th step, there are totally $2m_0$ possible states for (S_{k+1}, α_{k+1}), given by

$$(S_{k+1}, \alpha_{k+1}) = \begin{cases} (Su_i, i), & \text{with probability } p_i p_{ii}^{\alpha}, \\ (Su_i^{-1}, i), & \text{with probability } (1 - p_i) p_{ii}^{\alpha}, \\ (Su_i, j), & \text{with probability } p_i p_{ij}^{\alpha}, \ j \neq i, \\ (Su_i^{-1}, j), & \text{with probability } (1 - p_i) p_{ij}^{\alpha}, \ j \neq i. \end{cases} \tag{3}$$

An illustrative two-regime tree is depicted in Fig. 1.

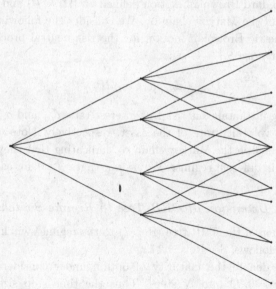

Fig. 1.   A non-completely recombining tree for two regimes

It can be seen that the state space for the two-dimensional Markov chain $\{(S_k, \alpha_k)\}_{k \geq 0}$ is the set

$$\left\{ (S_0 u_1^{2l_1 - k_1} u_2^{2l_2 - k_2} \cdots u_{m_0}^{2l_{m_0} - k_{m_0}}, i) \Big| 0 \leq l_i \leq k_i, 1 \leq i \leq m_0, \right.$$
$$\left. \sum_{i=1}^{m_0} k_i = k, 0 \leq k \leq N \right\}, \qquad (4)$$

where S_0 is the initial stock price. It was shown in Aingworth et al. [1, Proposition 1] that the number of distinct asset prices at the kth period is $\begin{pmatrix} k + 2m_0 - 1 \\ 2m_0 - 1 \end{pmatrix} = O(k^{2m_0 - 1})$. Thus the tree does not grow linearly and therefore is not computationally simple according to Ref. 21. Consequently, it works well for small number of regimes and for moderate number of tree steps. However, if m_0 and N are large, such an implementation will become computationally formidable due to node explosion. For example, consider a model with 20 regimes. If 100 time steps were used, then the last step of the tree would have $\begin{pmatrix} 139 \\ 39 \end{pmatrix} \approx 4.7 \times 10^{192}$ nodes. Apparently, most computers can not handle such a heavy computation requirement.

2.3. *A Regime-Switching Recombining Tree*

To have a tree structure that grows linearly, complete recombination of nodes needs to be achieved at each time step. For the generalized CRR tree discussed in the previous section, node recombination is only achieved partially, namely, nodes do combine if the regime stays the same. However, whenever a switch of regime occurs, the change of asset price will very likely produce new nodes, due to different change factor u_i in different regime. To resolve this issue, we adjust the change factor u_i in a suitable way. Meanwhile, it is necessary to match the local mean and variance calculated from the tree to that implied by the continuous regime-switching diffusion, in order for the discrete tree approximations to converge to its underlying continuous process (see next section). We present the tree design in this section. For numerical experiments and an application of the tree method to the Heston's stochastic volatility model, we refer the reader to Ref. 19.

First, let $X_t = \ln(S_t/S_0)$. Then $S_t = S_0 e^{X_t}$ where X_t is the solution of the stochastic differential equation:

$$dX_t = a_{\alpha_t} dt + \sigma_{\alpha_t} dB_t, \quad X_0 = 0, \qquad (5)$$

where

$$a_{\alpha_t} = r_{\alpha_t} - d_{\alpha_t} - \frac{1}{2}\sigma_{\alpha_t}^2. \qquad (6)$$

We design a completely recombined tree for the process X_t instead of the asset price S_t as done in Refs. 1,3. The corresponding approximations of asset price can be easily obtained from the approximations of X_t. Divide the option life $[0, T]$ into N steps and let $h = \frac{T}{N}$ be the step size. Let $(X_k, \alpha_k) := (X_t, \alpha_t)_{t=kh}$ be the state at the kth step of the tree, $k = 0, 1, \ldots, N$. We use three branches for each regime, a up move, a down move, and a middle stay (no move). The up and down moves are carefully chosen so that they must take values among $2b + 1$ evenly spaced points, where the specification of b will be discussed shortly. Let constant $\bar{\sigma} > 0$ and positive integer b be given. Assume $X_k \doteq x$, then at the next step, X_{k+1} must take the three values from the set of $2b + 1$ values given by $\{x + j\bar{\sigma}\sqrt{h}, j = -b, -b+1, \ldots, 0, \ldots, b-1, b\}$. Specifically, for regime i at step k, i.e., $(X_k, \alpha_k) = (x, i)$, let l_i (to be determined) be the number of upward moves of X_{k+1}. Then, the three branches associated with regime i are given by $x + l_i\bar{\sigma}\sqrt{h}$, x and $x - l_i\bar{\sigma}\sqrt{h}$. Next, let $p_{i,u}$, $p_{i,m}$ and $p_{i,d}$ be the conditional probabilities corresponding to the up, middle and down branches. By matching the mean and variance implied by the trinomial lattice to that implied by the SDE (5), we have,

$$
\begin{cases}
(l_i\bar{\sigma}\sqrt{h})p_{i,u} - (l_i\bar{\sigma}\sqrt{h})p_{i,d} = a_i h, \\
(l_i\bar{\sigma}\sqrt{h})^2 p_{i,u} + (l_i\bar{\sigma}\sqrt{h})^2 p_{i,d} = \sigma_i^2 h + a_i^2 h^2, \\
p_{i,u} + p_{i,m} + p_{i,d} = 1,
\end{cases}
\tag{7}
$$

where $a_i = r_i - d_i - \frac{1}{2}\sigma_i^2$. It follows that,

$$
\begin{aligned}
p_{i,u} &= \frac{\sigma_i^2 + a_i(l_i\bar{\sigma})\sqrt{h} + a_i^2 h}{2(l_i\bar{\sigma})^2}, \\
p_{i,d} &= \frac{\sigma_i^2 - a_i(l_i\bar{\sigma})\sqrt{h} + a_i^2 h}{2(l_i\bar{\sigma})^2}, \\
p_{i,m} &= 1 - \frac{\sigma_i^2 + a_i^2 h}{(l_i\bar{\sigma})^2}.
\end{aligned}
\tag{8}
$$

Consequently, emanating from the node (x, i), there are $3m_0$ states for (X_{k+1}, α_{k+1}), given by

$$
(X_{k+1}, \alpha_{k+1}) = \begin{cases}
(x + l_i\bar{\sigma}\sqrt{h}, j), & \text{with probability } p_{ij}^\alpha p_{i,u}, \\
(x, j) & \text{with probability } p_{ij}^\alpha p_{i,m}, \ j = 1, \ldots m_0, \\
(x - l_i\bar{\sigma}\sqrt{h}, j), & \text{with probability } p_{ij}^\alpha p_{i,d},
\end{cases}
\tag{9}
$$

where p_{ij}^α denote the one-step transition probabilities of the Markov chain α_k from state i to j (see (2)). It can be seen that the number of nodes at the kth step is $m_0(2bk + 1)$, linear in k. Hence, our design does yield a computationally simple tree. If we consider again the previous example, namely, $m_0 = 20$ and $N = 100$, then the last step of the new tree will have $20(200b+1) = 80020$ if $b = 20$ is chosen, a much smaller number comparing to 4.7×10^{192}.

Fig. 2 displays a recombining tree with 7 branches ($b = 3$) emanating from each node (a heptanomial tree structure).

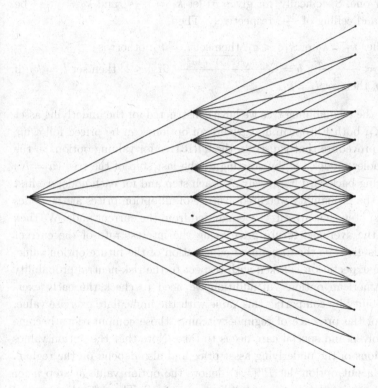

Fig. 2. A two-step recombining heptanomial tree

Note that if l_i and h are not chosen properly, it is possible that (8) results in negative branch probabilities. The following proposition gives conditions under which the branch probabilities are guaranteed to be non-negative.

Proposition 1. *The following assertions hold for the solutions given by (8).*

(1) If $a_i = 0$, then $0 \le p_{i,u}, p_{i,m}, p_{i,d} \le 1$ provided that $l_i \bar{\sigma} \ge \sigma_i$;

(2) If $a_i \ne 0$, then $0 \le p_{i,u}, p_{i,m}, p_{i,d} \le 1$ provided that (a) $\sigma_i < l_i \bar{\sigma} \le 2\sigma_i$ and $h \le \frac{(l_i \bar{\sigma})^2 - \sigma_i^2}{a_i^2}$, or (b) $l_i \bar{\sigma} > 2\sigma_i$ and $h \le \frac{[l_i \bar{\sigma} - \sqrt{(l_i \bar{\sigma})^2 - 4\sigma_i^2}\,]^2}{4a_i^2}$.

Remark 1. The numerical results in Ref. 19 show that the approximate option prices are insensitive to the choice of parameter $\bar{\sigma}$. Hence the choice of l_i can be flexible as long as the conditions in Proposition 1 are satisfied. One way is to choose the l_i value so that the upper constraint for h takes the larger one. Specifically, for given $\bar{\sigma}$, let $k_1 = \lfloor \frac{2\sigma_i}{\bar{\sigma}} \rfloor$ and $k_2 = \lceil \frac{2\sigma_i}{\bar{\sigma}} \rceil$ be the floor and ceiling of $\frac{2\sigma_i}{\bar{\sigma}}$, respectively. Then,

- if either $k_1 = k_2$ or $k_1 \bar{\sigma} < \sigma_i$, then set $l_i = k_2$; otherwise,
- let $a = \frac{(k_1 \bar{\sigma})^2 - \sigma_i^2}{a_i^2}$, $b = \frac{[k_2 \bar{\sigma} - \sqrt{(k_2 \bar{\sigma})^2 - 4\sigma_i^2}\,]^2}{4a_i^2}$, if $a \le b$, then set $l_i = k_2$; if $a > b$, then set $l_i = k_1$.

Using the recombining tree we have constructed for the underlying asset process X_t, both European and American options can be priced following a similar procedure that is used in the CRR tree for valuing options in the Black-Scholes model, i.e., by starting at the last step of the tree ($n = N$) and working backward iteratively. At each step and for each node, we first calculate the probability weighted average of all option prices at the nodes in the next step that are directly emanated from the current node. We then discount the averaged future price using the interest rate of the current node. This gives us the discounted expectation of the future option value, where the expectation is taken with respect to the risk-neutral probability $\tilde{P}$. For American options, in addition we need to check the early exercise possibility by comparing this value with the immediate exercise value. Because of the presence of regime-switching, those computations become more involved and special care needs to take. Note that the option values are functions of the underlying asset price and also depend on the regime. Consider a put option, let $P^n(x, i)$ denote the option value at step n for the node associated with the state $(X_n, \alpha_n) = (x, i)$. Then, for the terminal step $n = N$,

$$P^N(x, i) = \max\{K - S_0 e^x, 0\}, \quad i = 1, \ldots, m_0, \tag{10}$$

where K is the strike price of the option. At step $n < N$, for European put,

$$
\begin{aligned}
P^n(x, i) = e^{-r_i h} \sum_{j=1}^{m_0} p_{ij}^\alpha \Big[& P^{n+1}(x + l_i \bar{\sigma} \sqrt{h}, j) p_{i,u} \\
& + P^{n+1}(x - l_i \bar{\sigma} \sqrt{h}, j) p_{i,d} + P^{n+1}(x, j) p_{i,m} \Big],
\end{aligned}
\tag{11}
$$

and for American put,

$$P^n(x,i) = \max\left\{K - S_0 e^x, e^{-r_i h} \sum_{j=1}^{m_0} p_{ij}^\alpha \left[P^{n+1}(x + l_i \bar\sigma \sqrt{h}, j) p_{i,u}\right.\right.$$

$$\left.\left. + P^{n+1}(x - l_i \bar\sigma \sqrt{h}, j) p_{i,d} + P^{n+1}(x, j) p_{i,m}\right]\right\}. \tag{12}$$

Formulae for call options are obtained by changing $K - S_0 e^x$ to $S_0 e^x - K$ in (10)-(12).

3. Convergence of the Tree Approximations

In this section we establish the convergence results of the discrete approximation processes implied by the regime-switching trees to the continuous-time processes. We prove the convergence under a general framework that includes the completely recombining tree and the partially recombining CRR tree as two special cases.

We consider the following one-dimensional stochastic differential equation modulated by the Markov chain α_t,

$$dX_t = \mu(X_t, \alpha_t)dt + \sigma(X_t, \alpha_t)dB_t, \quad t \geq 0, \tag{13}$$

where $\mu(\cdot, \cdot) : \mathbb{R} \times \mathcal{M} \to \mathbb{R}$ and $\sigma(\cdot, \cdot) : \mathbb{R} \times \mathcal{M} \to \mathbb{R}$ are appropriate functions satisfying the Lipschitz condition

$$|\mu(x_1, i) - \mu(x_2, i)| \leq K|x_1 - x_2|, \quad |\sigma(x_1, i) - \sigma(x_2, i)| \leq K|x_1 - x_2|, \tag{14}$$

and the linear growth condition,

$$|\mu(x, i)| \leq K(1 + |x|), \quad |\sigma(x, i)| \leq K(1 + |x|), \tag{15}$$

for all $x, x_1, x_2 \in \mathbb{R}$ and for each $i \in \mathcal{M}$, where K is a positive constant (In what follows we will use K as a generic positive constant; that is, its values may vary at different places). In addition, we assume $\sigma(x, i) > 0$ for all $x \in \mathbb{R}$ and for each $i \in \mathcal{M}$. Note that for the regime-switching model (5) on which the tree construction is based, $\mu(x, i) = r_i - d_i - \frac{1}{2}\sigma_i^2$ and $\sigma(x, i) = \sigma_i$. Clearly, (14) and (15) are satisfied.

For each $i \in \mathcal{M}$, let function $V(x, i)$ be twice continuously differentiable in x. Define an operator $\mathcal{L}$ by

$$\mathcal{L}V(x, i) = \mu(x, i)\frac{dV}{dx}(x, i) + \frac{1}{2}\sigma^2(x, i)\frac{d^2V}{dx^2}(x, i) + QV(x, \cdot)(i) \tag{16}$$

where

$$QV(x, \cdot)(i) = \sum_{j \neq i} q_{ij}[V(x, j) - V(x, i)] = \sum_{j=1}^{m_0} q_{ij}V(x, j) \tag{17}$$

and $Q = (q_{ij})_{m_0 \times m_0}$ is the generator of α_t.

In what follows we will use $\varepsilon > 0$ (instead of h as in the previous section) for the time step size. Then $N = \lfloor \frac{T}{\varepsilon} \rfloor$ is the number of steps for a tree construction. Let the initial state X_0 and initial regime α_0 be given. For each fixed $\varepsilon > 0$, we construct a continuous-time process $(X^\varepsilon(t), \alpha^\varepsilon(t))$, $t \geq 0$ such that,

$$X^\varepsilon(0) = X_0, \ \alpha^\varepsilon(0) = \alpha_0, \tag{18}$$

$$X^\varepsilon(t) = X_n, \ \alpha^\varepsilon(t) = \alpha_n, \ \text{for } n\varepsilon \leq t < (n+1)\varepsilon, \ n = 0, 1, \ldots, N-1, \tag{19}$$

where for given $(X_n, \alpha_n) = (x, i)$, (X_{n+1}, α_{n+1}) is determined as following: First for the discrete Markov chain $\{\alpha_n, n \geq 0\}$, the transition probability matrix is given by

$$P^\varepsilon = I + \varepsilon Q, \tag{20}$$

that implies

$$P\{\alpha_{n+1} = j | \alpha_n = i\} = \delta_{ij} + \varepsilon q_{ij}, \ 1 \leq i, j \leq m_0, \ 0 \leq n \leq N-1, \tag{21}$$

where $\delta_{ij} = 1$ if $i = j$ and $\delta_{ij} = 0$ if $i \neq j$. It can be shown by applying Theorem 3.1 in Ref. 27 that the interpolated process $\alpha^\varepsilon(\cdot)$ converges weakly to the continuous Markov process $\alpha(\cdot)$ generated by Q.

Next, we assume that for given $(X_n, \alpha_n) = (x, i)$, X_{n+1} can take n_0 different values $X^{\varepsilon, l}(x, i)$, $l = 1, \ldots, n_0$ with conditional probabilities,

$$P\left\{ X_{n+1} = X^{\varepsilon, l}(x, i) \Big| X_n = x, \alpha_n = i \right\} = p^{\varepsilon, l}(x, i), \ l = 1, 2, \ldots n_0, \tag{22}$$

where functions $X^{\varepsilon, l}(\cdot, \cdot) : \mathbb{R} \times \mathcal{M} \to \mathbb{R}$ satisfy $-\infty < X^{\varepsilon, l}(x, i) < \infty$ for all $x \in \mathbb{R}$ and each $i \in \mathcal{M}$, and $p^{\varepsilon, l}(\cdot, \cdot) : \mathbb{R} \times \mathcal{M} \to \mathbb{R}$ satisfy

$$0 \leq p^{\varepsilon, l}(x, i) \leq 1, \ \sum_{l=1}^{n_0} p^{\varepsilon, l}(x, i) = 1, \ \text{for all } x \in \mathbb{R}, \ i \in \mathcal{M}. \tag{23}$$

Moreover, we assume throughout this section that, for $i \in \mathcal{M}$, $x \in \mathbb{R}$,

$$E\left[X_{n+1} \Big| X_n = x, \alpha_n = i \right] = \sum_{l=1}^{n_0} p^{\varepsilon, l}(x, i) X^{\varepsilon, l}(x, i) = x + \mu(x, i)\varepsilon, \tag{24}$$

and

$$\begin{aligned} E\left[(X_{n+1} - x)^2 \Big| X_n = x, \alpha_n = i \right] &= \sum_{l=1}^{n_0} p^{\varepsilon, l}(x, i) \left(X^{\varepsilon, l}(x, i) - x \right)^2 \\ &= \sigma^2(x, i)\varepsilon + \mu^2(x, i)\varepsilon^2. \end{aligned} \tag{25}$$

Note that these conditions are all satisfied by the regime-switching recombining tree design presented in Section 2 with $n_0 = 3$.

The following lemma presents an upper bound of the second moments of X_n, $0 \leq n \leq \lfloor \frac{T}{\varepsilon} \rfloor$ which will be used in the weak convergence proof.

Lemma 1. *Under Assumptions (14), (15), (24) and (25), for any fixed $T > 0$ and any $\varepsilon > 0$, we have*

$$\sup_{0 \leq n \leq \lfloor T/\varepsilon \rfloor} E\left[X_n\right]^2 \leq (X_0^2 + KT)e^{KT} < \infty. \tag{26}$$

Proof. Using (24) and (25) we have,

$$E\left[X_{n+1}^2 \Big| X_n, \alpha_n\right] = E\left[(X_n + (X_{n+1} - X_n))^2 \Big| X_n, \alpha_n\right]$$
$$= X_n^2 + 2X_n \mu\left(X_n, \alpha_n\right)\varepsilon + \sigma^2\left(X_n, \alpha_n\right)\varepsilon + \mu^2\left(X_n, \alpha_n\right)\varepsilon^2.$$

Using (15) we have,

$$\mu^2\left(X_n, \alpha_n\right) \leq K\left(1 + X_n^2\right) \text{ and } \sigma^2\left(X_n, \alpha_n\right) \leq K\left(1 + X_n^2\right).$$

It follows that

$$E\left[X_{n+1}^2 \Big| X_n, \alpha_n\right] \leq X_n^2 + K\varepsilon\left(1 + X_n^2\right). \tag{27}$$

Taking the expectation on both sides of (27) leads to

$$E\left[X_{n+1}^2\right] \leq E\left[X_n^2\right] + K\varepsilon + K\varepsilon E\left[X_n^2\right]. \tag{28}$$

Iterating (28), we obtain

$$E\left[X_n^2\right] \leq X_0^2 + Kn\varepsilon + K\varepsilon \sum_{j=0}^{n-1} E\left[X_j^2\right]. \tag{29}$$

Applying the Gronwall's inequality yields

$$E\left[X_n^2\right] \leq (X_0^2 + Kn\varepsilon)e^{KT} \leq (X_0^2 + KT)e^{KT},$$

and then (26) follows. $\square$

It follows immediately that

$$\sup_{0 \leq t \leq T} E\left[X^\varepsilon(t)\right]^2 \leq K < \infty. \tag{30}$$

Next, we establish the tightness of the continuous processes $\{X^\varepsilon(t) : 0 \leq t \leq T\}$ in a suitable function space. To this end, we rewrite the approximation (22) into the following iterative equation:

$$X_{n+1} = X_n + \mu\left(X_n, \alpha_n\right)\varepsilon + \xi\left(X_n, \alpha_n\right), \tag{31}$$

where the random functions $\xi(\cdot, \cdot)$ are defined as following: For each $x \in \mathbb{R}$ and each $i \in \mathcal{M}$, $\xi(x, i)$ takes values from the set of n_0 values $\{X^{\varepsilon,l}(x, i) - x - \mu(x, i)\varepsilon, l = 1, \ldots, n_0\}$ with probabilities $p^{\varepsilon,l}(x, i)$, $l = 1, 2, \ldots n_0$, as defined in (22) and (23). It follows from (24) and (25) that

$$
\begin{aligned}
E\left[\xi\left(X_n, \alpha_n\right) \Big| X_n = x, \alpha_n = i\right] &= 0, \\
E\left[\xi^2\left(X_n, \alpha_n\right) \Big| X_n = x, \alpha_n = i\right] &= \sigma^2(x, i)\varepsilon.
\end{aligned}
\tag{32}
$$

Lemma 2. *Under Assumptions (14), (15), (24) and (25), the interpolation process $\{X^\varepsilon(t) : 0 \leq t \leq T\}$ is tight in $D[0, T]$, the space of functions that are right continuous and have left limits, endowed with the Skorohod topology.*

Proof. For any $\delta > 0$, $t > 0$, $0 \leq s \leq \delta$ and $t + s \leq T$, we have,

$$
\begin{aligned}
E\left[X^\varepsilon(t + s) - X^\varepsilon(t)\right]^2 &= E\left[\sum_{k=t/\varepsilon}^{(t+s)/\varepsilon-1} (X_{k+1} - X_k)\right]^2 \\
&= E\left[\sum_{k=t/\varepsilon}^{(t+s)/\varepsilon-1} \left(\mu\left(X_k, \alpha_k\right)\varepsilon + \xi\left(X_k, \alpha_k\right)\right)\right]^2 \\
&\leq 2\varepsilon^2 E\left[\sum_{k=t/\varepsilon}^{(t+s)/\varepsilon-1} \mu\left(X_k, \alpha_k\right)\right]^2 \\
&\quad + 2E\left[\sum_{k=t/\varepsilon}^{(t+s)/\varepsilon-1} \xi\left(X_k, \alpha_k\right)\right]^2.
\end{aligned}
$$

Note that for simplicity, in the above we have used t/ε and $(t+s)/\varepsilon$ for the integer parts of t/ε and $(t+s)/\varepsilon$, respectively. We will adapt this convention for the rest of the paper.

Note that

$$
\begin{aligned}
\varepsilon^2 E\left[\sum_{k=t/\varepsilon}^{(t+s)/\varepsilon-1} \mu\left(X_k, \alpha_k\right)\right]^2 &\leq \varepsilon\delta \sum_{k=t/\varepsilon}^{(t+s)/\varepsilon-1} E\left[\mu^2\left(X_k, \alpha_k\right)\right] \\
&\leq \varepsilon\delta K \sum_{k=t/\varepsilon}^{(t+s)/\varepsilon-1} \left(1 + E[X_k^2]\right) \leq K\delta^2,
\end{aligned}
$$

and

$$E\left[\sum_{k=t/\varepsilon}^{(t+s)/\varepsilon-1} \xi(X_k,\alpha_k)\right]^2$$

$$= \sum_{k=t/\varepsilon}^{(t+s)/\varepsilon-1} E\left[\xi^2(X_k,\alpha_k)\right] + 2\sum_{k<l} E\left[\xi(X_k,\alpha_k)\xi(X_l,\alpha_l)\right]$$

$$= \sum_{k=t/\varepsilon}^{(t+s)/\varepsilon-1} E\left[E_k\left[\xi^2(X_k,\alpha_k)\right]\right] + 2\sum_{k<l} E\left[\xi(X_k,\alpha_k)E_l\left[\xi(X_l,\alpha_l)\right]\right]$$

$$= \varepsilon\sum_{k=t/\varepsilon}^{(t+s)/\varepsilon-1} E\left[\sigma^2(X_k,\alpha_k)\right] \le K\varepsilon\sum_{k=t/\varepsilon}^{(t+s)/\varepsilon-1}\left(1+E[X_k^2]\right) \le K\delta,$$

where we have used E_k for the conditional expectation given (X_k,α_k). We have

$$E\left[X^\varepsilon(t+s) - X^\varepsilon(t)\right]^2 \le K(\delta+\delta^2).$$

Then it follows that

$$\lim_{\delta\to 0}\limsup_{\varepsilon\to 0} E\left[X^\varepsilon(t+s) - X^\varepsilon(t)\right]^2 = 0.$$

Then the tightness criterion (see Ref. 18, P47) implies that $\{X^\varepsilon(\cdot)\}$ is tight in $D[0,T]$. $\square$

Note that it can be shown by applying Theorem 3.1 in Ref. 27 that the interpolated process $\{\alpha^\varepsilon(\cdot)\}$ is tight. It then follows that $\{X^\varepsilon(\cdot),\alpha^\varepsilon(\cdot)\}$ is tight in $D([0,T];\mathbb{R}\times\mathcal{M})$, the space of functions that are defined on $D[0,T]$ taking values in $\mathbb{R}\times\mathcal{M}$, and that are right continuous and have left limits, endowed with the Skorohod topology.

Since $\{X^\varepsilon(\cdot),\alpha^\varepsilon(\cdot)\}$ is tight, by the Prohorov's Theorem, it has convergent subsequences. We select such a subsequence and still denote it by $\{X^\varepsilon(\cdot),\alpha^\varepsilon(\cdot)\}$ for notational simplicity. Denote the limit by $\{X(\cdot),\alpha(\cdot)\}$. By Skorohod representation, we may assume without loss of generality that the subsequence $\{X^\varepsilon(\cdot),\alpha^\varepsilon(\cdot)\}$ converges to $\{X(\cdot),\alpha(\cdot)\}$ w.p.1, and that the convergence is uniform on any bounded interval. Next we apply the Martingale problem formulation method (Refs. 25,28) to characterize the limit process $\{X(\cdot),\alpha(\cdot)\}$.

Theorem 1. *Under Assumptions (14), (15), (24) and (25), the interpolation process $(X^\varepsilon(\cdot),\alpha^\varepsilon(\cdot))$ converges weakly to $(X(\cdot),\alpha(\cdot))$ as $\varepsilon\to 0$ such that $(X(\cdot),\alpha(\cdot))$ is the solution of the martingale problem associated with operator $\mathcal{L}$ defined by (16).*

Proof. We are to show that for each $i \in \mathcal{M}$ and for any function $\varphi(\cdot, i) \in C_0^2$,

$$M_\varphi(t) := \varphi(X(t), \alpha(t)) - \varphi(X_0, \alpha_0) - \int_0^t \mathcal{L}\varphi(X(s), \alpha(s)) \, ds \tag{33}$$

is a martingale. This is done via a series of approximations that average out the unwanted terms.

Following Refs. 25,28, the claim (33) will be verified if we can show that for each $i \in \mathcal{M}$, any bounded and continuous function $h_j(\cdot, i)$, any positive integer κ, any $0 < t_j \le t$ with $j \le \kappa$, $s > 0$, and $t + s \le T$,

$$E\left\{ \prod_{j=1}^{\kappa} h_j(X(t_j), \alpha(t_j)) \Big[\varphi(X(t+s), \alpha(t+s)) - \varphi(X(t), \alpha(t)) \right.$$
$$\left. - \int_t^{t+s} \mathcal{L}\varphi(X(u), \alpha(u)) \, du \Big] \right\} = 0. \tag{34}$$

In what follows we show that (34) holds for the pair $(X^\varepsilon(\cdot), \alpha^\varepsilon(\cdot))$ and hence for the pair $(X(\cdot), \alpha(\cdot))$. The proof is divided into four steps.

Step 1. We begin by choosing a sequence of positive integers $\{m_\varepsilon\}$ satisfying $m_\varepsilon \to \infty$ as $\varepsilon \to 0$ but $\Delta_\varepsilon := \varepsilon m_\varepsilon \to 0$. Again, we use t/Δ_ε and $(t+s)/\Delta_\varepsilon$ for their integer parts (to avoid the use of the floor function notation). Introduce the index set

$$\chi_l^\varepsilon = \{k : lm_\varepsilon \le k \le (l+1)m_\varepsilon - 1\}. \tag{35}$$

Then we can rewrite $\varphi(X^\varepsilon(t+s), \alpha^\varepsilon(t+s)) - \varphi(X^\varepsilon(t), \alpha^\varepsilon(t))$ as

$$\varphi(X^\varepsilon(t+s), \alpha^\varepsilon(t+s)) - \varphi(X^\varepsilon(t), \alpha^\varepsilon(t))$$
$$= \sum_{l=t/\Delta_\varepsilon}^{(t+s)/\Delta_\varepsilon - 1} \left[\varphi(X_{(l+1)m_\varepsilon}, \alpha_{(l+1)m_\varepsilon}) - \varphi(X_{(l+1)m_\varepsilon}, \alpha_{lm_\varepsilon}) \right]$$
$$+ \sum_{l=t/\Delta_\varepsilon}^{(t+s)/\Delta_\varepsilon - 1} \left[\varphi(X_{(l+1)m_\varepsilon}, \alpha_{lm_\varepsilon}) - \varphi(X_{lm_\varepsilon}, \alpha_{lm_\varepsilon}) \right]. \tag{36}$$

Step 2. We work on the second summation (the third line) of (36) in this step. Noting that

$$\sum_{l=t/\Delta_\varepsilon}^{(t+s)/\Delta_\varepsilon - 1} \left[\varphi(X_{(l+1)m_\varepsilon}, \alpha_{lm_\varepsilon}) - \varphi(X_{lm_\varepsilon}, \alpha_{lm_\varepsilon}) \right]$$
$$= \sum_{l=t/\Delta_\varepsilon}^{(t+s)/\Delta_\varepsilon - 1} \sum_{k:k \in \chi_l^\varepsilon} \left[\varphi(X_{k+1}, \alpha_{lm_\varepsilon}) - \varphi(X_k, \alpha_{lm_\varepsilon}) \right],$$

we have ,

$$
\lim_{\varepsilon \to 0} E\left\{ \prod_{j=1}^{\kappa} h_j(X^\varepsilon(t_j), \alpha^\varepsilon(t_j)) \sum_{l=t/\Delta_\varepsilon}^{(t+s)/\Delta_\varepsilon - 1} \left[\varphi(X_{(l+1)m_\varepsilon}, \alpha_{lm_\varepsilon}) - \varphi(X_{lm_\varepsilon}, \alpha_{lm_\varepsilon}) \right] \right\}
$$

$$
= \lim_{\varepsilon \to 0} E\left\{ \prod_{j=1}^{\kappa} h_j(X^\varepsilon(t_j), \alpha^\varepsilon(t_j)) \left[\sum_{l=t/\Delta_\varepsilon}^{(t+s)/\Delta_\varepsilon - 1} \sum_{k:k\in\chi_l^\varepsilon} \varphi_x'(X_k, \alpha_{lm_\varepsilon})\mu(X_k, \alpha_k)\varepsilon \right. \right.
$$

$$
+ \sum_{l=t/\Delta_\varepsilon}^{(t+s)/\Delta_\varepsilon - 1} \sum_{k:k\in\chi_l^\varepsilon} \varphi_x'(X_k, \alpha_{lm_\varepsilon})\xi(X_k, \alpha_k)
$$

$$
+ \frac{1}{2} \sum_{l=t/\Delta_\varepsilon}^{(t+s)/\Delta_\varepsilon - 1} \sum_{k:k\in\chi_l^\varepsilon} \varphi_{xx}''(X_k, \alpha_{lm_\varepsilon})\mu^2(X_k, \alpha_k)\varepsilon^2
$$

$$
+ \frac{1}{2} \sum_{l=t/\Delta_\varepsilon}^{(t+s)/\Delta_\varepsilon - 1} \sum_{k:k\in\chi_l^\varepsilon} \varphi_{xx}''(X_k, \alpha_{lm_\varepsilon})\xi^2(X_k, \alpha_k)
$$

$$
+ \sum_{l=t/\Delta_\varepsilon}^{(t+s)/\Delta_\varepsilon - 1} \sum_{k:k\in\chi_l^\varepsilon} \varphi_{xx}''(X_k, \alpha_{lm_\varepsilon})\mu(X_k, \alpha_k)\xi(X_k, \alpha_k)\varepsilon
$$

$$
+ \left. \left. \sum_{l=t/\Delta_\varepsilon}^{(t+s)/\Delta_\varepsilon - 1} \sum_{k:k\in\chi_l^\varepsilon} \int_0^1 \left[\varphi_{xx}''(X_k + u\widetilde{\Delta x_k}, \alpha_{lm_\varepsilon}) - \varphi_{xx}''(X_k, \alpha_{lm_\varepsilon}) \right] \widetilde{\Delta x_k}^2 \, du \right] \right\},
$$

$$
\tag{37}
$$

where $\widetilde{\Delta x_k} = \mu(X_k, \alpha_k)\varepsilon + \xi(X_k, \alpha_k)$. Since $\varphi_{xx}''(\cdot, i)$ is continuous with compact support, the weak convergence of $(X^\varepsilon(\cdot), \alpha^\varepsilon(\cdot))$, the Skorohod representation, and the boundedness of $h_j(\cdot, i)$ then yield

$$
\lim_{\varepsilon \to 0} E\left\{ \prod_{j=1}^{\kappa} h_j(X^\varepsilon(t_j), \alpha^\varepsilon(t_j)) \sum_{l=t/\Delta_\varepsilon}^{(t+s)/\Delta_\varepsilon - 1} \sum_{k:k\in\chi_l^\varepsilon} \int_0^1 \left[\varphi_{xx}''(X_k + u\widetilde{\Delta x_k}, \alpha_{lm_\varepsilon}) \right. \right.
$$

$$
\left. \left. - \varphi_{xx}''(X_k, \alpha_{lm_\varepsilon}) \right] \widetilde{\Delta x_k}^2 \, du \right\} = 0.
$$

$$
\tag{38}
$$

Since $\varphi_x'(\cdot, i)$ and $\mu(\cdot, i)$ are continuous for each $i \in \mathcal{M}$, we have

$$
\lim_{\varepsilon \to 0} E \left\{ \prod_{j=1}^{\kappa} h_j(X^\varepsilon(t_j), \alpha^\varepsilon(t_j)) \sum_{l=t/\Delta_\varepsilon}^{(t+s)/\Delta_\varepsilon - 1} \sum_{k:k \in \chi_l^\varepsilon} \varphi_x'(X_k, \alpha_{lm_\varepsilon}) \mu(X_k, \alpha_k) \varepsilon \right\}
$$
$$
= \lim_{\varepsilon \to 0} E \left\{ \prod_{j=1}^{\kappa} h_j(X^\varepsilon(t_j), \alpha^\varepsilon(t_j)) \sum_{l=t/\Delta_\varepsilon}^{(t+s)/\Delta_\varepsilon - 1} \varepsilon \sum_{k:k \in \chi_l^\varepsilon} \varphi_x'(X_{lm_\varepsilon}, \alpha_{lm_\varepsilon}) \mu(X_{lm_\varepsilon}, \alpha_k) \right\}
$$
$$
= \lim_{\varepsilon \to 0} E \left\{ \prod_{j=1}^{\kappa} h_j(X^\varepsilon(t_j), \alpha^\varepsilon(t_j)) \sum_{l=t/\Delta_\varepsilon}^{(t+s)/\Delta_\varepsilon - 1} \varepsilon \sum_{k:k \in \chi_l^\varepsilon} E_{lm_\varepsilon} \right.
$$
$$
\left. \times \left[\sum_{i=1}^{m_0} \varphi_x'(X_{lm_\varepsilon}, \alpha_{lm_\varepsilon}) \mu(X_{lm_\varepsilon}, i) I_{\{\alpha_k = i\}} \right] \right\}.
$$

Note that when $\varepsilon \to 0$ and $l\varepsilon m_\varepsilon = l\Delta_\varepsilon \to u$, $(l+1)\Delta_\varepsilon \to u$ and for any k satisfying $lm_\varepsilon \le k < (l+1)m_\varepsilon$, $\varepsilon k \to u$. In addition, in view of the weak convergence of $\alpha^\varepsilon(\cdot)$ to $\alpha(\cdot)$ and the Skorohod representation (without changing notation), we may assume that $I_{\{\alpha^\varepsilon(u)=i\}} \to I_{\{\alpha(u)=i\}}$ w.p.1. As a result, for the above limit, we have,

$$
\lim_{\varepsilon \to 0} E \left\{ \prod_{j=1}^{\kappa} h_j(X^\varepsilon(t_j), \alpha^\varepsilon(t_j)) \sum_{l=t/\Delta_\varepsilon}^{(t+s)/\Delta_\varepsilon - 1} \varepsilon \sum_{k:k \in \chi_l^\varepsilon} E_{lm_\varepsilon} \right.
$$
$$
\left. \times \left[\sum_{i=1}^{m_0} \varphi_x'(X_{lm_\varepsilon}, \alpha_{lm_\varepsilon}) \mu(X_{lm_\varepsilon}, i) I_{\{\alpha_k = i\}} \right] \right\}
$$
$$
= \lim_{\varepsilon \to 0} E \left\{ \prod_{j=1}^{\kappa} h_j(X^\varepsilon(t_j), \alpha^\varepsilon(t_j)) \sum_{i=1}^{m_0} \sum_{l=t/\Delta_\varepsilon}^{(t+s)/\Delta_\varepsilon - 1} \Delta_\varepsilon \varphi_x'(X^\varepsilon(l\Delta_\varepsilon), \alpha^\varepsilon(l\Delta_\varepsilon)) \right.
$$
$$
\mu(X^\varepsilon(l\Delta_\varepsilon), i) \times \frac{1}{m_\varepsilon} E_{lm_\varepsilon} \left[\sum_{k:k \in \chi_l^\varepsilon} I_{\{\alpha^\varepsilon(\varepsilon k)=i\}} \right] \Bigg\}
$$
$$
= E \left\{ \prod_{j=1}^{\kappa} h_j(X(t_j), \alpha(t_j)) \sum_{i=1}^{m_0} \int_t^{t+s} \varphi_x'(X(u), \alpha(u)) \mu(X(u), i) I_{\{\alpha(u)=i\}} \, du \right\}
$$
$$
= E \left\{ \prod_{j=1}^{\kappa} h_j(X(t_j), \alpha(t_j)) \int_t^{t+s} \varphi_x'(X(u), \alpha(u)) \mu(X(u), \alpha(u)) \, du \right\}.
$$

$$(39)$$

In view of (32), we have

$$
E\left\{\prod_{j=1}^{\kappa} h_j(X^\varepsilon(t_j), \alpha^\varepsilon(t_j)) \sum_{l=t/\Delta_\varepsilon}^{(t+s)/\Delta_\varepsilon-1} \sum_{k:k\in\chi_l^\varepsilon} \varphi'_x(X_k, \alpha_{lm_\varepsilon}) \xi(X_k, \alpha_k)\right\}
$$
$$
= E\left\{\prod_{j=1}^{\kappa} h_j(X^\varepsilon(t_j), \alpha^\varepsilon(t_j)) \sum_{l=t/\Delta_\varepsilon}^{(t+s)/\Delta_\varepsilon-1} \sum_{k:k\in\chi_l^\varepsilon} \varphi'_x(X_k, \alpha_{lm_\varepsilon}) E_k\left[\xi(X_k, \alpha_k)\right]\right\}
$$
$$
= 0,
$$

$$(40)$$

and similarly,

$$
E\left\{\prod_{j=1}^{\kappa} h_j(X^\varepsilon(t_j), \alpha^\varepsilon(t_j)) \sum_{l=t/\Delta_\varepsilon}^{(t+s)/\Delta_\varepsilon-1} \sum_{k:k\in\chi_l^\varepsilon} \varphi''_{xx}(X_k, \alpha_{lm_\varepsilon})\right.
$$
$$
\left. \times \mu(X_k, \alpha_k) \xi(X_k, \alpha_k)\varepsilon \right\} = 0.
$$

$$(41)$$

Using the boundedness of $\varphi''_{xx}(\cdot, i)$, $h_j(\cdot, i)$, the linear growth condition (15) for $\mu(\cdot, i)$, and the estimate (26), we have,

$$
E\left|\prod_{j=1}^{\kappa} h_j(X^\varepsilon(t_j), \alpha^\varepsilon(t_j)) \sum_{l=t/\Delta_\varepsilon}^{(t+s)/\Delta_\varepsilon-1} \sum_{k:k\in\chi_l^\varepsilon} \varphi''_{xx}(X_k, \alpha_{lm_\varepsilon})\mu^2(X_k, \alpha_k)\varepsilon^2\right|
$$
$$
\leq K \sum_{l=t/\Delta_\varepsilon}^{(t+s)/\Delta_\varepsilon-1} \sum_{k:k\in\chi_l^\varepsilon} \varepsilon^2 E\left[\mu^2(X_k, \alpha_k)\right] \leq K\varepsilon^2 \sum_{l=t/\Delta_\varepsilon}^{(t+s)/\Delta_\varepsilon-1} \sum_{k:k\in\chi_l^\varepsilon} \left(1 + E\left[X_k^2\right]\right)
$$
$$
\leq K\varepsilon^2 \cdot \frac{T}{\varepsilon} = O(\varepsilon) \to 0 \text{ as } \varepsilon \to 0.
$$

$$(42)$$

In view of (32), we have

$$
E\left\{\prod_{j=1}^{\kappa} h_j(X^\varepsilon(t_j), \alpha^\varepsilon(t_j)) \sum_{l=t/\Delta_\varepsilon}^{(t+s)/\Delta_\varepsilon-1} \sum_{k:k\in\chi_l^\varepsilon} \varphi''_{xx}(X_k, \alpha_{lm_\varepsilon}) \xi^2(X_k, \alpha_k)\right\}
$$
$$
= E\left\{\prod_{j=1}^{\kappa} h_j(X^\varepsilon(t_j), \alpha^\varepsilon(t_j)) \sum_{l=t/\Delta_\varepsilon}^{(t+s)/\Delta_\varepsilon-1} \sum_{k:k\in\chi_l^\varepsilon} \varphi''_{xx}(X_k, \alpha_{lm_\varepsilon}) E_k\left[\xi^2(X_k, \alpha_k)\right]\right\}
$$
$$
= E\left\{\prod_{j=1}^{\kappa} h_j(X^\varepsilon(t_j), \alpha^\varepsilon(t_j)) \sum_{l=t/\Delta_\varepsilon}^{(t+s)/\Delta_\varepsilon-1} \sum_{k:k\in\chi_l^\varepsilon} \varphi''_{xx}(X_k, \alpha_{lm_\varepsilon}) \sigma^2(X_k, \alpha_k)\varepsilon \right\}.
$$

By using a similar argument to the one leading to (39), we can show that

$$
\lim_{\varepsilon \to 0} E \left\{ \prod_{j=1}^{\kappa} h_j(X^\varepsilon(t_j), \alpha^\varepsilon(t_j)) \sum_{l=t/\Delta_\varepsilon}^{(t+s)/\Delta_\varepsilon - 1} \sum_{k: k \in \chi_l^\varepsilon} \varphi_{xx}''(X_k, \alpha_{lm_\varepsilon}) \xi^2(X_k, \alpha_k) \right\}
$$

$$
= E \left\{ \prod_{j=1}^{\kappa} h_j(X(t_j), \alpha(t_j)) \int_t^{t+s} \varphi_{xx}''(X(u), \alpha(u)) \sigma^2(X(u), \alpha(u)) \, du \right\}. \tag{43}
$$

Using (38)–(43) in (37), we obtain

$$
\lim_{\varepsilon \to 0} E \left\{ \prod_{j=1}^{\kappa} h_j(X^\varepsilon(t_j), \alpha^\varepsilon(t_j)) \sum_{l=t/\Delta_\varepsilon}^{(t+s)/\Delta_\varepsilon - 1} \left[\varphi(X_{(l+1)m_\varepsilon}, \alpha_{lm_\varepsilon}) - \varphi(X_{lm_\varepsilon}, \alpha_{lm_\varepsilon}) \right] \right\}
$$

$$
= E \left\{ \prod_{j=1}^{\kappa} h_j(X(t_j), \alpha(t_j)) \int_t^{t+s} \left[\varphi_x'(X(u), \alpha(u)) \mu(X(u), \alpha(u)) \right. \right.
$$

$$
\left. \left. + \ \varphi_{xx}''(X(u), \alpha(u)) \sigma^2(X(u), \alpha(u)) \right] du \right\}.
$$

$$\tag{44}$$

Step 3. Now we work on the first summation (the second line) of (36). Using the continuity of $\varphi(\cdot, i)$ for each $i \in \mathcal{M}$, we can argue along the same line as in Step 2 that the limit of

$$
\sum_{l=t/\Delta_\varepsilon}^{(t+s)/\Delta_\varepsilon - 1} \left[\varphi(X_{(l+1)m_\varepsilon}, \alpha_{(l+1)m_\varepsilon}) - \varphi(X_{(l+1)m_\varepsilon}, \alpha_{lm_\varepsilon}) \right]
$$

is the same as that of

$$
\sum_{l=t/\Delta_\varepsilon}^{(t+s)/\Delta_\varepsilon - 1} \left[\varphi(X_{lm_\varepsilon}, \alpha_{(l+1)m_\varepsilon}) - \varphi(X_{lm_\varepsilon}, \alpha_{lm_\varepsilon}) \right] + \eta_\varepsilon
$$

where $\eta_\varepsilon \to 0$ in probability uniformly in t, as $\varepsilon \to 0$. We then have,

$$
E \left\{ \prod_{j=1}^{\kappa} h_j(X^\varepsilon(t_j), \alpha^\varepsilon(t_j)) \sum_{l=t/\Delta_\varepsilon}^{(t+s)/\Delta_\varepsilon - 1} \left[\varphi(X_{lm_\varepsilon}, \alpha_{(l+1)m_\varepsilon}) - \varphi(X_{lm_\varepsilon}, \alpha_{lm_\varepsilon}) \right] \right\}
$$

$$
= E \left\{ \prod_{j=1}^{\kappa} h_j(X^\varepsilon(t_j), \alpha^\varepsilon(t_j)) \sum_{l=t/\Delta_\varepsilon}^{(t+s)/\Delta_\varepsilon - 1} \sum_{k: k \in \chi_l^\varepsilon} \left[\varphi(X_{lm_\varepsilon}, \alpha_{k+1}) - \varphi(X_{lm_\varepsilon}, \alpha_k) \right] \right\}
$$

$$= E\left\{ \prod_{j=1}^{\kappa} h_j(X^\varepsilon(t_j), \alpha^\varepsilon(t_j)) \sum_{l=t/\Delta_\varepsilon}^{(t+s)/\Delta_\varepsilon - 1} \sum_{k: k \in \chi_l^\varepsilon} \sum_{i=1}^{m_0} E_k\{[\varphi(X_{lm_\varepsilon}, \alpha_{k+1}) \right.$$

$$\left. - \varphi(X_{lm_\varepsilon}, \alpha_i)] I_{\{\alpha_k = i\}} \} \right\}$$

$$= E\left\{ \prod_{j=1}^{\kappa} h_j(X^\varepsilon(t_j), \alpha^\varepsilon(t_j)) \sum_{l=t/\Delta_\varepsilon}^{(t+s)/\Delta_\varepsilon - 1} \sum_{k: k \in \chi_l^\varepsilon} \sum_{i=1}^{m_0} \left[\sum_{j=1}^{m_0} \varphi(X_{lm_\varepsilon}, j) \right. \right.$$

$$\left. \left. \times P\{\alpha_{k+1} = j | \alpha_k = i\} - \varphi(X_{lm_\varepsilon}, i) \right] I_{\{\alpha_k = i\}} \right\}$$

$$= E\left\{ \prod_{j=1}^{\kappa} h_j(X^\varepsilon(t_j), \alpha^\varepsilon(t_j)) \sum_{l=t/\Delta_\varepsilon}^{(t+s)/\Delta_\varepsilon - 1} \sum_{k: k \in \chi_l^\varepsilon} \sum_{i=1}^{m_0} \left[\sum_{j=1}^{m_0} \varphi(X_{lm_\varepsilon}, j)(\delta_{ij} + \varepsilon q_{ij}) \right. \right.$$

$$\left. \left. - \varphi(X_{lm_\varepsilon}, i) \right] I_{\{\alpha_k = i\}} \right\}$$

$$= E\left\{ \prod_{j=1}^{\kappa} h_j(X^\varepsilon(t_j), \alpha^\varepsilon(t_j)) \sum_{l=t/\Delta_\varepsilon}^{(t+s)/\Delta_\varepsilon - 1} \sum_{k: k \in \chi_l^\varepsilon} \sum_{i=1}^{m_0} \sum_{j=1}^{m_0} \varepsilon q_{ij} \varphi(X_{lm_\varepsilon}, j) I_{\{\alpha_k = i\}} \right\}$$

$$\to E\left\{ \prod_{j=1}^{\kappa} h_j(X(t_j), \alpha(t_j)) \int_t^{t+s} Q\varphi(X(u), \cdot)(\alpha(u)) \, du \right. \quad \text{as } \varepsilon \to 0,$$

$$(45)$$

in which (21) is used for the one-step transition probability of $\{\alpha_n, n \geq 0\}$.

Step 4. By combining Step 1 to 3, we have

$$E\left\{ \prod_{j=1}^{\kappa} h_j(X^\varepsilon(t_j), \alpha^\varepsilon(t_j)) [\varphi(X^\varepsilon(t+s), \alpha^\varepsilon(t+s)) - \varphi(X^\varepsilon(t), \alpha^\varepsilon(t))] \right\}$$

$$\to E\left\{ \prod_{j=1}^{\kappa} h_j(X(t_j), \alpha(t_j)) \int_t^{t+s} \mathcal{L}\varphi(X(u), \alpha(u)) \, du \right\} \quad \text{as } \varepsilon \to 0.$$

On the other hand, by virtue of the weak convergence of $(X^\varepsilon(\cdot), \alpha^\varepsilon(\cdot))$, the Skorohod representation, and the continuity of $h_j(\cdot, i)$ and $\varphi(\cdot, i)$, we have,

$$E\left\{\prod_{j=1}^{\kappa} h_j(X^\varepsilon(t_j), \alpha^\varepsilon(t_j)) \left[\varphi(X^\varepsilon(t+s), \alpha^\varepsilon(t+s)) - \varphi(X^\varepsilon(t), \alpha^\varepsilon(t))\right]\right\}$$

$$\to E\left\{\prod_{j=1}^{\kappa} h_j(X(t_j), \alpha(t_j)) \left[\varphi(X(t+s), \alpha(t+s)) - \varphi(X(t), \alpha(t))\right]\right\} \quad \text{as } \varepsilon \to 0.$$

Hence Equation (34) is verified, and the desired result follows and the proof of Theorem 1 is completed. □

4. Concluding Remarks

In this paper we present a multinomial tree method for option pricing when the underlying asset price follows a regime-switching model. The new tree grows linearly as the number of time steps increases. We prove the weak convergence of the discrete tree approximations to the continuous regime-switching process by applying the method of martingale problem formulation.

References

1. D.D. Aingworth, S.R. Das and R. Motwani, A simple approach for pricing equlty options with Markov switching state variables, *Quantitative Finance*, **6** (2006), 95-105.
2. C. Albanese, Affine lattice models, *International Journal of Theoretical and Applied Finance*, **8** (2005), 223-238.
3. N.P.B. Bollen, Valuing options in regime-switching models, *Journal of Derivatives*, Vol. **6** (1998), 38-49.
4. P. Boyle and T. Draviam, Pricing exotic options under regime switching, *Insurance: Mathematics and Econimics*, Vol. **40** (2007), 267-282.
5. J. Buffington and R.J. Elliott, American options with regime switching, *Int. J. Theor. Appl. Finance*, **5** (2002), 497-514.
6. L.B. Chang and K. Palmer, Smooth convergence in the binomial model, *Finance and Stochastics*, Vol. **11**, No. **1** (2007), 91-105.
7. J. Cox, S. Ross, and M. Rubinstein, Option pricing, a simplified approach, *Journal of Financial Economics* **7** (1979), 229-263.
8. F. Diener and M. Diener, Asymptotics of the price oscillations of a european call option in a tree model, *Mathematical Finance*, **14** (2004), 271-293.
9. P. Eloe, R.H. Liu and J.Y. Sun, Double barrier option under regime-switching exponential mean-reverting process, *International Journal of Computer Mathematics*, Vol. **86**, No. **6** (2009), 964-981.

10. I. Florescu and F.G. Viens, Stochastic volatility: Option pricing using a multinomial recombining tree, *Applied Mathematical Finance*, **15** (2008), 151-181.
11. X. Guo, Information and option pricings, *Quant. Finance*, Vol. **1** (2000), 38-44.
12. X. Guo and Q. Zhang, Closed-form solutions for perpetual American put options with regime switching, *SIAM J. Appl. Math.*, **64**, No. **6** (2004), 2034-2049.
13. M.R. Hardy, A regime-switching model for long-term stock returns, *North American Actuarial Journal*, Vol. **5**, No. **2** (2001), 41-53.
14. H. He, Convergence from discrete to continuous time contingent claims prices, *The Review of Finanical Studies*, Vol. **3** (1990), 523-546.
15. S. Heston and G.F. Zhou, On the rate of convergence of discrete-time contingent claims, *Mathematical Finance*, **10** (2000), 53-75.
16. J.E. Hilliard and A. Schwartz, Binomial option pricing under stochastic volatility and correlated state variables, *Journal of Derivatives*, Fall (1996), 23-39
17. A.Q.M. Khaliq and R.H. Liu, New numerical scheme for pricing American option with regime-switching, *Int. J. Theor. Appl. Finance*, Vol. **12**, No. **3** (2009), 319-340.
18. H.J. Kushner, *Approximation and Weak Convergence Methods for Random Processes with Applications to Stochastic Systems Theory*, MIT Press, Cambridge, MA, 1984.
19. R.H. Liu, Regime-switching recombining tree for option pricing, *Int. J. Theor. Appl. Finance*, Vol. **13**, No. **3** (2010), forthcoming.
20. R.H. Liu, Q. Zhang and G. Yin, Option pricing in a regime switching model using the Fast Fourier Transform, *Journal of Applied Mathematics and Stochastic Analysis*, Vol. 2006, Article ID 18109, doi:10.1155/JAMSA/2006/18109.
21. D.B. Nelson and K. Ramaswamy, Simple binomial processes as diffusion approximations in financial models, *The Review of Finanical Studies*, Vol. **3**, No. **3** (1990), 393-430.
22. J.A. Primbs, M. Rathinam and Y. Yamada, Option pricing with a pentanomial lattice model that incorporates skewness and kurtosis, *Applied Mathematical Finance*, **14** (2007), 1-17.
23. J.B. Walsh, The rate of convergence of the binomial tree scheme, *Finance and Stochastics*, Vol. **7**, No. **3** (2003), 337-361.
24. D.D. Yao, Q. Zhang and X.Y. Zhou, A regime-switching model for European options, in *Stochastic Processes, Optimization, and Control Theory: Applications in Financial Engineering, Queueing Networks, and Manufacturing Systems* (2006), H.M. Yan, G. Yin, and Q. Zhang Eds., 281-300, Springer.
25. G. Yin, Q. Song, and Z. Zhang, Numerical solutions for jump-diffusions with regime switching, *Stochastics*, **77** (2005), 61-79.
26. G. Yin and Q. Zhang, *Continuous-Time Markov Chains and Applications: A Singular Perturbation Approach*, Springer, 1998.

27. G. Yin and Q. Zhang, and G. Badowski, Discrete-time singularly perturbed Markov chains: Aggregation, occupation measures, and switching diffusion limit, *Advances in Applied Probability*, 35 (2003), 449-476.
28. G. Yin and C. Zhu, *Hybrid Switching Diffusions: Properties and Applications*, Springer, 2010.

Optimal Reinsurance under a Jump Diffusion Model

Shangzhen Luo

Department of Mathematics,
University of Northern Iowa,
Cedar Falls, Iowa, 50614-0506, USA
Email: luos@uni.edu

We study an optimal dynamic control problem of an insurance company whose wealth process is modeled by a jump diffusion. The company can purchase proportional reinsurance to reduce its risk level and invest all surplus in a risk free asset. The objective is to maximize the expected value of an exponential utility of the terminal wealth. We derive explicit optimal value function and find its associated reinsurance strategy.

Keywords: Stochastic control; HJB equation; proportional reinsurance; exponential utility; jump diffusion.

1. Introduction

During last few decades, various applications of stochastic control theory in actuarial science have been appearing. See, e.g., Asmussen and Taksar,[1] Browne,[2] Browne,[3] Emanuel et. al,[5] Höjgaard and Taksar,[7] Höjgaard and Taksar,[8] Liang,[9] Luo et. al,[10] Schmidli,[11] Schmidli,[12] Taksar and Markussen[13] etc. Specifically, Browne[2] studied a diffusion model with investment under exponential utility maximization and ruin minimization criteria; Yang and Zhang[15] studied the same problem under more general jump diffusion settings. Noticing that in many recent actuarial models, e.g., Höjgaard and Taksar,[7] Höjgaard and Taksar,[8] and Luo et. al,[10] control of reinsurance purchase is incorporated as a part of the management of the insurance company. In this paper, we apply stochastic control theory to solve an optimization problem on reinsurance. We consider an exponential utility maximization as the optimization scenario; more exactly the objective is to seek an optimal proportional reinsurance policy to maximize the expected utility of the terminal wealth; here we assume that the surplus of the insurance company is modeled by a perturbed compound Poisson process, which is a jump diffusion process. Note that this class of risk processes has been studied extensively in the literature (see, e.g., Dufresne and Gerber,[4] Gerber and Yang,[6] and Wang and Wu[14]). A similar study to this paper has been conducted in Liang,[9] where the utility maximization problem was studied under both diffusion approximation and jump diffusion models. There investment was not considered. However in this paper (as in Taksar and Markussen[13]) we assume that there is a risk free asset in the market, and we then solve the optimization problem in the case with all surplus invested in the riskless asset. More explicit results are given when the claim size distribution is exponential or gamma. Economic analysis on relations between optimal reinsurance strategy and exogenous parameters is also provided.

We start with a perturbed Cramer-Lundberg risk process in differential form

$$dR_t = \mu dt + \sigma dw_t - dL_t,$$
$$R_0 = x, \tag{1}$$

where μ is the premium rate, x is the initial surplus level, σ is the volatility of the perturbing process w_t which is a standard Brownian motion independent of process L_t, and L_t is a compound Poisson process given by

$$L_t = \sum_{i=1}^{N_t} Y_i, \tag{2}$$

here Y_i's are i.i.d. non-negative claims with common distribution function F, moment generating function M_Y and finite expectation α, and $\{N_t\}_{t \geq 0}$ is a Poisson process (with intensity λ) that counts the number of claims. We suppose that the insurance company can buy *proportional reinsurance* to reduce the aggregate loss L_t; however, the risk of the perturbing noise σw_t cannot be controlled by reinsurance, and then the model is never perfectly hedged. The retention level of proportional reinsurance $u_t (> 0)$ at time t is determined by the management of the insurance company. We denote the reinsurance control policy by $\pi = \{u_t\}_{t \geq 0}$. For a fixed retention level u_t at time t, with proportion of u_t, the claims are paid by the insurance company, and the rest proportion of $(1 - u_t)$ by reinsurance company. In return, the insurance company diverts its premium at rate of β. In this paper, we call the reinsurance is *expensive* if $\beta > \lambda \alpha$, and *cheap* if $\beta \leq \lambda \alpha$; we note here that the quantity $\lambda \alpha$ is the expected loss per unit time. We also assume that the surplus is invested in a risk free asset with interest rate r, then the dynamics of surplus under policy π become

$$dR_t^\pi = (\mu - \beta(1 - u_t) + rR_t^\pi)dt + \sigma dw_t - u_t dL_t,$$
$$R_0 = x. \tag{3}$$

The rest of the paper is organized as follows: in Section 2 we formulate the optimization problem and give its corresponding HJB equation; in Section 3 we provide a general solution to the HJB equation; and we consider the cases when the claim size follows exponential distribution and gamma distribution in Section 4.

2. Formulation and HJB Equation

Suppose a complete probability space $(\Omega, \mathcal{F}, P)$ endowed with filtration $\mathcal{G} = \{\mathcal{G}_t\}_{t \geq 0}$ and the standard Brownian motion $\{w_t\}_{t \geq 0}$ is adapted to $\mathcal{G}$. A policy $\pi = \{u_t\}_{t \geq 0}$ is said to be *admissible* if $\{u_t\}_{t \geq 0}$ is a predictable process with respect to the filtration $\mathcal{G}$ and for each $t \geq 0$, $0 \leq u_t \leq 1$. The set of all admissible policies is denoted by Π.

Now consider an exponential utility function

$$U(x) = c - \frac{\delta}{\gamma}e^{-\gamma x}, \tag{4}$$

where c, $\gamma(> 0)$, and $\delta(> 0)$ are constants. For any admissible policy $\pi \in \Pi$, define a value function as

$$V_\pi(t, x) = E[U(R_T^\pi)|R_t^\pi = x], \tag{5}$$

where $0 \leq t \leq T$ and $-\infty < x < \infty$. The objective is to find the optimal value function

$$V(t, x) = \sup_{\pi \in \Pi} V_\pi(x), \tag{6}$$

and an optimal control policy π^* such that $V(t, x) = V_{\pi^*}(t, x)$.

For any admissible control policy π, applying Ito's lemma to a function $g(t, x)$ of the jump diffusion process governed by (3), we obtain the following:

$$g(t, R_t^\pi) - g(0, x)$$

$$= \int_0^t \{g_s(s, R_s^\pi) + [\mu - \beta(1 - u_s) + rR_s^\pi]g_x(s, R_s^\pi) + \frac{1}{2}\sigma^2 g_{xx}(s, R_s^\pi)\}ds$$

$$+ \int_0^t g_x(s, R_s^\pi)\sigma dw_s + \sum_{s \in \mathcal{J}_\mathcal{L}; s \leq t}[g(s, R_s^\pi) - g(s, R_{s-}^\pi)],$$

$$\tag{7}$$

where g_s, g_x, and g_{xx} are partial derivatives, and J_L is the set of jumping times of the claim process $L := \{L_s\}_{s \geq 0}$. Thus the generator of the controlled process under a fixed policy $u_t \equiv u$ is given by

$$
\begin{aligned}
\mathcal{A}^u g(t,x) = & g_t(t,x) + [\mu - \beta(1-u) + rx]g_x(t,x) + \frac{1}{2}\sigma^2 g_{xx}(t,x) \\
& + \lambda E[g(t, x - uY) - g(t,x)].
\end{aligned}
\tag{8}
$$

Suppose the function V has continuous partial derivatives V_t, V_x and V_{xx}, and all limits and expectations can be interchanged, then one can show that the function V solves an HJB equation given as follows

$$
\sup_{0 \leq u \leq 1} \mathcal{A}^u V(t,x) = 0,
\tag{9}
$$

with boundary condition

$$
V(T,x) = U(x).
\tag{10}
$$

We shall find a classical solution V with $V_x > 0$ and $V_{xx} < 0$ to the HJB equation on $\mathcal{I}$ where

$$
\mathcal{I} = [0,T] \times (-\infty, \infty).
\tag{11}
$$

3. Solution to the HJB Equation

In this section, we solve the HJB equation. Motivated by Browne[3] and Yang and Zhang,[15] we try to fit a solution of the following form

$$
V(t,x) = c - \frac{\delta}{\gamma} \exp\{-\gamma x e^{r(T-t)} + h(T-t)\},
\tag{12}
$$

with boundary condition $V(T,x) = u(x)$ or $h(0) = 0$, where function h is to be determined later. Now we compute partial derivatives:

$$
\begin{aligned}
V_t(t,x) = & (V(t,x) - c)[-h'(T-t) + \gamma rx e^{r(T-t)}], \\
V_x(t,x) = & (V(t,x) - c)(-\gamma e^{r(T-t)}), \\
V_{xx}(t,x) = & (V(t,x) - c)\gamma^2 e^{2r(T-t)}.
\end{aligned}
\tag{13}
$$

Plug-in the expressions in (13) into HJB equation (9) and simplify, then it becomes

$$
\begin{aligned}
\sup_{0 \leq u \leq 1} \big\{ & -h'(T-t) - \gamma(\mu - \beta(1-u))e^{r(T-t)} \\
& + \frac{1}{2}\gamma^2 \sigma^2 e^{2r(T-t)} + \lambda M_Y(\gamma e^{r(T-t)}u) - \lambda \big\} = 0.
\end{aligned}
\tag{14}
$$

By differentiation, the maximizer over $u \in (-\infty, \infty)$ in the HJB equation, denoted by $\tilde{u}(t)$ (it does not depend on x), is the unique solution of the following equation

$$-\frac{\beta}{\lambda} + E(Y \exp\{u\gamma e^{r(T-t)}Y\}) = 0. \tag{15}$$

Thus the maximizer over $u \in [0,1]$ in the HJB equation, denoted by $u^*(t)$ is given as follows:

$$u^*(t) = \begin{cases} 0 & if \ \ \tilde{u}(t) \leq 0 \\ \tilde{u}(t) & if \ \ 0 < \tilde{u}(t) < 1 \\ 1 & if \ \ \tilde{u}(t) \geq 1 \end{cases}. \tag{16}$$

We note that $u^*(t)$ and $\tilde{u}(t)$ are both continuous and increasing functions in t.

Theorem 3.1. *The optimal value function V in (6) admits form (12), its associated optimal investment feedback function $u^*(t)$ is given by (16), and function h in (12) solves the following equation*

$$h'(T-t) = -\gamma(\mu - \beta(1 - u^*(t)))e^{r(T-t)}$$
$$+ \frac{1}{2}\gamma^2\sigma^2 e^{2r(T-t)} + \lambda M_Y(\gamma e^{r(T-t)} u^*(t)) - \lambda, \tag{17}$$

with boundary condition $h(0) = 0$.

Lemma 3.1. *Under the optimal reinsurance policy $\pi^* = \{u^*(t)\}$, the process $\{V(t, R_t^{\pi^*})\}_{0 \leq t \geq T}$ is a martingale.*

Proof. Apply Ito's formula to $V(s, R_s^{\pi^*})$ and take expectation, then we have

$$E[V(t, R_t^{\pi^*}) - V(0, x)] = \int_0^t \mathcal{A}^{u^*(s)} V(s, R_s^{\pi^*})ds + E[\int_0^t \sigma V_x(s, R_s^{\pi^*})dw_s],$$
$$= 0,$$

since $\mathcal{A}^{u^*(s)} V(s, R_s^{\pi^*}) = 0$ and $V_x(s, R_s^{\pi^*})$ is squarely integrable. $\quad\square$

One can also show that $\{V(t, R_t^{\pi})\}_{0 \leq t \geq T}$ is a super-martingale for any admissible control π, and the theorem follows immediately.

Remark 3.1. In the case of cheap reinsurance, i.e., $\beta \leq \lambda\alpha$, it always holds that $\tilde{u}(t) \leq 0$. Thus the optimal reinsurance strategy is

$$u^*(t) \equiv 0;$$

it implies that the optimal control policy for the insurance company is to buy 100% of reinsurance, and hence the risk on claims is eliminated. Consequently, the optimal value function is given as follows

$$V(t,x) = c - \frac{\delta}{\gamma} \exp\{-\gamma x e^{r(T-t)} + \frac{\sigma^2 \gamma^2}{4r} e^{2r(T-t)}$$

$$- \frac{\gamma}{r}(\mu - \beta)e^{r(T-t)} - \frac{\sigma^2 \gamma^2}{4r} + \frac{\gamma}{r}(\mu - \beta)\},$$

which does not depend on the claim size distribution.

In the following section, we shall consider the case of expensive reinsurance, i.e., $\beta > \lambda\alpha$, for two special claim distributions - exponential distribution and Gamma distribution.

4. Special Claim Distributions

In this section, we study the cases with exponential claims and gamma claims. Some economic analysis and financial interpretations are provided. Certain insights on the interplay between the optimal reinsurance strategy and the exogenous parameters are also given.

4.1. *The Case with Exponential Claims*

In this subsection, we study the special case when the claim size distribution is exponential with mean α. Immediately, Equation (15) becomes

$$\frac{\beta}{\lambda} = E[Y \exp\{u\gamma e^{r(T-t)}Y\}]$$

$$= \frac{1}{\alpha} \int_0^\infty y e^{-\frac{y}{\alpha}} \exp\{u\gamma e^{r(T-t)}y\}dy$$

$$= \frac{1}{\alpha[\frac{1}{\alpha} - u\gamma e^{r(T-t)}]^2},$$

thus

$$\tilde{u}(t) = \frac{\frac{1}{\alpha} - \sqrt{\frac{\lambda}{\alpha\beta}}}{\gamma e^{r(T-t)}}. \tag{18}$$

Note that when the reinsurance is expensive $\beta > \lambda\alpha$, it always holds $\tilde{u}(t) > 0$.

In the sequel, we give three theorems according to the following three possible parameter cases:

(1) $\frac{\frac{1}{\alpha} - \sqrt{\frac{\lambda}{\alpha\beta}}}{\gamma e^{rT}} \geq 1$,

(2) $0 < \dfrac{\frac{1}{\alpha} - \sqrt{\frac{\lambda}{\alpha\beta}}}{\gamma} < 1$, and

(3) $\dfrac{\frac{1}{\alpha} - \sqrt{\frac{\lambda}{\alpha\beta}}}{\gamma} \geq 1 > \dfrac{\frac{1}{\alpha} - \sqrt{\frac{\lambda}{\alpha\beta}}}{\gamma e^{rT}}$.

The theorems are readily derived from Theorem 3.1, and we omit their proofs. We note that if $\dfrac{\frac{1}{\alpha} - \sqrt{\frac{\lambda}{\alpha\beta}}}{\gamma e^{rT}} \geq 1$, then $\tilde{u}(t) \geq 1$ and $u^*(t) = 1$; further, when $u^*(x) \equiv 1$, the function h in (12) solves

$$h'(T - t) = g_1(T - t),$$

where

$$g_1(T - t) = -\gamma\mu e^{r(T-t)} + \frac{1}{2}\sigma^2\gamma^2 e^{2r(T-t)} + \frac{\lambda}{\alpha(\frac{1}{\alpha} - \gamma e^{r(T-t)})} - \lambda. \quad (19)$$

Thus we obtain:

Theorem 4.1. *If* $\dfrac{\frac{1}{\alpha} - \sqrt{\frac{\lambda}{\alpha\beta}}}{\gamma e^{rT}} \geq 1$, *then the optimal reinsurance retention level is*

$$u^*(t) \equiv 1,$$

and the optimal value function is given by

$$V(t,x) = c - \frac{\delta}{\gamma}\left(\frac{1 - \alpha\gamma}{1 - \alpha\gamma e^{r(T-t)}}\right)^{\lambda/r} \exp\{-\gamma x e^{r(T-t)} + \frac{\sigma^2\gamma^2}{4r}e^{2r(T-t)}$$
$$- \frac{\gamma}{r}\mu e^{r(T-t)} + \lambda(T - t) - \frac{\sigma^2\gamma^2}{4r} + \frac{\gamma}{r}\mu\}.$$

Remark 4.1. In Theorem 4.1, the assumption $\dfrac{\frac{1}{\alpha} - \sqrt{\frac{\lambda}{\alpha\beta}}}{\gamma e^{rT}} \geq 1$ is equivalent to $\beta \geq \dfrac{\lambda}{\alpha(\frac{1}{\alpha} - \gamma e^{rT})^2}$ and $\frac{1}{\alpha} - \gamma e^{rT} > 0$. Noticing

$$\beta \geq \frac{\lambda}{\alpha(\frac{1}{\alpha} - \gamma e^{rT})^2} > \frac{\lambda}{\alpha(\frac{1}{\alpha})^2} = \lambda\alpha,$$

it indicates that the reinsurance rate β is very expensive. As a result, the insurance company never purchases reinsurance under such an expensive rate.

Notice that

$$M_Y(\gamma e^{r(T-t)}\tilde{u}(t)) = M_Y(\frac{1}{\alpha} - \sqrt{\frac{\lambda}{\alpha\beta}})$$

$$= \frac{1}{\alpha[\frac{1}{\alpha} - (\frac{1}{\alpha} - \sqrt{\frac{\lambda}{\alpha\beta}})]}$$

$$= \sqrt{\frac{\beta}{\lambda\alpha}},$$

thus when $u^*(t) \equiv \tilde{u}(t)$, the function h in (12) solves

$$h'(T - t) = g_2(T - t),$$

where

$$g_2(T - t) = -\gamma(\mu - \beta)e^{r(T-t)} + \frac{1}{2}\sigma^2\gamma^2 e^{2r(T-t)} - \frac{\beta}{\alpha} + 2\sqrt{\frac{\lambda\beta}{\alpha}} - \lambda. \quad (20)$$

We note here that under constraint $0 < \frac{\frac{1}{\alpha} - \sqrt{\frac{\lambda}{\alpha\beta}}}{\gamma} < 1$, it always holds that $0 < \tilde{u}(t) < 1$. These discussions lead to the following theorem:

Theorem 4.2. *If* $0 < \frac{\frac{1}{\alpha} - \sqrt{\frac{\lambda}{\alpha\beta}}}{\gamma} < 1$, *then the optimal reinsurance retention level is*

$$u^*(t) = \tilde{u}(t) = \frac{\frac{1}{\alpha} - \sqrt{\frac{\lambda}{\alpha\beta}}}{\gamma e^{r(T-t)}},$$

and the optimal value function is given by

$$V(t, x) = c - \frac{\delta}{\gamma}\exp\{-\gamma x e^{r(T-t)} + \frac{\sigma^2\gamma^2}{4r}e^{2r(T-t)} - \frac{\gamma}{r}(\mu - \beta)e^{r(T-t)}$$

$$+ (\frac{\beta}{\alpha} - 2\sqrt{\frac{\lambda\beta}{\alpha}} + \lambda)(T - t) - \frac{\sigma^2\gamma^2}{4r} + \frac{\gamma}{r}(\mu - \beta)\}.$$

Remark 4.2. Notice that the condition $0 < \frac{\frac{1}{\alpha} - \sqrt{\frac{\lambda}{\alpha\beta}}}{\gamma} < 1$ is equivalent to the union of the following two cases:

(1) $\alpha \geq \frac{1}{\gamma}$ and $\lambda\alpha < \beta$ (expensive reinsurance and high expectation of the claim distribution); and

(2) $\alpha < \frac{1}{\gamma}$ and $\lambda\alpha < \beta < \frac{\lambda}{\alpha(\frac{1}{\alpha} - \gamma)^2}$ (expensive reinsurance but not too expensive and low expectation of the claim distribution).

Under these conditions (high claim size expectation, or low claim size expectation with not too expensive reinsurance), the optimal policy always involves certain levels of reinsurance purchase, and it is never 0% or 100%.

Now we consider the last case of expensive reinsurance on parameters: $\frac{\frac{1}{\alpha}-\sqrt{\frac{\lambda}{\alpha\beta}}}{\gamma} \geq 1 > \frac{\frac{1}{\alpha}-\sqrt{\frac{\lambda}{\alpha\beta}}}{\gamma e^{rT}}$; under this constraint, we compute t_0, with $0 \leq t_0 < T$, given as the following,

$$t_0 = \frac{1}{r} \ln \frac{\frac{1}{\alpha} - \sqrt{\frac{\lambda}{\alpha\beta}}}{\gamma}, \tag{21}$$

so that

$$\tilde{u}(T - t_0) = 1.$$

Hence $\tilde{u}(t) \leq 1$ and $u^*(t) = \tilde{u}(t)$ when $0 < t \leq T - t_0$; $\tilde{u}(t) > 1$ and $u^*(t) = 1$ when $T - t_0 < t \leq T$. Now we conclude that the function h in (12) solves the following equation:

$$h'(T - t) = \begin{cases} g_1(T - t) & if \ 0 \leq T - t \leq t_0 \\ g_2(T - t) & if \ t_0 < T - t \leq T \end{cases},$$

where g_1 and g_2 are defined by (19) and (20) respectively. One checks $g_1(t_0) = g_2(t_0)$; thus the piece-wisely defined function above is continuous in t. The discussions above lead to the following theorem:

Theorem 4.3. *If* $\frac{\frac{1}{\alpha}-\sqrt{\frac{\lambda}{\alpha\beta}}}{\gamma} \geq 1 > \frac{\frac{1}{\alpha}-\sqrt{\frac{\lambda}{\alpha\beta}}}{\gamma e^{rT}}$, *then the optimal reinsurance retention level is*

$$u^*(t) = \begin{cases} 1 & if \ 0 \leq T - t \leq t_0 \\ \tilde{u}(t) & if \ t_0 < T - t \leq T \end{cases}, \tag{22}$$

and the optimal value function is given by

$$V(t, x) = c - \frac{\delta}{\gamma} \left(\frac{1 - \alpha\gamma}{1 - \alpha\gamma e^{r(T-t)}} \right)^{\lambda/r} \exp\{ -\gamma x e^{r(T-t)} + \frac{\sigma^2 \gamma^2}{4r} e^{2r(T-t)}$$

$$- \frac{\gamma}{r} \mu e^{r(T-t)} + \lambda(T - t) - \frac{\sigma^2 \gamma^2}{4r} + \frac{\gamma}{r} \mu \},$$

for $0 \leq T - t \leq t_0$, *and*

$$V(t,x) = c - \frac{\delta}{\gamma} \exp\{-\gamma x e^{r(T-t)} + \frac{\sigma^2\gamma^2}{4r} e^{2r(T-t)} - \frac{\gamma}{r}(\mu-\beta)e^{r(T-t)}$$

$$+ (\frac{\beta}{\alpha} - 2\sqrt{\frac{\lambda\beta}{\alpha}} + \lambda)(T-t) - \frac{\sigma^2\gamma^2}{4r} - (\frac{\beta}{\alpha} - 2\sqrt{\frac{\lambda\beta}{\alpha}})t_0 - \frac{\beta\gamma}{r}e^{rt_0}$$

$$- \frac{\lambda}{r}\ln\frac{\frac{1}{\alpha} - \gamma}{\sqrt{\frac{\lambda}{\alpha\beta}}} + \frac{\gamma}{r}\mu\},$$

for $t_0 < T - t \leq T$.

Remark 4.3. Under the condition in Theorem 4.3, the retention level increases to 100% as time evolves; the optimal strategy is to buy less reinsurance gradually down to 0% near the terminal time.

To this end, we have considered all the possible cases of the parameters.

4.2. The Case with Gamma Claims

In this subsection, we assume that claim size distribution F is Gamma with parameters a and b, and mean $\alpha = a/b$. Under this assumption, equation (15) becomes

$$\frac{\beta}{\lambda} = E[Y \exp\{u\gamma e^{r(T-t)}Y\}]$$

$$= \frac{1}{\alpha} \int_0^\infty \frac{b^a y^a \exp\{-by + u\gamma e^{r(T-t)}y\}}{\Gamma(a)} dy$$

$$= \frac{b^a \Gamma(a+1)}{\Gamma(a)(b - u\gamma e^{r(T-t)})^{(a+1)}}$$

$$\times \int_0^\infty \frac{(b - u\gamma e^{r(T-t)})^{(a+1)} y^{(a+1)-1} \exp\{-(b - u\gamma e^{r(T-t)})y\}}{\Gamma(a+1)} dy$$

$$= \frac{ab^a}{(b - u\gamma e^{r(T-t)})^{(a+1)}},$$

thus

$$\tilde{u}(t) = \frac{b - \left(\frac{\lambda a b^a}{\beta}\right)^{\frac{1}{a+1}}}{\gamma e^{r(T-t)}}. \tag{23}$$

We note that under expensive reinsurance $\beta > \lambda\alpha = \lambda\frac{a}{b}$, $\tilde{u}$ is always positive.

In the following, we give three theorems with respect to three possible conditions on the parameters characterized by:

(1) $\dfrac{b-\left(\frac{\lambda a b^a}{\beta}\right)^{\frac{1}{a+1}}}{\gamma e^{rT}} \geq 1,$

(2) $0 < \dfrac{b-\left(\frac{\lambda a b^a}{\beta}\right)^{\frac{1}{a+1}}}{\gamma} < 1,$ and

(3) $\dfrac{b-\left(\frac{\lambda a b^a}{\beta}\right)^{\frac{1}{a+1}}}{\gamma} \geq 1 > \dfrac{b-\left(\frac{\lambda a b^a}{\beta}\right)^{\frac{1}{a+1}}}{\gamma e^{rT}}.$

Theorem 4.4. *If* $\dfrac{b-\left(\frac{\lambda a b^a}{\beta}\right)^{\frac{1}{a+1}}}{\gamma e^{rT}} \geq 1$, *then the optimal reinsurance retention level is*

$$u^*(t) \equiv 1,$$

and the optimal value function is given by (12) *where h solves the following equation*

$$h'(T-t) = f_1(T-t),$$

where

$$f_1(T-t) = -\gamma \mu e^{r(T-t)} + \frac{1}{2}\sigma^2\gamma^2 e^{2r(T-t)} + \lambda\left(\frac{b}{b-\gamma e^{r(T-t)}}\right)^a - \lambda, \quad (24)$$

with boundary condition $h(0) = 0$.

Remark 4.4. In Theorem 4.4, the assumption on parameters $\dfrac{b-\left(\frac{\lambda a b^a}{\beta}\right)^{\frac{1}{a+1}}}{\gamma e^{rT}} \geq 1$ is equivalent to $\beta \geq \dfrac{\lambda a b^a}{(b-\gamma e^{rT})^{(a+1)}}$ and $b - \gamma e^{rT} > 0$. We observe that

$$\beta \geq \frac{\lambda a b^a}{(b-\gamma e^{rT})^{(a+1)}} > \frac{\lambda a b^a}{b^{(a+1)}} = \lambda \alpha,$$

and it implies that the reinsurance is very expensive. Consequently, the insurance company should never purchase reinsurance.

Now we compute

$$M_Y(\gamma e^{r(T-t)}\tilde{u}(t)) = M_Y\left(b - \left(\frac{\lambda a b^a}{\beta}\right)^{\frac{1}{a+1}}\right)$$

$$= \left[\frac{b}{\left(\frac{\lambda a b^a}{\beta}\right)^{\frac{1}{a+1}}}\right]^a.$$

Further we notice that under the condition $0 < \dfrac{b-\left(\frac{\lambda a b^a}{\beta}\right)^{\frac{1}{a+1}}}{\gamma} < 1$, it always holds that $0 < \tilde{u}(t) < 1$, and this leads to the following theorem:

Theorem 4.5. *If* $0 < \dfrac{b - \left(\frac{\lambda a b^a}{\beta}\right)^{\frac{1}{a+1}}}{\gamma} < 1,$ *then the optimal reinsurance retention level is*

$$u^*(t) = \tilde{u}(t) = \frac{b - \left(\frac{\lambda a b^a}{\beta}\right)^{\frac{1}{a+1}}}{\gamma e^{r(T-t)}},$$

and the optimal value function is given by (12) *and function* h *solves*

$$h'(T - t) = f_2(T - t),$$

where

$$f_2(T - t) = -\gamma(\mu - \beta)e^{r(T-t)} + \frac{1}{2}\sigma^2\gamma^2 e^{2r(T-t)} - \beta\left(b - \left(\frac{\lambda a b^a}{\beta}\right)^{\frac{1}{a+1}}\right)$$
$$+ \lambda\frac{b^a}{\left(\frac{\lambda a b^a}{\beta}\right)^{\frac{a}{a+1}}} - \lambda, \tag{25}$$

with boundary condition $h(0) = 0.$

Remark 4.5. The condition in Theorem 4.5, $0 < \dfrac{b - \left(\frac{\lambda a b^a}{\beta}\right)^{\frac{1}{a+1}}}{\gamma} < 1$, is equivalent to the union of the following two cases:

(1) $\alpha \geq \frac{a}{\gamma}$ and $\lambda\alpha < \beta$ (expensive reinsurance and high claim size expectation); and
(2) $\alpha < \frac{a}{\gamma}$ and $\lambda\alpha < \beta < \frac{\lambda a b^a}{\alpha(b-\gamma)^{a+1}}$ (expensive reinsurance but not too expensive and low claim size expectation).

Under these conditions, the optimal policy always involves certain positive level of reinsurance purchase that is never 0% or 100%.

Now we consider the last case of expensive reinsurance on parameters: $\dfrac{b - \left(\frac{\lambda a b^a}{\beta}\right)^{\frac{1}{a+1}}}{\gamma} \geq 1 > \dfrac{b - \left(\frac{\lambda a b^a}{\beta}\right)^{\frac{1}{a+1}}}{\gamma e^{rT}}$; under this constraint, there exists t_1, $0 \leq t_1 < T$, given as the following,

$$t_1 = \frac{1}{r}\ln\frac{b - \left(\frac{\lambda a b^a}{\beta}\right)^{\frac{1}{a+1}}}{\gamma}, \tag{26}$$

such that

$$\tilde{u}(T - t_1) = 1.$$

Consequently, $\tilde{u}(t) \leq 1$ and $u^*(t) = \tilde{u}(t)$ when $0 < t \leq T - t_1$; $\tilde{u}(t) > 1$ and $u^*(t) = 1$ when $T - t_1 < t \leq T$. Now we come to the following theorem:

Theorem 4.6. *If* $\dfrac{b-\left(\frac{\lambda ab^a}{\beta}\right)^{\frac{1}{a+1}}}{\gamma} \geq 1 > \dfrac{b-\left(\frac{\lambda ab^a}{\beta}\right)^{\frac{1}{a+1}}}{\gamma e^{rT}}$, *then the optimal rein-*
surance retention level is

$$u^*(t) = \begin{cases} 1 & if \ \ 0 \leq T-t \leq t_1 \\ \tilde{u}(t) & if \ \ t_1 < T-t \leq T \end{cases}, \tag{27}$$

and the optimal value function is given by (12), *where the function h solves*
the following equation:

$$h'(T-t) = \begin{cases} f_1(T-t) & if \ \ 0 \leq T-t \leq t_1 \\ f_2(T-t) & if \ \ t_1 < T-t \leq T \end{cases},$$

here f_1 and f_2 are defined by (24) *and* (25) *respectively, and the function*
above is continuous.

Under the paramerter condition in Theorem 4.6, the optimal strategy suggests to buy less reinsurance gradually down to 0% to the final time.

The solution of the case with gamma claims is now complete.

4.3. *Numerical Examples*

In this subsection, we give two examples with plots to demonstrate the results we obtained in the previous subsections.

Example 4.1. In this example we assume that the claim distribution is exponential with mean $\alpha = 0.4$, and the other model parameters are given as the following: $\beta = 1$, $\lambda = 2$, $\gamma = 0.2$, $r = 0.15$, $\sigma = 0.5$, $\mu = 0.2$, $c = 1$, and $\delta = 0.15$. Notice that the reinsurance premium rate β is higher than the expected loss per unit time $\lambda\alpha$, and the reinsurance is expensive. Further assume the surplus level is $x = 1$ and expiration time is $T = 4$; and compute $\dfrac{\frac{1}{\alpha}-\sqrt{\frac{\lambda}{\alpha\beta}}}{\gamma} \approx 1.3197 \geq 1$ and $\dfrac{\frac{1}{\alpha}-\sqrt{\frac{\lambda}{\alpha\beta}}}{\gamma e^{rT}} \approx 0.7242 < 1$; one can apply Theorem 4.3 to obtain the optimal risk exposure (reinsurance intention level) $u^*(t)$ and the optimal value function $V(t,x)$. There is a unique threshold point in time $T - t_0 \approx 2.1508$ at which the form of the optimal risk exposure function changes. The plot is given in Figure 1, where $u^*(t)$ is an increasing function in t; this means that as time approaches to the expiration date, less reinsurance is bought (down to 0% at the end).

Example 4.2. In this example we assume a Gamma claim distribution, whose parameters are $a = 2$ and $b = 3$. Note that its mean is $\alpha = \frac{a}{b} = 2/3$. The other parameters are given as below: $\beta = 1$, $\lambda = 1$, $\gamma = 0.3$, $r = 0.15$,

$\sigma = 0.5$, $\mu = 0.2$, $c = 1$, and $\delta = 0.15$. Further assume $T = 4$ and $x = 1$. Now compute $\dfrac{b-\left(\frac{\lambda_a b^a}{\beta}\right)^{\frac{1}{a+1}}}{\gamma} \approx 1.2642$ and $\dfrac{b-\left(\frac{\lambda_a b^a}{\beta}\right)^{\frac{1}{a+1}}}{\gamma e^{rT}} \approx 0.6938$. Thus we can apply Theorem 4.6 to obtain the optimal risk exposure function and the optimal value function. The plot is drawn in Figure 2. The unique threshold point is at $T - t_1 \approx 2.4371$.

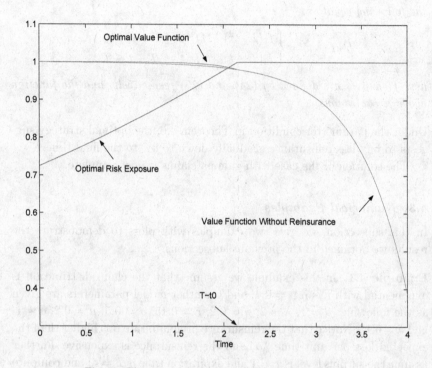

Fig. 1.   The case with exponential claims (Theorem 4.3)

Acknowledgments

This research is supported by a UNI summer research fellowship.

References

1. Asmussen, S. and Taksar, M.: Controlled Diffusion Models for Optimal Dividend Payout, *Insurance: Math. Econ.* **20**, 1–15 (1997)

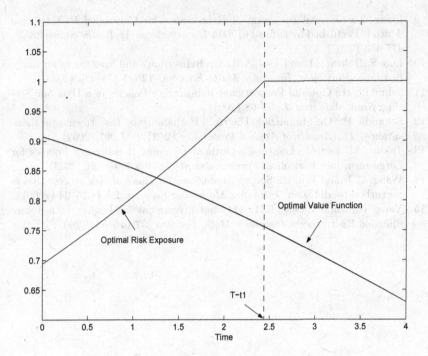

Fig. 2.   The case with Gamma claims (Theorem 4.6)

2. Browne, S.: Optimal Investment Policies for a Firm with a Random Risk Process: Exponential Utility and Minimizing the Probaiblity of Ruin, *Math. Ope. Res* **20**(4), 937–958 (1995)

3. Browne, S.: Survival and Growth with Liability: Optimal Portfolio Strategies in Continuous Time, *Math. Ope. Res* **22**, 468–492 (1997)

4. Dufresne, F. and Gerber, H. U. Risk theory for a compound Poisson process that is perturbed by diffusion. Insurance Math. Econom. **10** 51–59 (1991)

5. Emanuel, D.C., Harrison J.M. and Taylor, A.J.: A Diffusion Approximation for the Ruin Probability with Compounding Assets, *Scan. Actuarial J.* 37–45 (1975)

6. Gerber, H.U. and Yang, H.L.: Absolute Ruin Probabilities in a Jump Diffusion Risk Model with Investment, *North American Actuarial Journal* **11**(3), 159–169 (2007)

7. Höjgaard., B. and Taksar, M.: Optimal Proportional Reinsurance Policies for Diffusion Models with Transaction Costs, *Insurance Math. Econom.* **22**, 41–51 (1998)

8. Höjgaard., B. and Taksar, M.: Optimal Proportional Reinsurance Policies for Diffusion Models, *Scan. Actuarial J.* 166–180 (1998)

9. Liang, Z.: Optimal Proportional Reinsurance for Controlled Risk Process which is Perturbed by Diffusion, *Acta Mathematicae Applicatae Sinica* **23**(3), 477-488 (2007)
10. Luo, S., Taksar, M. and Tsoi, A.H.: On Reinsurance and Investment for Large Insurance Portfolios, *Insurance Math. Econom.* **42** (1) 434–444 (2008)
11. Schmidli, H.: Optimal Proportional Reinsurance Policies in a Dynamic Setting, *Scan. Actuarial J.* 55–68 (2001)
12. Schmidli, H.: On Minimizing The Ruin Probability by Investment and Reinsurance, *The Annals of Applied Probability* **12(3)**, 890–907 (2002)
13. Taksar, M. and Markussen C.: Optimal Dynamic Reinsurance Policies for Large Insurance Portfolios, *Finance and Stochastics* **7** 97–121 (2003)
14. Wang, G.J. and Wu, R.: Some distributions for classical risk process that is perturbed by diffusion, *Insurance Math. and Econom.* **26**(1) 15–24 (2000)
15. Yang, H.L. and Zhang, L.H.: Optimal Investment for Insurer with Jump-diffusion Risk Process *Insurance Math. Econom.* **37** 615–634 (2005)

Applications of Counting Processes and Martingales in Survival Analysis

Jianguo Sun

Department of Statistics
University of Missouri, USA
Email: sunj@missouri.edu

The counting process and martingale have been the research topic in probability for a long time and there exists a great deal of literature for both their studies and their applications. This article discusses their applications in survival analysis, a field that plays an essential and fundamental role in biostatistics and deals with experiments concerning times to events. In addition to being a key component in most of medical and public health studies, survival analysis is applied to many other fields including demographical studies, economics, reliability studies and social studies. Although survival analysis has existed for a long time, the modern survival analysis really started about 30 years ago when the counting process and martingale tools were applied for the advance of the field (Aalen, 1975, 1980; Andersen et al., 1993).

Keywords: Counting processes; Cox model; martingales; partial likelihood; right-censored failure time data; semiparametric regression analysis.

1. Introduction

Survival analysis refers to the statistical field that deals with data concerning positive random variables representing times to certain events (Kalbfleisch and Prentice, 2002). Examples of the event, often referred to as the failure or survival event, include death, the onset of a disease or certain milestone, the failure of a mechanical component of a machine, or learning something. The occurrence of the event is usually referred to as a failure and often we also use the terminology failure time data analysis and refer to the variable of interest as failure or survival time or variable. Failure time or survival data arise extensively in medical studies, but there are many other investigations that also produce survival data. These include biological studies, demographical studies, economic and financial studies, epidemiological studies, psychological experiments, reliability experiments, and sociological studies.

Table 1.   Remission times in weeks for acute leukemia patients

Group	Survival times in weeks
6-MP	6, 6, 6, 6*, 7, 9*, 10, 10*, 11*, 13, 16, 17*, 19*, 20*, 22, 23, 25* 32*, 32*, 34*, 35*
Placebo	1, 1, 2, 2, 3, 4, 4, 5, 5, 8, 8, 8, 8, 11, 11, 12, 12, 15, 17, 22, 23

Survival analysis differs from other statistical fields due to many reasons. One major reason, also the feature that distinguishes the analysis of failure time data from other statistical problems, is the existence of censoring such as right censoring that will be introduced below. Censoring mechanisms can be quite complicated and thus necessitate special methods of treatment. The methods available for other types of data are usually simply not appropriate for censored data. Truncation is another feature of some failure time data that requires special treatments. Because of censoring and/or truncation, failure time data are always incomplete and the goal of survival analysis is to develop statistical approaches that can handle such incompleteness.

To give an example, Table 1 presents a typical set of failure time data, which arose from a clinical trial on acute leukemia patients. The table gives remission times in weeks for 42 patients in two treatment groups. One treatment is the drug 6-mercaptopurine (6-MP) and the other is the placebo treatment. The study was performed over a one-year period and the patients were enrolled into the study at different times. In the table, the starred numbers are right-censored remission times, meaning that the true remission times were known only to be greater than these censored times. This happened because these patients were still in the state of remission at the end of the trial and the times observed are the amount of time from when the patient entered the study to the end of the study. For other patients, their remission times were observed exactly.

In survival analysis, it is common to use T to denote the nonnegative random variable under study or the failure time of interest. For inference about T, unlike other statistical fields, it is more common and convenient to consider or model the so-called survival and hazard functions defined as

$$S(t) = P(T > t), \ 0 < t < \infty$$

and

$$\lambda(t) = \lim_{\Delta t \to 0^+} \frac{P(t \le T < t + \Delta t \,|\, T \ge t)}{\Delta t},$$

respectively, assuming that T is a continuous variable. Here we will focus on continuous survival variables and the hazard function is defined differently for discrete survival variables (Kalbfleisch and Prentice, 2002). It can be easily shown that

$$\lambda(t) = \frac{f(t)}{S(t)} = -\frac{d \log S(t)}{dt}, \; S(t) = \exp\left[-\int_0^t \lambda(s)\, ds\right] = \exp\left[-\Lambda(t)\right],$$

where $f(t)$ denotes the density function of T and $\Lambda(t) = \int_0^t \lambda(s)\, ds$, usually referred to as the cumulative hazard function.

A counting process, usually denoted by $N(t)$, is a stochastic process that can be thought of as registering the occurrences in time of certain recurrent events. That is, $N(t)$ is a step function jumping by one each time when the event occurs. In survival analysis, more often one will face a multivariate counting process, a vector of counting processes all being zero at time zero with paths which are piecewise constant and nondecreasing, having jumps of size one only with no two processes jumping simultaneously. A martingale is also a stochastic process often regarded as a kind of pure random noise process. Many stochastic processes can be written as the sum of a martingale and a finite variation predictable stochastic process with the latter called the compensator of the process. In the following, we will show that many inference problems in survival analysis can be conveniently formulated using the counting process and martingale and thus one can apply the existing, rich theory about them to solve the problems much easily.

The remainder of the paper is organized as follows. In Section 2, we will consider regression analysis of right-censored failure time data using the Cox or proportional hazards model. Section 3 discusses regression analysis of current status failure time data using the additive hazards model. Current status failure time data mean that the failure time of interest is observed only to be either smaller or larger than a number, usually the observation time. In this case, each study subject is observed only once and one area that often produces such data is cross-sectional studies. In Section 4, regression analysis of bivariate current status failure time data will be described under the proportional hazards model. For all situations, the focus will be on estimation of regression parameters. Section 5 contains some concluding remarks.

2. Regression Analysis of Right-censored Failure Time Data Using the Cox Model

Consider a survival study that consists of n independent subjects. For subject i, let T_i denote the failure time of interest and Z_i a vector of covariates. In reality, it is usually the case that there exists a so-called censoring variable C_i such that one may only observe C_i and $T_i > C_i$. This is usually referred to as right-censoring and one common situation where this happens is that C_i represents the end of the study and by then, the subject is still alive or the event of interest has not occurred yet. In other words, the observed data have the form

$$\{ (X_i = \min(T_i, C_i), \delta_i = I(X_i = T_i), Z_i \; ; \; i = 1, ..., n \}.$$

Suppose that the goal of the study is to assess the effect of covariates on T_i or estimate the covariate effect. For this, a common approach is to specify a regression model and in survival analysis, perhaps the most commonly used regression model is the Cox or proportional hazards model defined as

$$\lambda(t|Z_i) = \lambda_0(t) \exp(Z_i'\beta) \tag{1}$$

(Cox, 1972). Here $\lambda(t|Z_i)$ denotes the hazard function of T_i given Z_i, $\lambda_0(t)$ is an unknown baseline hazard function, and β denotes the vector of regression parameters.

To estimate or make inference about β, the common approach is to maximize the partial likelihood

$$L(\beta) = \prod_{i=1}^{n} \left[\frac{\exp(Z_i'\beta)}{\sum_{X_j \geq X_i} \exp(Z_j'\beta)} \right]^{\delta_i}$$

or to solve the partial score function $U(\beta) = \partial \log L(\beta)/\partial\beta = 0$. This approach was originally proposed by Cox (1975) and has been studied by many authors (Andersen et al., 1993; Kalbfleisch and Prentice, 2002) using the formulation and theory of the counting process and martingale described below. A key advantage of this partial likelihood approach is that $L(\beta)$ is independent of the baseline hazard function $\lambda_0(t)$. In other words, one does not have to deal with or estimate $\lambda_0(t)$.

To express $L(\beta)$ or $U(\beta)$ using the counting process and martingale, define

$$N_{ri}(t) = I(X_i \leq t, \delta_i = 1)$$

and $Y_{ri}(t) = I(X_i \geq t)$, $i = 1, ..., n$. It is easy to see that $N_{ri}(t)$ is a counting process jumping from 0 to 1 at $t = T_i$ if the survival time T_i is observed instead of the censoring time C_i. $Y_{ri}(t)$ is usually referred to as the indicator process indicating if subject i is under study at time t. Also define

$$M_{ri}(t) = N_{ri}(t) - \int_0^t Y_{ri}(u) \exp(Z_i'\beta)\lambda_0(u)du, \ i = 1, ..., n,$$

$$S_r^{(j)}(\beta, t) = n^{-1} \sum_{i=1}^n Z_i^j Y_{ri}(t) \exp(Z_j'\beta), \ j = 0, 1,$$

and $E_r(\beta, t) = S_r^{(1)}(\beta, t)/S_r^{(0)}(\beta, t)$. Then $\{M_{ri}(t)\}$ are martingales and the partial likelihood function can be rewritten as $L(\beta) = L(\beta, \infty)$ with

$$L(\beta, t) = \prod_{i=1}^n \prod_{0 \le u \le t} \left[\frac{Y_{ri}(u) \exp(Z_i'\beta)}{\sum_{j=1}^n Y_{rj}(u) \exp(Z_j'\beta)} \right]^{\Delta N_{ri}(u)}.$$

Furthermore, we have $U(\beta) = U_r(\beta, \infty)$ with

$$U_r(\beta, t) = \frac{\partial \log L(\beta, t)}{\partial \beta} = \sum_{i=1}^n \int_0^t [Z_i - E_r(\beta, u)] \, dN_{ri}(u)$$

$$= \sum_{i=1}^n \int_0^t [Z_i - E_r(\beta, u)] \, dM_{ri}(u). \tag{2}$$

Let $\hat{\beta}_r$ denote the solution to $U(\beta, \infty) = 0$. It is clear that the integrand with respect to each $M_{ri}(t)$ is a predictable process and thus $U_r(\beta, t)$ is the sum of martingales. By applying the martingale theory and using expression (2), one can show that $\hat{\beta}_r$ is a consistent estimate of β in the proportional hazards model (1). Furthermore, $U_r(\beta, t)$ converges to a Gaussian process and $\hat{\beta}_r$ converges in distribution to a normal random vector. The details can be found in Andersen and Gill (1982).

3. Regression Analysis of Current Status Failure Time Data Using the Additive Hazards Model

In this section, instead of right censoring, we will discuss failure time data with a different type of censoring. First we will introduce current status failure time data and then consider their regression analysis.

Table 2.   Death times in days for 144 male RFM mice with lung tumors

Group	Tumor status	Death times
CE	With tumor	381, 477, 485, 515, 539, 563, 565, 582, 603, 616, 624, 650
		651, 656, 659, 672, 679, 698, 702, 709, 723, 731, 775, 779
		795, 811, 839
	No tumor	45, 198, 215, 217, 257, 262, 266, 371, 431, 447, 454, 459
		475, 479, 484, 500, 502, 503, 505, 508, 516, 531, 541, 553
		556, 570, 572, 575, 577, 585, 588, 594, 600, 601, 608, 614
		616, 632, 632, 638, 642, 642, 642, 644, 644, 647, 647, 653
		659, 660, 662, 663, 667, 667, 673, 673, 677, 689, 693, 718
		720, 721, 728, 760, 762, 773, 777, 815, 886
GE	With tumor	546, 609, 692, 692, 710, 752, 773, 781, 782, 789, 808, 810
		814, 842, 846, 851, 871, 873, 876, 888, 888, 890, 894, 896
		911, 913, 914, 914, 916, 921, 921, 926, 936, 945, 1008
	No tumor	412, 524, 647, 648, 695, 785, 814, 817, 851, 880, 913, 942
		986

3.1. Current Status Failure Time Data

As mentioned above, current status failure time data refer to the failure time data in which each failure time is known or observed to be either smaller or larger than a number, usually the observation time. The term current status data originates from demographical studies and such data are also sometimes referred to case I interval-censored data (Sun, 2006). Current status data occur when each study subject is observed only once and the only observed information for the survival event of interest is whether the event has occurred before or after the observation time, which in survival analysis is also referred to as either left- or right-censored. In addition to cross-sectional studies, tumorigenicity experiments are another field where one often faces current status failure time data.

Table 2 gives a standard set of current status data arising from a tu-morigenicity experiment of male RFM mice concerning lung tumors (Hoel and Walberg, 1972). In the experiment, the animals were put in one of two environments, conventional environment (CE) and germ-free environment (GE), and one of the goals was to assess the environment effects on the growth rate of lung tumors. In other words, the failure time of interest is the time to tumor onset. Unfortunately theses times or the occurrences of the tumors cannot be observed while they were alive. The information about these times that can observed is the death or sacrifice time and if lung tumor was present or absent at the death or sacrifice. The sacrifice here means that the animals were killed, which is often performed to obtain better information about the tumor growth.

3.2. *Regression Analysis using the Additive Hazards Model*

Consider a survival study that consists of n independent subjects and in which each study subject is observed or examined only at one time point for the survival event of interest. For subject i, let T_i and Z_i be defined as before and C_i denote the observation time. Then the observed data have the form

$$\{ C_i, \delta_i = I(T_i \leq C_i), Z_i ; i = 1, ..., n \}.$$

For regression analysis, as mentioned before, the proportional hazards model is the most commonly used but it may not fit data well sometimes. Also depending on the nature of the study, different models may be preferred that describe different aspects rather than those specified by the proportional hazards model. One such model is the additive hazards model defined as

$$\lambda(t|Z_i) = \lambda_0(t) + Z_i'\beta \tag{3}$$

(Lin and Ying, 1994, 1995), where $\lambda_0(t)$ and β are defined as in model (1). It can be easily seen that in model (1), β represents the multiplicative effect of covariates on the hazard function, while β in model (3) gives the additive effect of covariates, which are often more interesting in fields such as social studies.

To estimate β in model (3), define $Y_{ci}(t) = I(C_i \leq t)$ and the counting process

$$N_{ci}(t) = I\{C_i \leq \min(T_i, t)\}$$

for subject i, $i = 1, ..., n$. Then one can show that $N_{ci}(t)$ has the intensity

$$Y_{ci}(t) \exp(-tZ_i'\beta)\lambda_c(t)e^{-\Lambda_0(t)},$$

where $\lambda_c(t)$ denotes the hazard function of C_i. Furthermore, the martingale corresponding to $N_{ci}(t)$ can be defined as

$$M_{ci}(t) = N_{ci}(t) - \int_0^t Y_{ci}(u) \exp(-uZ_i'\beta)\lambda_c(u)e^{-\Lambda_0(u)}du.$$

Now we can develop the estimating equation for estimation of β using $N_{ci}(t)$ and $M_{ci}(t)$. For this, define

$$S_c^{(j)}(\beta, t) = n^{-1}\sum_{i=1}^n Y_{ci}(t)(tZ_i)^j \exp(-tZ_i'\beta) \quad j = 0, 1.$$

Then motivated by $U_r(\beta, t)$ given in Section 2, one can construct the following estimating function

$$U_c(\beta, t) = \sum_{i=1}^{n} \int_0^t \left[t\dot{Z}_i - \frac{S_c^{(1)}(\beta, u)}{S_c^{(0)}(\beta, u)} \right] dN_{ci}(u)$$

$$= \sum_{i=1}^{n} \int_0^t \left[tZ_i - \frac{S_c^{(1)}(\beta, u)}{S_c^{(0)}(\beta, u)} \right] dM_{ci}(u)$$

and estimate β by the solution to $U_c(\beta, \infty) = 0$. Let $\hat{\beta}_c$ denote the such defined estimate. Then as $\hat{\beta}_r$, by applying the counting and martingale theory, one can prove that $\hat{\beta}_c$ is consistent and converges in distribution to a normal random vector. The detailed derivation and proofs of the results given above can be found in Lin et al. (1998).

4. Regression Analysis of Bivariate Current Status Failure Time Data Using the Cox Model

4.1. *Assumptions and the Likelihood Function*

In many survival studies, there may exist two related failure time variables of interest instead of only one variable and in this case, it is obvious that one has to conduct bivariate or joint analysis of the variables or the observed data about them. Let T_1 and T_2 denote two failure time variables of interest on a study subject and Z the associated vector of covariates. Suppose that during the study, each study subject is examined or observed only once at time C. That is, only current status data are available for both T_1 and T_2. Note that in general, there may exist different observation times for T_1 and T_2, but for the simplicity, here we assume that the observation times for them are the same. Define

$$\delta_1 = I(T_1 \leq C), \; \delta_2 = I(T_2 \leq C),$$

indicating if the survival events represented by T_1 and T_2 have occurred before C. Then the observed data have the form

$$(C, \delta_1, \delta_2, Z)$$

for one subject.

For the analysis, instead of the partial likelihood and estimating equation approaches applied in Sections 2 and 3, respectively, we will consider the full likelihood approach here. For this, we will assume that T_1 and T_2 follow the copula model specified by

$$S(s,t) = P(T_1 > s, T_2 > t) = C_\alpha(S_1(s), S_2(t)) \tag{4}$$

for the joint survival function of T_1 and T_2, where C_α is a distribution function on the unit square, $\alpha \in R$ is a global association parameter, and S_1 and S_2 denote the marginal survival functions of T_1 and T_2, respectively. The copula model has been widely used in modeling bivariate variables in probability and statistics (Hougaard, 2000) and one attractive feature of model (4) is its flexibility in that it includes as special cases many useful bivariate failure time models. For example, it gives the Archimedean copula family by taking

$$C_\alpha(u,v) = \phi_\alpha\{\phi_\alpha^{-1}(u) + \phi_\alpha^{-1}(v)\}, \, 0 \le u, v \le 1,$$

where $0 \le \phi_\alpha \le 1$, $\phi_\alpha(0) = 1$, $\phi_\alpha' < 0$, $\phi_\alpha'' > 0$. More specifically, if letting $\phi_\alpha(u) = (1 + u)^{1/(1-\alpha)}$, the Laplace transformation of a gamma distribution, one has

$$C_\alpha(u,v) = (u^{1-\alpha} + v^{1-\alpha} - 1)^{1/(1-\alpha)}, \, \alpha > 1,$$

which is commonly referred to as the Clayton family (Clayton, 1978). Another attractive feature of copula models is that the marginal distributions do not depend on the choice of the association structure. Thus one can model the marginal distributions and the association separately.

To derive the full likelihood function, define the following counting processes

$$N_{11}(t) = \delta_1 \delta_2 I(C \le t), \; N_{10}(t) = \delta_1(1 - \delta_2) I(C \le t),$$

$$N_{01}(t) = (1 - \delta_1)\delta_2 I(C \le t), \; N_{00}(t) = (1 - \delta_1)(1 - \delta_2) I(C \le t).$$

Then the log likelihood function is proportional to

$$l(\alpha, S_1, S_2) = \sum_{j=0}^{1} \sum_{m=0}^{1} \int_0^\infty \log S_{jm}(\theta, t) \, dN_{jm}(t)$$

based on one subject, where

$$S_{11}(\theta, t) = P(T_1 \le t, T_2 \le t), \; S_{01}(\theta, t) = P(T_1 > t, T_2 \le t),$$

$$S_{10}(\theta, t) = P(T_1 \le t, T_2 > t), S_{00}(\theta, t) = P(T_1 > t, T_2 > t).$$

4.2. Efficient Estimation of Regression Parameters

For inference about covariate effects, we need to specify a regression model and for this, one way is to assume that both T_1 and T_2 follow the proportional hazards model (1) with the same regression parameter β and different baseline hazard functions $\lambda_1(t)$ and $\lambda_2(t)$, respectively. It is straightforward to generalize the idea and method discussed here to the situation where the regression parameters for T_1 and T_2 are different. Under the proportional hazards model, we have

$$S_1(t) = \exp(-\Lambda_1(t) \exp(Z'\beta)) \,, \; S_2(t) = \exp(-\Lambda_2(t) \exp(Z'\beta)) \,,$$

where $\Lambda_1(t) = \int_0^t \lambda_1(s)ds$ and $\Lambda_2(t) = \int_0^t \lambda_2(s)ds$ are the cumulative baseline hazard functions of T_1 and T_2, respectively. The log likelihood function $l(\alpha, S_1, S_2)$ then becomes a function of $\theta = (\beta', \alpha)'$ and $\Lambda_1(t)$ and $\Lambda_2(t)$.

For estimation of θ based on the log likelihood function $l(\theta, \Lambda_1, \Lambda_2)$, a common approach is to derive the efficient score function for θ. For this, let D_α, D_u and D_v denote the derivatives of $C_\alpha(u, v)$ with respect to α, u and v and at $u = S_1$ and $v = S_2$, respectively. For $j, m = 0, 1$, define

$$a_{jm}^{(1)} = -(-1)^j \exp(Z'\beta) \Lambda_1 S_1 [m + (-1)^m D_u] \,,$$

$$a_{jm}^{(2)} = -(-1)^m \exp(Z'\beta) \Lambda_2 S_2 [j + (-1)^j D_v] \,,$$

and $B_{jm} = (a_{jm}^{(1)} + a_{jm}^{(2)})^{-1} (-1)^{m+j} D_\alpha$. Also for $j, m = 0, 1$, let $a_{jm} = (a_{jm}^{(1)}, a_{jm}^{(2)})'$ and define the matrix

$$Z_{jm} = \begin{pmatrix} Z & Z \\ B_{jm} & B_{jm} \end{pmatrix}$$

and $b^* = H * G^{-1}$, where,

$$G = \sum_{j=0}^{1} \sum_{m=0}^{1} E\left[a_{jm}^{\otimes 2} Y \lambda_c / S_{jm} \right]$$

and

$$H = \sum_{j=0}^{1} \sum_{m=0}^{1} E\left[Z_{jm} a_{jm}^{\otimes 2} Y \lambda_c / S_{jm} \right] \,.$$

Then one can show that the efficient score function for θ has the form

$$\dot{l}_{\theta^*} = \sum_{j=0}^{1} \sum_{m=0}^{1} \int_0^\infty \frac{(Z_{jm} - b^*) a_{jm}}{S_{jm}} dN_{jm}(t) \tag{5}$$

based on a single study subject.

Suppose that the study involves n independent subjects and let N_{jmi}, Z_{jmi}, a_{jmi} and S_{jmi} be defined as N_{jm}, Z_{jm}, a_{jm} and S_{jm} with respect to subject i. Then for estimation of β or θ, it is natural to apply the empirical version of the efficient score function given in (5), which has the form

$$l_{\theta^*, n}(\theta, \Lambda_1, \Lambda_2) = \sum_{i=1}^{n} \sum_{j=0}^{1} \sum_{m=0}^{1} \int_0^\infty \frac{(Z_{jmi} - b_n^*) \, a_{jmi}}{S_{jmi}} \, dN_{jmi}(t).$$

That is, one can estimate θ by the solution to

$$l_{\theta^*, n}(\theta, \hat{\Lambda}_1, \hat{\Lambda}_2) = 0,$$

where $\hat{\Lambda}_1$ and $\hat{\Lambda}_2$ denote some consistent estimates of $\Lambda_1(t)$ and $\Lambda_2(t)$. As in Sections 2 and 3, one can apply the counting process and martingale theory to establish the asymptotic properties of the such defined estimate. Wang et al. (2008) provided more details about the discussion above.

5. Concluding Remarks

In the preceding sections, we discussed regression analysis of failure time data under three different censoring mechanisms and presented the counting process and martingale connection to these problems. As it can be seen, for all three situations, the estimation problems can be expressed using counting processes and martingales and thus one could conveniently employ the existing theory about them, especially various types of central limit results, to investigate or establish the asymptotic properties of the resulting estimates commonly expected and needed for inference. It is worth noting that without the use of the connection to the counting process and martingale and their theory, many of the asymptotic results could still be obtained by different approaches but would be much more difficult and complicated. In contrast, the application of the counting process and martingale makes these results much easy and straightforward and more importantly, some of the other results possible and available.

There are many other types of problems in survival analysis in addition to regression analysis or the assessment of covariate effects and most of these can be formulated and investigated using the counting process and martingale. For example, consider the right-censored failure time data discussed in Section 2 and suppose that one is interested in estimation of the overall survival function $S(t)$, which is usually the first thing that one does in analyzing failure time data. Suppose that there do not exist covariates and let N_{ri} and Y_{ri} be defined as in Section 2. Define

$$N_r(t) = \sum_{i=1}^{n} N_{ri}(t)\,,\ Y_r(t) = \sum_{i=1}^{n} Y_{ri}(t)\,.$$

It is apparent that $N_r(t)$ is still a counting process. For estimation of $S(t)$, the most commonly used approach is to use the Kaplan-Meier estimator (Kaplan and Meier, 1958) that can be expressed as

$$\hat{S}(t) = \prod_{s \le t} \left(1 - \frac{\Delta N(s)}{Y(s)} \right)\,.$$

As with estimation of regression parameters, one can easily apply the counting process and martingale theory to investigate the statistical properties of $\hat{S}(t)$ such as deriving the variance estimate. If there is no right censoring or all $\delta_i = 1$, then the Kaplan-Meier estimator reduces to one minus the empirical distribution function.

As discussed above, the Cox or proportional hazards model (1) is the most commonly used regression model in survival analysis and the additive hazards model (3) is also a popular choice. There exist many other models in survival analysis and the selection of an appropriate model for a given problem usually depends on the nature and prior knowledge of the problem but could be very difficult sometimes. Note that in both models (1) and (3), we assumed that covariates are time-independent and actually one can apply them even if the covariates are time-dependent. Furthermore, both models assume the linear effects of covariates and one can easily generalize them to include nonlinear covariate effects. In addition to these two models, another popular semiparametric model, motivated by the common linear regression model, is the accelerated failure time model given by

$$\log T = Z'\beta + \epsilon$$

(Jin et al., 2003), where ϵ denotes a random error with unknown distribution function. Note that this model directly models the effects of covariates on the failure time of interest instead of the hazard function as in models (1) and (3). Sometimes one may be more interested in the direct effects of covariates on survival function and in this case, one could apply the proportional odds model given by

$$\frac{S_Z(t)}{1 - S_Z(t)} = \exp(Z'\beta)\,\frac{S_0(t)}{1 - S_0(t)}$$

(Yang and Prentice, 1999), where $S_Z(t)$ denotes the survival function for a subject with covariates Z and $S_0(t)$ is an unknown baseline survival function.

In addition to right censoring and current status censoring, there exist other types of censorings or mechanisms that cause incompleteness in survival analysis. For example, instead of being observed only once as with current status censoring, a study subject may be examined or observed a few times during the study period, which is often the situation in medical or public health follow-up studies. In these situations, the failure time of interest could be only observed to belong to a window given by the last observation time at which the survival event has not occurred and the first observation time at which the survival event has occurred. This is commonly referred to as interval censoring (Sun, 2006). Truncation is another feature that often exists in failure time data and it also causes incompleteness and needs special treatments as censoring although the mechanism behind it is usually different from that behind censoring. For example, left-truncation means that the survival event could be observed or known only if it occurs after certain time or another event. More discussion on this and all types of censoring can be found in, for example, Klein and Moeschberger (2003) and Lawless (2003).

References

Aalen, O. O. (1975). Statistical inference for a family of counting processes. *Ph.D Thesis*, University of California, Berkeley.

Aalen, O. O. (1980). A model for nonparametric regression analysis of counting processes. *Spring Lect. Notes Statist.* **2**, 1-25. Mathematical Statistics and Probability Theory. W. Klonecki, A. Kozek, and J. Rosiński, editors.

Andersen, P. K., Borgan, O., Gill, R. D. and Keiding, N. (1993). *Statistical models based on counting processes*. Springer-Verlag: New York.

Andersen, P. K. and Gill, R. D. (1982). Cox regression model for counting processes: a large sample study. *The Annals of Statistics*, **10**, 1100-1120.

Clayton, D. G. (1978). A model for association in bivariate life tables and its application in epidemiological studies of familial tendency in chronic disease incidence. *Biometrika*, **65**, 141-151.

Cox, D. R. (1972). Regression analysis and life tables (with discussion). *J. R. Statist. Soc. B*, 34, 187-220.

Cox, D. R. (1975). Partial likelihood. *Biometrika*, **62**, 269-276.

Hoel, D. G. and Walburg, H. E. (1972). Statistical analysis of survival experiments. *Journal of National Cancer Institute*, **49**, 361-372.

Hougaard, P. (2000). *Analysis of multivariate survival data*. Springer-Verlag: New York.

Jin, Z., Lin, D. Y., Wei, L. J. and Ying, Z. (2003). Rank-based inference for the accelerated failure time model. *Biometrika*, **90**, 341-353.

Kalbfleisch, J. D., and Prentice, R. L. (2002). *The statistical analysis of failure time data*. Second edition. Wiley: New York

Kaplan, E. L. and Meier, P. (1958). Nonparametric estimation from incomplete observations. *Journal of the American Statistical Association*, **53**, 457-481.

Klein, J. P. and Moeschberger, M. L. (2003). *Survival analysis*, Springer-Verlag: New York.

Lawless, J. F. (2003). *Statistical models and methods for lifetime data*. John Wiley: New York.

Lin, D. Y., Oakes, D. and Ying, Z. (1998). Additive hazards regression with current status data. *Biometrika*, 85, 289-298.

Lin, D. Y. and Ying, Z. (1994). Semiparametric analysis of the additive risk model. *Biometrika*, **81**, 61-71.

Lin, D. Y. and Ying, Z. (1995). Semiparametric analysis of general additive-multiplicative hazard models for counting processes. *The Annals of Statistics*, **23**, 1712-1734.

Sun, J. (2006). *The statistical analysis of interval-censored failure time data*. Springer.

Wang, L., Sun, J. and Tong, X. (2008). Efficient estimation for the proportional hazards model with bivariate current status data. *Lifetime Data Analysis*, accepted.

Yang, S. and Prentice, R. L. (1999). Semiparametric inference in the proportional odds regression model. *Journal of the American Statistical Association*, **94**, 125-136.

Stochastic Algorithms and Numerics for Mean-Reverting Asset Trading

Q. Zhang

Department of Mathematics
University of Georgia
Athens, GA 30602, USA
Email: qingz@math.uga.edu

C. Zhuang

Marshall School of Business
University of Southern California
Los Angeles, CA 90089, USA
Email: czhuang@usc.edu

G. Yin

Department of Mathematics
Wayne State University
Detroit, MI 48202, USA
Email: gyin@math.wayne.edu

This work considers trading a mean-reverting asset. The strategy is to determine a low price to buy and a high price to sell so that the expected return is maximized. Slippage cost is imposed on each transaction. Our effort is devoted to developing a recursive stochastic approximation type algorithm to estimate the desired selling and buying prices. The advantage of this approach is that the underlying asset is model free. Only observed stock prices are required, so it can be performed on line. After briefly presenting the asymptotic results, simulations and real market data are used to demonstrate the performance of the proposed algorithm.

Keywords: Stochastic approximation; stochastic optimization; trading strategy.

1. Introduction

This work is concerned with developing a systematic numerical procedure for trading a mean-reverting asset. The trading strategy consists of two ingredients, buy and sell. One wishes to buy low and sell high. Nevertheless, it is challenging to be able to correctly identify these low and high prices in practice. The purpose of this paper is to develop and implement an easy while systematical procedure to determine the buying and selling prices when the underlying asset price is subject to a mean-reverting process.

A mean-reverting model is often used in financial and energy markets to capture the price movements that will eventually move back to an "equilibrium" level. Empirical studies on mean-reverting stock prices can be traced back to the 1930s (see Ref. 10). The research was carried further in many studies. Among them, Ref. 13 and Ref. 25 were the first to provide direct empirical evidence that mean reversion occurs in U.S. stock market over long horizon. Ref. 2 provided international evidence to support mean-reverting stock prices in 18 countries during the period 1969 to 1996.

Mean reversion is also found in short-horizon expected returns (Ref. 7). Other than stock prices, mean-reverting models are also used to characterize stochastic volatility (Ref. 16) and asset price in energy market (Ref. 1).

There is a large body of literature studying trading rule in financial markets for many years, especially on the sell side. For example, in Ref. 29 a selling rule is studied when the stock price evolves according to a series of geometric Brownian motions (GBM) coupled with a continuous-time finite state Markov chain. An investor makes a selling decision whenever the stock price exceeds the target price or hit the stop-loss price. The objective is to determine these threshold prices by maximizing a discounted expected reward function, and the optimal threshold values are obtained by solving a set of two-point boundary value problems. In Ref. 14, the trailing stop strategy is considered for two models for stock prices: a discrete-time random walk and continuous-time Brownian motion. A trailing stop order maintains a stop price at a precise percentage below the market price. For both models, they discuss the question of optimizing the percentage from the current price to the stop. In Ref. 15, the optimal selling rule was considered for stock price under a switching GBM model and the optimal stopping problem is solved by using a smooth-fit technique. In Ref. 24 the liquidation problem of a large block of stocks was considered. Other than these analytical results, various numerical methods have been developed to compute these threshold. In Ref. 28, a stochastic approximation technique was used to obtain the optimal selling rule. Further numerical and asymptotic results were obtained in Ref. 27. In addition, a liner programming approach was developed in Ref. 17 and fast Fourier transformation was used in Ref. 23. Furthermore, capital gain taxes and transaction costs in stock selling were considered in Ref. 5, 9 and 11, among others.

On the other hand, most work on the buying side of trading is qualitative. For example, contrarian and momentum strategies were studied in Ref. 4, Ref. 8, and Ref. 18–20. Not until recently was a rigorous mathematical analysis on the buying side provided in Ref. 30, in which they considered buying and selling for assets governed by mean-reverting processes. The objective is to buy and sell the underlying asset sequentially to maximize the discounted reward function when the slippage cost is taken into account. [Slippage cost usually refers to the difference between the estimated price and the actual price paid.] In Ref. 30, the optimal buying and selling prices were obtained using dynamic programming approach and the associated HJB equations for the value functions. One makes a buy decision when the market price hits the buying price and makes a sell decision when the market price exceeds the selling price. In order to implement the strategy in Ref. 30, one needs to know the values of parameters in the mean-reverting processes in order to compute the optimal threshold values. In practice, it is difficulty to determine those values. Taking this point into consideration, in Ref. 26, a stochastic approximation algorithm was proposed to solve the problem for buying and selling the asset once. Instead of solving two quasi-algebraic equations, the problem is formulated as a stochastic optimization procedure. The algorithm is model free and uses observed stock prices only. In this paper, we further develop the algorithm that allows buying and selling to take place a multiple number of times.

The essential feature of our approach is the use of stochastic approximation methods (see Ref. 22 and Ref. 6 for up-to-date development of stochastic approximation algorithms). The proposed stochastic approximation algorithm allows us to deal with general model free case and use only observed stock prices to determine the optimal buying and selling prices. Therefore the proposed method can be easily implemented in practice.

The rest of the paper is arranged as follows. Section 2 offers a precise formulation of the problem and the description of algorithm. Section 3 proceeds with the convergence and rates of convergence of the proposed algorithm. To demonstrate the feasibility and efficiency of the algorithm, numerical experiments using simulations and real market data are given in Section 4. We demonstrate that the proposed algorithms provide sound estimated optimal threshold values; they can be easily implemented in real time and provide guidelines for stock trading. Finally we conclude this paper with some further remarks in Section 5. Since the main concerns in this paper are to develop sound numerical algorithms, and since the theoretical development has been setup in Ref. 26, the paper is concentrated on the numerical aspects. The detailed proofs are omitted for brevity.

2. Problem Formulation

In Ref. 30, they assume that $X(t) \in \mathbb{R}$ is a mean-reverting process governed by

$$dX(t) = a(b - X(t))dt + \sigma dW(t), \quad X(0) = x, \tag{1}$$

where $a > 0$ is the rate of reversion, b is the equilibrium level, $\sigma > 0$ is the volatility, and $W(t)$ is a standard Brownian motion. Then the asset price is given by

$$S(t) = \exp(X(t)). \tag{2}$$

In our formulation, we do not require asset price $S(t)$ to be any specific stochastic process or follow any specific distribution. We only assume that asset price can be observed. Based on the observed stock price, two sequences of stopping times $\tau^{\{b_i\}}$ and $\tau^{\{s_i\}}$ with

$$0 \le \tau^{\{b_1\}} \le \tau^{\{s_1\}} \le \tau^{\{b_2\}} \le \tau^{\{s_2\}} \le \cdots$$

are considered. One makes a buying decision at time $\tau^{\{b_i\}}$ and makes a selling decision at time $\tau^{\{s_i\}}$, with $i = 1, 2, \ldots$. Suppose that $0 < K < 1$ is the percentage of slippage per transaction and $\rho > 0$ is the discount factor. We aim to find the optimal buying price and selling price that maximize a suitable reward function. Thus the formulation is

$$\text{Problem } \mathcal{P}: \begin{cases} \text{Find argmax } \Phi(\theta) = E[J(\theta)], \\ \theta = (\theta^1, \theta^2)' \in (0, \infty) \times (0, \infty), \\ J(\theta) = \sum_{i=1}^{\infty} [\exp(-\rho\tau^{\{s_i\}})S(\tau^{\{s_i\}})(1 - K) \\ \qquad - \exp(-\rho\tau^{\{b_i\}})S(\tau^{\{b_i\}})(1 + K)], \end{cases} \tag{3}$$

where

$$\tau^{\{b_1\}} = \inf\{t > 0, S(t) \le \exp(\theta^1)\},$$
$$\tau^{\{b_i\}} = \inf\{t > \tau^{\{s_{i-1}\}}, S(t) \le \exp(\theta^1)\}, \text{ for } i \ge 2,$$
$$\tau^{\{s_i\}} = \inf\{t > \tau^{\{b_i\}}, S(t) \ge \exp(\theta^2)\}, \text{ for } i \ge 1.$$

Note that $\tau^{\{b_i\}}$ and $\tau^{\{s_i\}}$ denote the stopping times for buying and selling respectively, θ^1 and θ^2 denote the buying and selling threshold values respectively, and $S(t)$ is the stock price at time t.

The analytic solution is obtained in Ref. 30 when $S(t)$ is governed by (2). However, the solution depends on the values of a and b in (1), which are difficult to be determined in practice. Our contribution is to devise a optimization procedure that estimates the optimal threshold value θ and only requires observed stock prices. We will use a stochastic approximation procedure (SA) to resolve the problem by constructing a sequence of estimates of optimal threshold value θ, using

$$\theta_{n+1} = \theta_n + \{\text{step size}\}\{\text{gradient estimate of } \Phi(\theta)\}. \tag{4}$$

Let us begin with a simple noisy finite difference scheme. The only provision is that $S(t)$ can be observed. Associated with the iteration number n, denote the threshold value by θ_n. Let us begin with an arbitrary initial guess θ_0, we construct a sequence of estimates $\{\theta_n\}$ recursively as follows. Then we can determine stopping times $\tau_n^{\{b_i\}}$ and $\tau_n^{\{s_i\}}$, buying times and selling times as

$$\tau_n^{\{b_1\}} = \inf\{t > 0, S(t) \leq \exp(\theta_n^1)\},$$
$$\tau_n^{\{b_i\}} = \inf\{t > \tau_n^{\{s_{i-1}\}}, S(t) \leq \exp(\theta_n^1)\}, \text{ for } i \geq 2,$$
$$\tau_n^{\{s_i\}} = \inf\{t > \tau_n^{\{b_i\}}, S(t) \geq \exp(\theta_n^2)\}, \text{ for } i \geq 1.$$

Define a combined process ξ_n that includes the random effect from $S(t)$ and the stopping times $\tau_n^{\{b_i\}}$ and $\tau_n^{\{s_i\}}$ as

$$\xi_n = (S(\tau_n^{\{b_1\}}), S(\tau_n^{\{s_1\}}), S(\tau_n^{\{b_2\}}), S(\tau_n^{\{s_2\}}), ..., \tau_n^{\{b_1\}}, \tau_n^{\{s_1\}}, \tau_n^{\{b_2\}}, \tau_n^{\{s_2\}}, ...)'. \tag{5}$$

We call ξ_n the sequence of collective noise. Let $\widetilde{\Phi}(\theta, \xi)$ be the observed value of the objective function $\Phi(\theta)$ with collective noise ξ. When the threshold values is set at θ, take random samples of size n_0 with sequence $\{\xi_{n,l}^{\pm}\}_{l=1}^{n_0}$ such that

$$\widehat{\Phi}(\theta, \xi_n^{\pm}) \stackrel{\text{def}}{=} \frac{\widetilde{\Phi}(\theta, \xi_{n,1}^{\pm}) + \cdots + \widetilde{\Phi}(\theta, \xi_{n,n_0}^{\pm})}{n_0}. \tag{6}$$

We assume that

$$E\widehat{\Phi}(\theta, \xi_n^{\pm}) = \Phi(\theta) \text{ for each } \theta. \tag{7}$$

Then for each θ, $\widehat{\Phi}(\theta, \xi_n^{\pm})$ is an estimate of $\Phi(\theta)$. In the simulation study, we can use independent random samples to estimate the expected value of $\Phi(\theta_n)$. The law of large numbers implies that $\widehat{\Phi}(\theta, \xi_n^{\pm})$ converges to $\Phi(\theta)$ w.p.1 as $n_0 \to \infty$. We will not assume the independence in the proof of convergence theorem. In lieu of using (6) with $\widehat{\Phi}(\theta, \xi_{n,l}^{\pm})$, we will use the form $\widehat{\Phi}(\theta, \xi_n)$.

To obtain the desire estimate, we construct a stochastic approximation procedure with finite difference gradient estimates. Define $Y_n^{\pm} = (Y_n^{\pm,1}, Y_n^{\pm,2})$ as

$$Y_n^{\pm,\iota}(\theta, \xi_n^{\pm}) = \widehat{\Phi}(\theta \pm \delta_n e_\iota, \xi_n^{\pm}), \text{ for, } \iota = 1, 2, \tag{8}$$

where e_ι is the standard unit vectors with $e_1 = (1,0)'$ and $e_2 = (0,1)'$, $\xi_n^\pm$ are two different collective noises taken at threshold values $\theta \pm \delta_n e_\iota$, respectively, and δ_n is the finite difference sequence satisfying $\delta_n \to 0$ as $n \to \infty$. We write $Y_n^\pm = Y_n^\pm(\theta, \xi_n^\pm)$. For simplicity, henceforth, we often use ξ_n to represent ξ_n^+ and ξ_n^- if there is no confusion. The gradient estimate at iteration n is given by

$$D\widehat{\Phi}(\theta_n, \xi_n) \stackrel{\text{def}}{=} (Y_n^+ - Y_n^-)/(2\delta_n). \tag{9}$$

Then the recursive algorithm is

$$\theta_{n+1} = \theta_n + \varepsilon_n D\widehat{\Phi}(\theta_n, \xi_n), \tag{10}$$

where ε_n is a sequence of real numbers known as step size. A frequently used choice of step size and finite difference sequences is $\varepsilon_n = O(1/n)$ and $\delta_n = O(1/n^{1/6})$. Throughout this paper, this is our default choice of step size and finite difference sequences.

To proceed, we define

$$
\begin{aligned}
\lambda_n &= (Y_n^+ - Y_n^-) - E_n(Y_n^+ - Y_n^-), \\
\eta_n^\iota &= [E_n Y_n^{+,\iota} - \Phi(\theta_n + \delta_n e_\iota)] - [E_n Y_n^{-,\iota} - \Phi(\theta_n - \delta_n e_\iota)], \quad \iota = 1,2, \\
\beta_n^i &= \frac{\Phi(\theta_n + \delta_n e_\iota) - \Phi(\theta_n - \delta_n e_\iota)}{2\delta_n} - \Phi_{\theta^\iota}(\theta_n), \quad \iota = 1,2,
\end{aligned}
\tag{11}
$$

where E_n denotes the conditional expectation with respect to $\mathcal{F}_n$, the σ-algebra generated by $\{\theta_1, \xi_j^\pm : j < n\}$, $\Phi_{\theta^\iota}(\theta) = (\partial/\partial\theta^\iota)\Phi(\theta)$, and $\Phi_\theta = (\Phi_{\theta^1}(\cdot), \Phi_{\theta^2}(\cdot))'$ denotes the gradient of $\Phi(\cdot)$. In the above, η_n^ι and β_n^ι for $\iota = 1, 2$ represent the noise and bias, and λ_n is a martingale difference sequence. It is reasonable to assume that after taking the conditional expectations, the resulting function is smooth. Thus we separate the noise into two parts, uncorrelated noise λ_n and correlated noise η_n. In what follows, we write $\eta_n = (\eta_n^1, \eta_n^2)'$ and $\beta_n = (\beta_n^1, \beta_n^2)'$, and note that $\eta_n = \eta_n(\theta_n, \xi_n)$. With the above definitions, algorithm (10) becomes

$$\theta_{n+1} = \theta_n + \varepsilon_n \Phi_\theta(\theta_n) + \varepsilon_n \frac{\lambda_n}{2\delta_n} + \varepsilon_n \beta_n + \varepsilon_n \frac{\eta_n(\theta_n, \xi_n)}{2\delta_n}. \tag{12}$$

3. Convergence and Rates of Convergence

This section studies the convergence and rates of convergence of the recursive algorithm. We will show that θ_n defined in (10) is closely related to an ordinary differential equation (ODE). The stationary points of ODE are the optimal buying and selling prices that we are seeking.

To carry out the study of convergence, we define the following:

$$
\begin{cases}
t_n = \displaystyle\sum_{i=1}^{n-1} \varepsilon_i, \quad m(t) = \max\{n : t_n \le t\}, \\
\theta^0(t) = \theta_n \ \text{ for } \ t \in [t_n, t_{n+1}), \ \theta^n(t) = \theta^0(t + t_n), \\
N_n = \min\{i : t_{n+i} - t_n \ge T\}, \ \text{ for an arbitrary } \ T > 0.
\end{cases}
\tag{1}
$$

Note that $\theta^0(\cdot)$ is a piecewise constant process and $\theta^n(\cdot)$ is its shift. With the above definition, the interpolated process $h^n(\cdot)$ becomes

$$
\begin{aligned}
\theta^n(t) = \theta_n &+ \sum_{j=n}^{m(t_n+t)-1} \varepsilon_j \Phi_\theta(\theta_j) + \sum_{j=n}^{m(t_n+t)-1} \varepsilon_j \frac{\lambda_j}{2\delta_j} \\
&+ \sum_{j=n}^{m(t_n+t)-1} \varepsilon_j \beta_j + \sum_{j=n}^{m(t_n+t)-1} \frac{\varepsilon_j}{2\delta_j} \eta_j(\theta_j, \xi_j).
\end{aligned}
\tag{2}
$$

We need the following conditions:

(A1) The second derivative $\Phi_{\theta\theta}(\cdot)$ is continuous.

(A2) For each compact set G,

(a) $\sup_n E|Y_n^\pm I_{\{\theta_n \in G\}}|^2 < \infty$.

(b) For each θ belonging to a bounded set,

$$
\sup_n \sum_{j=n}^{n+N_n-1} E^{\frac{1}{2}} |E_n \eta_j(\theta, \xi_j)|^2 < \infty, \quad \lim_n \sup_{0 \le i < N_n} E|\widetilde{\gamma}_i^n| = 0, \tag{3}
$$

where

$$
\widetilde{\gamma}_i^n = \frac{1}{\varepsilon_{n+i}} \sum_{j=n+i}^{n+N_n-1} \frac{\varepsilon_j}{2\delta_j} E_{n+i}[\eta_j(\theta_{n+i+1}, \xi_j) - \eta_j(\theta_{n+i}, \xi_j)], \ i \le N_n - 1.
$$

Remark 3.1. Our default choice of step size and finite difference sequences is $\varepsilon_n = O(1/n)$ and $\delta_n = O(1/n^{1/6})$, it follows that the sequences $\{\varepsilon_n\}$ and $\{\delta_n\}$ satisfy $0 < \varepsilon_n \to 0$, $\sum_n \varepsilon_n = \infty$, $0 < \delta_n \to 0$, and $\varepsilon_n/\delta_n^2 \to 0$ as $n \to \infty$. Moreover,

$$
\limsup_n \sup_{0 \le i \le N_n - 1} \frac{\varepsilon_{n+i}}{\varepsilon_n} < \infty, \quad \limsup_n \frac{\delta_{n+i}}{\delta_n} < \infty,
$$

$$
\limsup_n \left[\frac{(\varepsilon_{n+i}/\delta_{n+i}^2)}{(\varepsilon_n/\delta_n^2)} \right] < \infty.
$$

For simplicity, we use a mixing condition. Assume that $\xi_n^\pm = g_0(\zeta_n^\pm)$ where $g_0(\cdot)$ is a real-valued function, $\{\zeta_{n,\ell}^\pm\}$ are homogeneous finite-state

Markov chains whose transition matrices are irreducible and aperiodic. Thus the noise is bounded since the Markov chain takes only finite values. Then $\xi_{n,\ell}^{\pm}$ are ϕ-mixing sequences with exponential mixing rates (Ref. 3 [p.167]), i.e., $\varpi(j) = c_0 \varpi^j$ for some $c_0 > 0$ and some $0 < \varpi < 1$. Using the exponential mixing rates, conditions (A2)(a) and (A2)(b) are easily verified.

Theorem 3.2. *Assume that (A1)–(A2) and that $\{\theta_n\}$ is tight in $\mathbb{R}^2$. Then $\theta^n(\cdot)$ converges weakly to $\theta(\cdot)$, the solution to the differential equation*

$$\dot{\theta} = \Phi_\theta(\theta), \tag{4}$$

which has a unique solution for each initial condition.

Corollary 3.3. *Suppose that (4) has a unique stationary point θ_* that is globally asymptotically stable in the sense of Liapunov, and that $\{s_n\}$ is a sequence of real numbers such that $s_n \to \infty$. Then $\theta^n(s_n + \cdot)$ converges weakly to θ_* as $n \to \infty$.*

To proceed, for simplicity, take $\varepsilon_n = 1/n$ and $\delta_n = \delta/n^{1/6}$. We assume all the conditions of Theorem 3.2 holds. Define $u_n = n^{1/3}(\theta_n - \theta_*)$. The rate of convergence study aims to exploit the asymptotic properties of this scaled sequence. We shall show that the weak limit of the interpolation of u_n is a diffusion process. To proceed, let us state conditions needed in what follows.

(A3) Assume $\theta_n \to \theta_*$ in probability, and $\Phi_{\theta\theta\theta}(\cdot)$ exists and is continuous in a neighborhood of θ_*. In addition,

   (a) $\{u_n\}$ is tight;
   (b) all eigenvalues of $\Phi_{\theta\theta}(\theta_*) + (1/3)I$ have negative real parts;
   (c) for each θ,

   $$\eta_n(\theta, \xi) = \eta_n(\theta_*, \xi) + \eta_{n,\theta}(\theta_*, \xi)(\theta - \theta_*)$$
   $$+ \left(\int_0^1 [\eta_{n,\theta}(\theta_* + (\theta_n - \theta_*)s, \xi) - \eta_{n,\theta}(\theta_*, \xi)]ds \right)(\theta - \theta_*);$$

   (d) the sequence $\{\eta_n(\theta_*, \xi_n)\}$ is stationary ϕ-mixing such that $E|\eta_n(\theta_*, \xi_n)|^{2+\Delta} < \infty$ for some $\Delta > 0$ and $E\eta_n(\theta_*, \xi_n) = 0$ and that the mixing measure $\varpi(\cdot)$ is given by $\varpi(j) = \sup_{A \in \mathcal{F}^{n+j}} E^{(1+\Delta)/(2+\Delta)} |P(A|\mathcal{F}_n) - P(A)|^{(2+\Delta)/(1+\Delta)}$, satisfying $\sum_{j=1}^{\infty} (\varpi(j))^{\Delta/(1+\Delta)} < \infty$.

Remark 3.4. Applying perturbed Liapunov function methods, (A3)(a) can be verified. (see Ref. 22 [Section 10]). The existence and continuity of $\Phi_{\theta\theta\theta}(\cdot)$ in a neighborhood of θ_* allows us to linearize $\Phi(\cdot)$ about θ_*. Conditions (A3) (c) and (d) concern about the sequence $\eta_n(\theta, \xi)$. Let us examine these conditions in conjunction with (6) and using independent samples and noises $\{\xi_{n,l}^{\pm}\}$. It is important to note that due to the use of the stopping times $\tau^{\{b_i\}}$ and $\tau^{\{s_i\}}$, $\widehat{\Phi}(\theta, \xi)$ defined in (6) may not be continuous in θ. However, we can assume that its expectation is smooth.

Take for instance, $\widetilde{\Phi}(\theta, \xi) = \Phi(\theta) + f_0(\theta)\xi$, where $f_0(\theta)$ is a bounded and continuous function. Suppose that for a positive integer m_0, $\{\xi_{n,l}^{\pm}\}$ are m_0-dependent sequences (see Ref. 3 [p.167]). For example, $\xi_{n,l}^{\pm} = \sum_{j=0}^{m_0} c_j \zeta_{n-j,l}^{\pm}$, where $\{\zeta_{n,l}^{\pm}\}$ are martingale difference sequences satisfying $E|\zeta_{n,l}^{\pm}|^2 < \infty$. Then they are mixing processes and the mixing measures satisfy $\varpi(j) = 0$ for all $j > m_0$. For each θ belonging to a bounded set, it is easily verified that $E\left|\frac{1}{n_0}\sum_{l=1}^{n_0}[\widetilde{\Phi}(\theta + \delta_n e_\iota, \xi_{n,l}^{\pm})]\right|^2 < \infty$. Thus $E|Y_n^{\pm,\iota}|^2 < \infty$. In addition, for each $j > n$, $E_n[\eta_j(\theta, \xi_j)] = E_n\{[E_j Y_j^{+,\iota}(\theta, \xi_j) - \Phi(\theta + \delta_n)] - [E_j Y^{-,\iota}(\theta, \xi_j)_j - \Phi(\theta - \delta_n)]\} = 0$. It is easy to see that $\eta_n(\theta_*, \xi_n)$ is a zero mean sequence and is ϕ-mixing (in fact n_0-dependent). Thus (A3)(d) is verified.

Lemma 3.5. *Assume (A1)–(A3).*

(a) *The following inequalities hold:*

$$|E\eta_j(\theta_*, \xi_j)\eta_k(\theta_*, \xi_k)| \leq K\left(\varpi(j)\right)^{\Delta/(1+\Delta)},$$
$$E\left|E(\eta_{n+j}(\theta_*, \xi_{n+j})|\mathcal{F}_n)\right| \leq K\left(\varpi(j)\right)^{\Delta/(1+\Delta)}; \tag{5}$$

(b) *The weak limit of the sequence $\sum_{j=n}^{m(t_n+t)-1}(\eta_j(\theta_*, \xi_j) + \lambda_j^*)/\sqrt{j}$ is an $\mathbb{R}^2$-valued Brownian motion $\widetilde{w}(\cdot)$ with covariance Σt,*

Define a piecewise constant interpolation

$$u^0(t) = u_n, \ t \in [t_n, t_{n+1}), \ \text{and} \ u^n(t) = u^0(t_n + t).$$

Theorem 3.6. *Assume that (A3) holds. Then $u^n(\cdot)$ converges weakly to a diffusion process $u(\cdot)$ that is a solution of the stochastic differential equation*

$$du = \left\{\left(\Phi_{\theta\theta}(\theta_*) + \frac{I}{3}\right)u + \frac{\delta^2}{3!}\begin{pmatrix}\Phi_{\theta^1,\theta^1,\theta^1}(\theta_*)\\ \Phi_{\theta^2,\theta^2,\theta^2}(\theta_*)\end{pmatrix}\right\}dt + \frac{1}{2\delta}d\widetilde{w}, \tag{6}$$

where $\widetilde{w}(\cdot)$ is the Brownian motion with covariance $\Sigma^{1/2}(\Sigma^{1/2})'t = \Sigma t$ given by Lemma 3.5.

Remark 3.7. To easily handle real market data, we do not take a sample average as in (6). Instead of that, we set $n_0 = 1$ and use $\widehat{\Phi}(\theta, \xi_n^\pm) = \widetilde{\Phi}(\theta, \xi_n^\pm)$ without using sample averages. The resulting estimate is not expected to be a smooth and the bias will be larger. Nevertheless, it does provide us a reasonable estimate.

Define $Y_n^\pm$ as in (8) and use algorithm (10). Since conditions concerning the noise given in Sections 3 and 4 all refer to the function $\widehat{\Phi}(\theta, \xi_n)$, we obtain the convergence and rate of convergence of the algorithm just as in the previous sections. We summarize the above as the following proposition.

Proposition 3.1. *Let $n_0 = 1$, under the conditions of Theorem 3.2 and Theorem 3.6, the conclusions of these theorems continue to hold.*

4. Numerical Demonstration

In this section, we report our simulation results and numerical experiments. We first use Monte Carlo simulation and compare our algorithm with the analytic solution obtained in Ref. 30. Then we test our algorithm using real market data.

4.1. *Simulation Study*

In this section, we compare the results of stochastic approximation algorithm with the closed-form solution in Ref. 30. We find that the proposed algorithm indeed provides good approximation results. Recall that the mean-reverting SDE follows

$$dX(t) = a(b - X(t))dt + \sigma dW(t), \quad X(0) = x, \tag{1}$$

and the stock price is given by $S(t) = \exp(X(t))$. First, we take $a = 0.8$, $b = 2$, $\sigma = 0.5$, $x = 0$, let the slippage rate $K = 0.01$, and the discount rate $\rho = 0.5$. In this case, the analytic solution in Ref. 30 gives $(\theta_*^1, \theta_*^2) = (1.331, 1.631)$.

We let $n_0 = 5000$, where n_0 is the number of random samples used in each iteration; see (6). Then we use (1) to simulate the stock prices and the recursive algorithm (10) is applied for 200 iterations. The sequence of ξ_n and δ_n are chosen to be $\xi_n = 1/(n + k_0)$ and $\delta_n = 1/(n^{1/6} + k_1)$, respectively, where k_0 and k_1 are some positive constants. The proposed algorithm yields the optimal estimation of $(\theta^1, \theta^2) = (1.3225, 1.6292)$. The (absolute) error $= \sqrt{(\theta^1 - \theta_*^1)^2 + (\theta^2 - \theta_*^2)^2} = 0.0087$, and the relative error $= \text{error}/\sqrt{(\theta_*^1)^2 + (\theta_*^2)^2} = 0.0041$.

Note that the analytic solutions depend on the knowledge of various parameters: a, b, σ, K and ρ. We next vary one of the parameters at a time and compare the approximation results with analytic results.

We first vary the values of b, the equilibrium levels of stock price. Then we compute the threshold values (θ^1, θ^2) associated with varying b. Intuitively, larger b would result in larger threshold values (θ^1, θ^2). Our approximation results confirm this. The detail results are reported in Table 1, where θ_*^1 and θ_*^2 are threshold values calculated by analytic solution and θ^1 and θ^2 are threshold values calculated by SA method; see (10).

Table 1. (θ^1, θ^2) with varying b: average error $= 0.01511$, average relative error $= 0.0102$

b	1	1.5	2	2.5	3
θ_*^1	0.331	0.831	1.331	1.831	2.331
θ_*^2	0.631	1.131	1.631	2.131	2.631
θ^1	0.3303	0.8211	1.3225	1.8199	2.3127
θ^2	0.6504	1.1384	1.6292	2.1403	2.6407
error	0.0194	0.0124	0.0087	0.0145	0.0207
relative error	0.0272	0.0088	0.0041	0.0052	0.0059

Next, we vary a. A larger a means fast reversion rate for X_t to reach the equilibrium level b and thus results in larger reward in short time. Consistently, Table 2 shows that the values of (θ^1, θ^2) increase in a.

Table 2. (θ^1, θ^2) with varying a:average error $= 0.0413$, average relative error $= 0.0216$

a	0.6	0.7	0.8	0.9	1
θ_*^1	1.175	1.264	1.331	1.383	1.425
θ_*^2	1.375	1.564	1.631	1.683	1.725
θ^1	1.146	1.238	1.300	1.367	1.425
θ^2	1.497	1.559	1.621	1.685	1.731
error	0.1254	0.0265	0.0326	0.016	0.006
relative error	0.0693	0.0132	0.0155	0.0074	0.0027

In Table 3, we vary the volatility σ. The larger the σ, the greater the range for the stock price $S_t = \exp(X_t)$. To avoid closing positions due to normal price fluctuations, one needs to increase the values of (θ^1, θ^2). Consistent with this, Table 3 shows the values (θ^1, θ^2) increase in σ.

In Table 4, we vary the discount rate ρ. A larger discount rate ρ implies smaller return and smaller threshold values (θ^1, θ^2). These are confirmed in Table 4.

Table 3. (θ^1, θ^2) with varying σ: average error = 0.0339, average relative error = 0.0158

σ	0.3	0.4	0.5	0.6	0.7
θ_*^1	1.231	1.275	1.331	1.400	1.481
θ_*^2	1.531	1.575	1.631	1.700	1.781
θ^1	1.213	1.274	1.316	1.360	1.469
θ^2	1.496	1.564	1.624	1.732	1.831
error	0.0394	0.0110	0.0166	0.0512	0.0514
relative error	0.0200	0.0055	0.0079	0.0232	0.0222

Table 4. (θ^1, θ^2) with varying ρ: average error = 0.0638, average relative error = 0.0310

ρ	0.3	0.4	0.5	0.6	0.7
θ_*^1	1.681	1.456	1.331	1.206	1.081
θ_*^2	1.781	1.756	1.631	1.506	1.281
θ^1	1.604	1.440	1.320	1.181	1.060
θ^2	1.901	1.782	1.627	1.521	1.384
error	0.1426	0.0305	0.0117	0.0292	0.1051
relative error	0.0582	0.0134	0.0056	0.0151	0.0627

Finally, we choose different slippage rates K. The results in Tables 5 suggest that θ_1 is decreasing slightly in K and θ_2 is almost flat. The possible explanation is that larger slippage cost discourages stock transactions and thus has to be compensate by smaller θ_1.

Table 5. (θ^1, θ^2) with varying K: average error = 0.0656, average relative error = 0.0316

K	0.001	0.005	0.01	0.015	0.02
θ_*^1	1.431	1.431	1.331	1.331	1.231
θ_*^2	1.631	1.531	1.631	1.531	1.631
θ^1	1.413	1.351	1.316	1.264	.1252
θ^2	1.585	1.619	1.624	1.633	1.634
error	0.0494	0.1189	0.0166	0.1220	0.0212
relative error	0.0228	0.0568	0.0079	0.0602	0.0104

It is clear to see from the above tables that the stochastic approximation constructed in this paper indeed provide sound estimates of optimal threshold values $(\theta^1, \theta^2)'$. The average error is 0.0440 and the average relative error is only 0.0220, or 2.20%.

4.2. *Tests with Market Data*

In this section, we study the performance of the algorithm using real market data. The proposed algorithm are employed as follows:

Step 1. We collect stock prices during the period January 2, 2002 to December 31, 2007;

Step 2. We divide the whole period into three sub-periods: Period 1, the first 500 trading days beginning at January 2, 2002; Period 2, the subsequent 500 trading day following Period 1; and Period 3, the next 500 trading days following Period 2;

Step 3. We run 600 iterations of stochastic approximation algorithm (10) on stock prices in the Period 1 and compute the threshold values (θ^1, θ^2);

Step 4. We use the threshold values (θ^1, θ^2) obtained in Step 3 to simulate trading the same stock in Period 2 and compute the overall dollar return. Recall that we buy stocks whenever stock price $S(t) < \exp(\theta^1)$ and sell stocks whenever stock price $S(t) > \exp(\theta^2)$;

Step 5. Again, we run 600 iterations of algorithm (10) on stock prices in the Period 2 and compute the threshold values (θ^1, θ^2);

Step 6. We use the threshold values (θ^1, θ^2) calculated in Step 5 to simulate trading the same stock in Period 3 and compute the overall dollar return.

In the above procedure, if a stock is bought in any period but the stock price doesn't exceed the selling price after buying, we will sell stock at the end of the period regardless of selling price. Now we choose different stocks to conduct above experiment. For example, Figure 1 is the graph of historical prices of Wal-Mart Stores Inc. during the period January 2, 2002 to December 31, 2007.

Applying stochastic approximation algorithm to Period 1, the computed buying price is \$44.6088 and the selling price is \$53.3786. Using these threshold values in Period 2, we buy stock at \$44.46 on Aug 23, 2005 and sell it at \$47.12 at the end of Period 2. The dollar return is \$1.7442 in Period 2. Using stock prices during Period 2, the calculated buying price is \$42.2969 and the selling price is \$47.8489. Applying these threshold values in Period 3, we buy stock at \$41.73 on Jul 14, 2006 and sell it at \$47.92 on Sep 26, 2006, and we buy it again at \$42.06 on Sep 5, 2007 and finally sell it at \$48.45 on Dec 5, 2007. The total dollar return in Period 3 is \$10.7784.

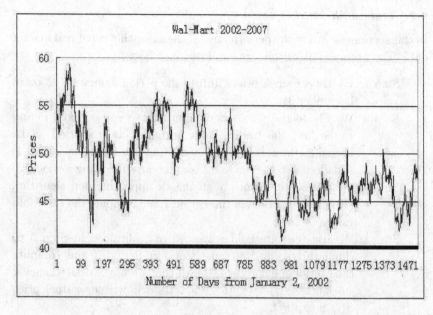

Fig. 1.   The Prices of Wal-Mart from Jan 2, 2002 to Dec 31, 2007

We apply the same procedure to Home Depot's stock, see Figure 2 for the daily stock price during the period January 2, 2002 to December 31, 2007.

Based on the prices of Home Depot in Period 1, after 600 iteration of (10), the calculated buying price is $34.0484 and selling price is $38.9126. Using the threshold values above, we trade Home Depot in Period 2. We first buy it at $33.29 on Sep 2, 2004 and sell it at $39.45 on Nov 10, 2004. We buy it again at $33.96 on May 10, 2005. However, before the ending of Period 2, the price doesn't raise above the selling price $38.9126 again. Thus we have to sell it at the end of Period 2 for $32.77. The total dollar return in Period 2 is $3.5753. Using the prices in Period 2, the computed selling price is $32.7166 and $38.7353. Therefore, in Period 3, we buy it at $32.61 on Dec 27, 2005 and sell it at $39.87 on Aug 9, 2007, resulting a total profit of $6.5352.

The same procedure are also applied to other stocks. The detail trading results are shown in Table 6. As can be seen, using the threshold values computed by the proposed algorithm does not necessarily trigger transactions in every period. However, overall speaking, the proposed algorithm in this paper may provide trading guidelines in practice.

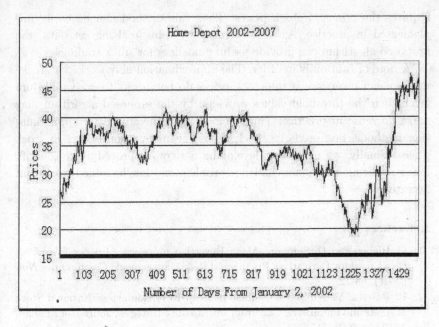

Fig. 2.   The Prices of Home Depot from Jan 2, 2002 to Dec 31, 2007

Table 6.   Trading Results

Stock	Period 2			Period 3		
	Selling Price	Buying Price	Total Profit	Selling Price	Buying Price	Total Profit
AIG	$52.15	$63.65	$13.95	$52.08	$59.94	$9.04
Du Pont	$30.30	$39.50	$0.00	$36.23	$46.00	$9.07
Home Depot	$34.05	$38.91	$3.58	$32.72	$38.74	$6.54
IBM	$54.30	$82.13	$0.00	$71.82	$83.38	$10.98
Intel	$12.63	$28.15	$0.00	$19.45	$27.00	$7.51
Microsoft	$19.81	$22.24	$0.00	$21.03	$26.94	$5.90
3M	$46.99	$69.21	$0.00	$69.90	$80.09	$20.06
Verizon	$25.66	$31.58	$0.07	$28.83	$34.50	$7.66
Wal-Mart	$44.61	$53.38	$1.74	$42.30	$47.85	$10.78

5. Further Remarks

A stochastic approximation algorithm has been developed for buying low and selling high strategy in stock trading. Compared with the analytic solution, the simulation results indicate that this algorithm provides sound estimates for optimal buying and selling prices. One advantage of the proposed method is its simple recursive form. In addition, this method only requires the observed stock prices. Thus this method can be easily implemented in practice. As demonstrated by using real market data, the proposed algorithm can provide useful guidelines for stock tradings.

A note of caution is in order. The approximation only works for stocks under mean reversion. If the stock prices do not revert to an equilibrium level, then the threshold values provided by the proposed algorithm may make no sense in practice. Thus, before using the stochastic approximation methods, one needs to check if the mean reversion occurs in stock prices. Finally, we note that developing a rigorous procedure to identify mean-reverting assets is a challenge problem and maybe added to current literatures.

References

1. C. Blanco and D. Soronow, Mean Reverting processes - Energy Price Processes used for Derivatives Pricing and Risk Management, *Commodities Now* (2001), 68-72.
2. R. Balvers, Y. Wu and E. Gilliland, Mean Reversion across National Stock Markets and Parametric Constrain Investment Strategies, *Journal of Finance*, **55(2)** (2000), 745-772.
3. P. Billingsley, *Convergence of Probability Measures*, J. Wiley, New York, 1968.
4. W.F.M.De Bondt and R. Thaler, Does the Stock Market Overreact? *Journal of Finance*, **40**, (1985), 793-805.
5. A. Cadenillas and S.R. Pliska, Optimal Trading of a Security when There are Taxes and Transaction Costs, *Finance and Stochastics*, **3**, (1999), 137-165.
6. H.F. Chen, *Stochastic Approximation and Its Applications*, Kluwer Academic, Dordrecht, Netherlands, 2002.
7. J. Conrad and G. Kaul, Mean Reversion in Short-Horizon Expected Returns, *Review of Financial Studies*, **2**, (1989), 225-240.
8. J. Conrad and G. Kaul, An Anatomy of Trading Strategies, *Review of Financial Studies*, **11**, (1998), 489-519.
9. G.M. Constantinides, Capital Market Equilibrium with Personal Tax, *Econometrica*, **51**, (1983), 611-636.
10. A. Cowles and H. Jones, Some Posteriori Probability in Stock Market Action, *Econometrica* , **5**, (1937), 611-636.
11. R.M. Dammon and C.S. Spat, The Optimal Trading and Pricing of Securities with Asymmetric Capital Gains Tax and Transaction Costs, *Review of Financial Studies*, **9**, 1996, 921-952.
12. S. N. Ethier and T. G. Kurtz, *Markov Processes: Characterization and Convergence*, J. Wiley, New York, 1986.
13. E. Fama and K. French, Permanent and Temporary Components of Stock Prices, *Journal of Political Economy*, **96** (1988), 246-273.

14. P.W. Glynn and D.L. Iglehart, Trading Securities Using Trailing Stops, *Management Science*, **41**, (1995), 1096-1106.
15. X. Guo and Q. Zhang, Optimal Selling Rules in a Regime Switching Model, *IEEE Trans. Automatic Control*, **50**, (2005), 1450-1455.
16. C.M. Hafner and H. Herwartz, Option Pricing under Linear Autoregressive Dynamics, Heteroskedasticity, and Conditional Leptokurtosis, *Journal of Empirical Finance*, **8**, (2001), 1-34.
17. K. Helmes, Computing optimal Selling Rules for Stocks using Linear Programming, *Mathematics of Finance*, G. Yin and Q. Zhang, (Eds), 187-198, Contemporary Mathematics, American Mathematical Society, 2004.
18. N. Jagadeesh and S. Titman, Returns to Buying Winners and Selling Losers: Implication for Stock Market Efficiency, *Journal of Finance*, **48**, (1993), 65-91.
19. N. Jagadeesh and S. Titman, Overreaction, Delayed Reaction, and Contrarian Profits, *Review of Financial Studies*, **8**, (1995), 973-993.
20. N. Jagadeesh and S. Titman, Short-Horizon Return Reversals and the Bid-Ask Spread, *Journal of Financial Intermediation*, **4**, (1995), 116-132.
21. H.J. Kushner, *Approximation and Weak Convergence Methods for Random Processes, with applications to Stochastic Systems Theory*, MIT Press, Cambridge, MA, 1984.
22. H.J. Kushner and G. Yin, *Stochastic Approximation and Recursive Algorithms and Applications*, 2nd Ed., Springer, New York, 2003.
23. R.H. Liu, Q. Zhang, and G. Yin, Option Pricing in a Regime-Switching Model using the Fast Fourier Transform, *Journal of Applied Mathematics and Stochastic Analysis*, Vol. 2006, Article ID 18109, 22 pages, 2006. doi:10.1155/JAMSA/2006/18109
24. M. Pemy, G. Yin and Q. Zhang, Liquidation of a Large Block of Stock, *Journal of Banking and Finance*, **31**, (2007), 1295-1305.
25. J. Poterba and L. Summers, Mean Reversion in Stock Prices: Evidence and Implications, *Journal of Financial Economics*, **22** (1988), 27-59.
26. Q.S. Song, G. Yin, and Q. Zhang, Stochastic optimization methods for buying-low-and-selling-high strategies, *Stochastic Anal. Appl.*, **27** (2009), 523-542.
27. G. Yin, Q. Zhang, F. Liu, R.H. Liu and Y. Chang, Stock Liquidation via Stochastic Approximation using NASDAQ Daily and Intra-day Data, *Mathematical Finance*, **16**, (2006), 217-236.
28. G. Yin, R.H. Liu and Q. Zhang, Recursive Algorithms for Stock Liquidation: A Stochastic Optimization Approach, *SIAM J. Optim*, **13**, (2002), 240-263.
29. Q. Zhang, Stock Trading: An Optimal Selling Rule, *SIAM J. Control Optim.*, **40**, (2001), 64-87.
30. H. Zhang and Q. Zhang, Trading A Mean-Reverting Asset: Buy Low and Sell High, *Automatica*, **44**, (2008), 1511-1518.